DEDICATION

For Stacy, Stefani, and Rob

ABOUT THE AUTHORS

Dr. Steven M. Gerson (PhD, Texas Tech University) has taught English and technical writing at Johnson County Community College, Overland Park, KS, since 1978. Sharon J. Gerson (MA, Texas Tech University) has taught English, technical writing, and professional writing at DeVry University, Kansas City, since 1987. Sharon and Steve have owned and operated Steve Gerson Consulting since 1980. During that time, the two have provided business writing services and facilitated over 500 technical/business writing workshops, focusing on writing and grammar, to over 10,000 participants. They have worked with employees, managers, and supervisors from corporations including Sprint, H&R Block, Commerce Bank, Honeywell/Bendix-Allied Aerospace, Avon, Puritan—Bennett, General Electric, J.C. Penney, the General Services Administration, the Missouri Department of Transportation, the Missouri Department of Natural Resources, and the Federal Highway Department.

Steve and Sharon have published many articles on technical/business writing and given numerous professional presentations on technical/business writing at regional, national, and international conferences.

Steve and Sharon are the authors of *Technical Writing: Process and Product* (5th edition, Prentice Hall). Steve is the author of *Writing That Works: A Teacher's Guide to Technical Writing* (2nd edition, Kansas State Department of Education). Sharon is the coauthor of *The Red Bridge Reader* (3rd edition, Prentice Hall). Both Sharon and Steve are senior members of the Society for Technical Communication, where Steve is an STC Fellow.

PREFACE

Welcome to *Workplace Communication: Process and Product*, a concise version of our highly successful technical writing textbook *Technical Writing: Process and Product*, currently in its 5th edition.

This compact yet thoroughly developed textbook, geared toward technical writing, business communication, and professional communication courses, gives students and instructors an accessible and lively introduction to workplace communication. We engage students through real-world business scenarios stitched throughout every chapter, easy-to-follow criteria, clear and relevant examples, a process approach to writing with before/after examples, and diverse end-of-chapter activities involving multiple learning styles.

Workplace Communication: Process and Product will appeal to a general, diverse, and inclusive audience of students in career programs, the sciences, business, and technology. Our book is appropriate for community colleges, technical schools, and universities.

What You Will Find in *Workplace Communication*

Communication at Work. Each chapter begins with a workplace scenario featuring a real person in business and industry facing a communication challenge. The "Communication at Work" scenarios provide students insight into actual workplace communication issues, allowing students to understand writing in a business context. These scenarios, stitched throughout the chapters, are complete with samples of writing and writer's reflections on how communication decisions are made.

Writer's Insight. Each chapter includes boxed writer's insights based on the chapter's opening scenarios. Unique to this textbook, every chapter provides insight into why writers make certain communication decisions regarding word usage, organization, layout, and tone. These reflections emphasize the importance of self-assessment, decision making, and techniques for solving writing-related problems.

The Writing Process at Work. As with our 5th edition textbook *Technical Writing: Process and Product*, this concise version focuses on the writing process. The goal—lifelong learning! We want to give students skills that they can transfer from the classroom into applicable communication challenges on the job.

To accomplish this goal, each chapter focuses on prewriting options, drafting techniques, and revising and rewriting skills, complete with samples and reflections (commentaries by writers regarding the communication decisions they have made). Every chapter also includes a process log. The process logs highlight professional and student-written examples of prewriting, rough drafts, and revisions to guide students through the writing process.

By providing students a process approach to writing, complete with options, examples, reflections, and process logs, the text offers a resource for successful communication during a semester and, more importantly, a resource for continued learning and application on the job.

Before and After Examples. To emphasize the importance of revision and to show students how to revise successfully, we provide numerous "before" and "after" examples. These examples clarify why and how to achieve changes in tone, style, layout, content, sensitivity to diverse audiences, and grammar.

Checklists. Every chapter includes easy-to-access checklists for each communication channel (letters, reports, memos, proposals, Web sites, e-mail, descriptions, instructions, etc.). Checklists benefit teachers and students. The checklists are an easy teaching tool for instruction and post-instruction assessment of assignments. The checklists benefit students through peer evaluations and self-assessment.

Developing Workplace Skills. Creative and diverse assignments—Case Studies, Individual and Team Projects, Problem Solving Think Pieces, and Web Workshops—allow for teamwork, individual learning, decision making, and practical application. Each end-of-chapter assignment allows students multiple opportunities for

- Self-assessment (applying the skills they've learned in the chapters).
- Teamwork (by working with others in team projects).
- Critical thinking/problem solving. Workplace communication requires choices rather than memorized responses. Our unique "Problem Solving Think Pieces" and "Case Studies" encourage the types of challenging decision making that students will encounter on the job.

Electronic Communication. *Workplace Communication: Process and Product* is the only current textbook that provides detailed discussions of twenty-first-century electronic communication channels. Some textbooks cover e-mail and Web design. In our textbook, in addition to discussing and providing examples of e-mail and Web sites, we also provide in-depth coverage of Web logs (blogs), instant messaging, and online help.

Real-World Examples. The book's many professional writing samples provide models for success in the workplace. The examples, drawn from actual businesses, industries, and governmental agencies, focus on multiple disciplines relevant to students. These disciplines include banking, insurance, marketing, education, health care, biomedical, computer information systems, information technology, automotive, electronics, hospitality management, HVAC, health information management, and more.

Thorough Coverage. *Workplace Communication: Process and Product* covers all major communication channels, such as memos, letters, short reports, long reports/proposals, instructions, technical descriptions, resumes, and emerging electronic communication options. In addition, we also thoroughly discuss issues of audience diversity; multiculturalism; strategies for everyday oral communication, informal oral presentations, and formal oral presentations; PowerPoint usage; and standards for APA and MLA documentation. We provide criteria for and examples of ways to incorporate graphics (tables and figures) in workplace communication.

The text also offers a grammar handbook in Appendix A. Appendix B covers works cited and references for researched information. Our unit on ethics is unique in that it focuses on ethical guidelines supported by the two leading organizations affiliated with workplace communication: The Society for Technical Communication and The Association for Business Communication.

Workplace Communication's Instructional Supplements

Companion Website: A Wealth of Online Materials.

We are especially excited about the wealth of new cases, exercises, activities and documents that have been developed for each chapter and are available for your use at our Companion Website located at *www.prenhall.com/gerson*. Online materials for each chapter in the text include the following:

- Chapter Learning Objectives. Overview of major chapter concepts.
- Writing Process Exercises. Prewriting/Writing/Rewriting assignments.
- Interactive Editing and Revision Exercises. Interactive documents allow students to see poorly done and corrected versions of documents with additional document revision exercises for assignments.
- Communication Cases. Students encounter real-world situations with links to outside content. A student response box is provided for students to send answers to the professor.
- Activities and Exercises. Activities specific to a variety of technical and career fields allow students to practice producing communication relevant to their interests.
- Collaboration Exercises. Assignments designed to provide practice writing and communicating in teams.
- Web Resources. Links to helpful online resources related to chapter content.
- Document Library. Additional documents and forms.
- Chapter Quizzes. Self-grading multiple-choice quizzes help students master chapter concepts and prepare for tests.

OneKey Distance Learning Solutions.

Ready-made Blackboard, WebCT, and CourseCompass online courses are available. If you adopt the text with a OneKey course, the student access cards will be packaged with the textbook at no additional charge to the student.

Instructor's Resources.

- Instructor's Manual (ISBN: 0-13-228809-5) The online Instructor's Manual is loaded with helpful teaching notes for your classroom. Included in the manual are answers to the chapter quiz questions, a test bank, and instructor notes for assignments and activities located on the Companion Website.
- Instructor's Resource CD (ISBN: 0-13-228811-7) The IRCD includes the following components:
    - TestGen (a computerized test generator)
    - PowerPoint Lecture Presentation Package
    - Instructor's Manual (in Microsoft Word)

All Instructor Resources can also be downloaded from the Instructor Resource Center at *www.prenhall.com*

To access supplementary materials online, instructors need to request an instructor access code. Go to *www.prenhall.com*, click the Instructor Resource Center link, and then click Register Today for an instructor access code. Within 48 hours after registering you will receive a confirming e-mail including an instructor access code. Once you have received your code, go to the site and log on for full instructions on downloading the materials you wish to use.

Acknowledgments

We would like to acknowledge the reviewers of this text for their constructive comments: Thomas Treffinger, Greenville Technical College; Susan Berston, City College of San Francisco; Cathie Cline, East Arkansas Community College; Arthur Khaw, Kirkwood Community College; Kathleen Fowler, Surry Community College; Suzy Barile, Wake Technical Community College; Stuart Brown, New Mexico State University; Anne Lehman, Milwaukee Area Technical College; and Kathryn Roberson, Gaston College.

CONTENTS

An Introduction to Workplace Communication

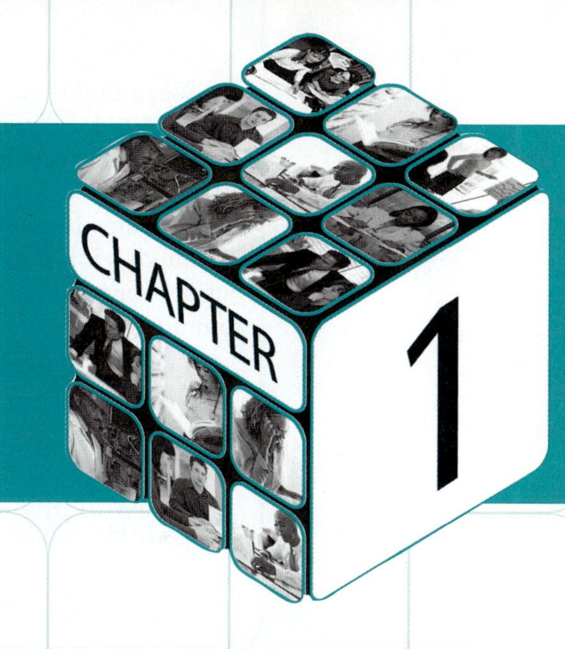

OBJECTIVES

When you complete this chapter, you will be able to do the following:

1. Define workplace communication.
2. Distinguish among the communication channels.
3. Recognize the importance of workplace communication in your career.
4. Collaborate effectively in teams.
5. Apply the process approach—prewriting, writing, and rewriting—to communicate effectively.
6. Test your knowledge of workplace writing through end-of-chapter activities:
   - Case Studies
   - Individual and Team Projects
   - Problem Solving Think Pieces
   - Web Workshop

COMMUNICATION AT WORK

Manuel "Buddy" Ramos uses many different communication channels when writing and speaking to customers, vendors, and colleagues.

Manuel "Buddy" Ramos is a client executive for ComputeToday, a company that provides technology applications (hardware and software) for customers in education, communications, financial services, healthcare, manufacturing, transportation, and utilities.

Buddy doesn't work in an office. He telecommutes, working from his home, airports, and hotels in cities throughout the world. Buddy spends **95 to 98 percent of his workday communicating with clients or internally with account teams.**

To accomplish his job duties, Buddy uses numerous communication channels. He spends about four to five hours a day speaking to customers and colleagues. He uses his cell phone for teleconference calls with his client teams and clients. Simultaneously, Buddy uses ComputeToday's intranet system to e-mail his team members. At the same time, Buddy and his coworkers are also online, using Internet metasearch options to find answers to their clients' questions. That's multitasking to achieve workplace communication.

When Buddy isn't on the phone, he is in face-to-face meetings with customers—locally, regionally, nationally, and internationally. Once Buddy has completed his meetings, either in person or on the phone, he has to supervise the 40-page proposals that will be sent to customers and the reports that will be sent to the home office.

Buddy's job is to help people implement new software solutions for their business operations. To accomplish this task, he must be an expert communicator,

- using multiple communication channels (Internet research tools, intranet e-mail systems, cell phone teleconferencing abilities, and face-to-face meetings);
- writing long proposals to customers, follow-up letters, and short reports to coworkers;
- working collaboratively with management, customers, vendors, and team members.

As Buddy says, "Communication is everything in my job."

What Is Workplace Communication?

Employers and employees, customers and vendors—if you buy, sell, or work within an organization, you will be involved with workplace communication. You will write business correspondence and speak to colleagues, clients, or sales representatives. Knowing how to communicate successfully in a work environment will help you to express your point of view and influence people. As Buddy Ramos at ComputeToday says, "Communication is everything" on the job.

What are the purposes of workplace communication? When will you be writing or speaking on the job? Consider these possibilities:

- As a computer information systems employee, you work at a 1-800 hotline help desk. A call comes through from a concerned client. Your job not only is to speak politely and professionally to the customer but also to follow up with a **one-page e-mail** documenting your responses.
- As a trust officer in a bank, one of your jobs is to make proposals to potential clients. In doing so, you will write a **20- to 40-page proposal** about your bank's services, and you will give an **oral presentation** to this client.
- As the manager of human resources, one of your major responsibilities is to document your training staff's accomplishments. To do so, you must write year-end **progress reports** for the employees, which will be used to justify their raises.
- As an entrepreneur, you want to advertise your new catering business. To do so, you plan to write **brochures** (to be distributed locally) and create a **Web site** (to expand your business opportunities globally).
- As a new graduate with an accounting degree, it is time to get a job. You need to write an effective **resume** and **letter of application** to show corporations what an outstanding asset you will be to their company. Then, you will need to interview well.

Using Different Communication Channels

Workplace communication takes many different forms. Not only will you communicate both orally and in writing, but also you will rely on various types of correspondence and technology, dependent upon the audience, purpose, and situation. To communicate successfully in the workplace, you must adapt to many different channels of communication.

For example, sometimes you might need to make an informal request for information. If you need an update from a vendor on delivery status, a brief e-mail message should suffice. Maybe you will need to make a telephone call to ensure that a client has received an order. Chatting with a coworker face to face about a team project is another common channel of informal business communication.

In other instances, formal reports, letters, or proposals are more appropriate communication channels. Your supervisor needs a written status report on the remodeling of your new showroom. You need to send a letter to a lawyer representing your company in a lawsuit. You are writing a long proposal to market a new product line. You are interviewing for a job.

Table 1.1 gives you examples of different communication channels, both oral and written.

Table 1.2 illustrates how different writers and speakers might use various channels to communicate effectively to both internal and external audiences. Internal audiences consist of the coworkers, subordinates, and supervisors in your workplace; external audiences consist of vendors, customers, and other workplace professionals.

Table 1.1 Communication Channels

Written Communication Channels	Oral Communication Channels
• E-mail	• Leading meetings
• Memos	• Conducting interviews
• Letters	• Making sales calls
• Reports	• Managing others
• Proposals	• Participating in teleconferences and videoconferences
• Fliers	• Facilitating training sessions
• Brochures	• Participating in collaborative team projects
• Newsletters	• Providing customer service
• Faxes	• Making telephone calls
• Internet Web sites	• Leaving voice-mail messages
• Intranet Web sites	• Making presentations at conferences or to civic organizations
• Extranet Web sites	• Participating in interpersonal communication at work
• Instant messaging	• Conducting performance reviews
• Blogging	
• Job information (resumes, letters of application, follow-up letters, interviews)	

TABLE 1.2 COMMUNICATION CHANNELS—AUDIENCE AND PURPOSE

Writers/ Speakers	Type of Communication and Communication Channel	Purposes	Internal or External
Human resources/ Training	Instructions—either hard copy or online (Internet, intranet, extranet)	Help employees and staff perform tasks	Internal
Marketing personnel	Brochures, sales letters, and phone calls	Promote new products or services	External
Customers	Inquiry or complaint letters or phone calls	Ask about/complain about products or services	External
Quality assurance	Investigative, incident, or progress reports	Report to regulatory agencies about events	External
Vendors	Newsletters, phone calls, and e-mail	Update clients on new prices, products, or services	External
Management	Oral presentations to department staff	Update employees on mergers, acquisitions, layoffs, raises, or site relocations	Internal
Everyone (management, employees, clients, vendors, governmental agencies)	E-mail, blogging, Web sites, and instant messaging	All conceivable purposes	Internal and External

Many communication channels overlap in terms of purpose and audience. If you are requesting information from a vendor, for example, you could write a letter, send an e-mail message, or make a telephone call. However, in other instances, communication channels are more exclusive. You would not want to communicate bad news—such as layoffs, loss of benefits, or corporate closings—to employees by way of mass e-mail messages or televised reports. In these instances, face-to-face meetings would be more appropriate. A key to successful workplace communication is choosing the right channel.

To clarify the use of different workplace communication channels, look at Figure 1.1. In a 2004 survey of approximately 120 companies employing over eight million people, the National Commission on Writing found that employees "almost always" use different forms of writing, including e-mail messages, PowerPoint, memos, letters, and reports ("Writing: A Ticket to Work" 11).

WHY IS WORKPLACE COMMUNICATION IMPORTANT?

At work, your primary job is not necessarily writing, or is it? You might be an accountant, computer information systems technician, training facilitator, salesperson, health information manager, supervisor, or paramedic. Your employers will expect

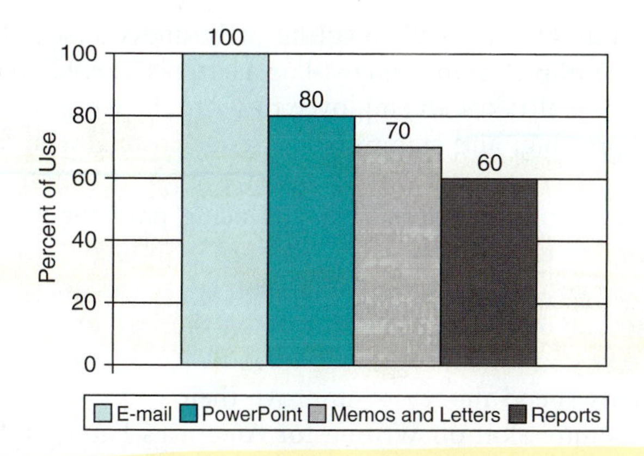

Figure 1.1 Channels "Almost Always" Used in Workplace Communication

expertise from you in those areas of specialization. In addition, to succeed on the job, you need to write and speak effectively to others—constantly. Workplace communication is important to you for many reasons.

The Importance of Workplace Communication When Operating a Business

Workplace communication is not a frill or an occasional occurrence. It is a major component of your job. Workplace communication (written and oral) allows you to manage your products, services, employees, and customers. Through letters, reports, e-mail, and teleconferences, you will communicate about how you manufacture your products, market your services, manage your staff, deliver your goods, meet deadlines, and explain to employees and customers how to follow procedures correctly. Through verbal communication, you will represent your company to civic leaders, clients, and vendors.

The National Commission on Writing's 2004 survey concluded that "two-thirds or more of . . . salaried employees have some responsibility for writing" and that "writing is almost a universal professional skill required in service industries as well as finance, insurance, and real estate" ("Writing: A Ticket to Work" 7). Other professions, such as automotive mechanics, surgical technicians, information technologists, firefighters, police officers, and health information management employees, also have on-the-job writing requirements.

The Amount of Time Communicating on the Job

In addition to serving valuable purposes on the job, getting a job, or meeting your needs as a customer, workplace communication is important because it is time-consuming. A North Carolina State University study concluded that employees spend approximately 31 percent of their time writing on the job. On the basis of a 40-hour workweek, 31 percent equals 12½ hours of writing each week (Miller et al. 10).

When asked how important writing was on the job, 90 percent of respondents to the university's survey stated that their on-the-job writing was "essential" to their performance and to their potential for advancement (Miller et al. 12).

The Cost of Communicating

You have heard it before—time is money. Here are four ways of looking at the cost of your workplace communication.

1. A recent study by Dartnell's Institute of Business Research tells us that the "average cost of producing and mailing a letter is $19.92" (Crane 1). That factors in the time it takes an employee to write the letter as well as the cost of the paper, printing, and stamp. If one letter costs almost $20, imagine how much an entire company's correspondence might cost annually.
2. Consider how much of your salary is being paid for your communication skills. If you make $35,000 a year and are spending 31 percent of your time writing, then your company is paying you approximately $10,850 to write. If you are not communicating effectively on the job, then you are asking your bosses to pay you for substandard work.
3. Corporations spend money to improve their employees' writing skills. The National Commission on Writing for America's Families, Schools, and Colleges reported that "More than 40 percent of responding firms offer or require training for salaried employees with writing deficiencies. 'We're likely to send out 200–300 people annually for skills upgrade courses like "business writing" or "technical writing," said one respondent.' Based on survey responses, the Commission estimates that remedying deficiencies in writing costs American corporations as much as $3.1 billion annually" ("Writing Skills").
4. Your communication skills do not just cost the company money; these talents can earn you and the company money. A well-written sales letter, brochure, proposal, or Web site can generate corporate income. Effectively written newsletters to clients and stakeholders can keep customers happy and bring in new clients. Good oral and written communication is not just part of your salary; good communication helps pay your wages.

The Importance of Building Interpersonal and Business Relationships

Your workplace does not just focus on money—the bottom line. A major component of a successful company is the environment it develops, the tone it expresses, the atmosphere it creates. Successful companies know that effective communication, both written and oral, creates a better workplace. These "soft skills" make customers want to shop with you and employees work for you.

Your workplace communication reflects something about you. E-mail messages, letters, memos, or telephone skills are a photograph of you and your company. If you write well, you are telling your audience that you can think logically and communicate your thoughts clearly. When your writing is grammatically correct or when your telephone tone of voice is calm and knowledgeable, you exhibit professionalism to your audience. Workplace communication is an extension of your interpersonal communication skills. Coworkers or customers will judge your competence based on what you say and how you say it.

The Use of Workplace Communication to Develop a Corporate Image and Accountability

There has never been a more important time in the history of international business for corporations to improve their images and to prove their accountability. The early years of the twenty-first century have been bad times for many stockholders, employees, and corporate managers. Newspapers and televisions report on the indictments of former executives from Tyco, Enron, WorldCom, Adelphia, RiteAid, and ImClone.

To combat or avoid negative perceptions, a company must use effective workplace communication. Look at how several companies promote their corporate image to employees and stakeholders:

Microsoft's Web site advertises its "Integrity and honesty." The Web site says that Microsoft cares about "Accountability for commitments, results, and quality to customers, shareholders, partners, and employees." Ultimately, Microsoft says that "Customer trust [is achieved] through the quality of our products and our responsiveness and accountability to customers and partners" ("Mission and Values").

Black & Veatch, an engineering, construction, and consulting firm, states in its Web site that it "strives to be sincere, fair and forthright, treating others with dignity and respecting their individual differences, feelings and contributions" ("About Us").

The Importance of Collaboration in the Workplace

In school and in business, often you will work in teams. Making matters more challenging, these teams will be diverse, consisting of people from many different areas of expertise, as well as different ages, sexes, cultures, languages, and races (see Chapter 2 for more discussion of audience).

Many reports, proposals, presentations, and Web sites are collaborative. Teams consist of engineers, graphic artists, marketing specialists, subject matter experts, and corporate employees in legal, delivery, production, sales, accounting, and management. These team projects extend beyond the company. A corporate team will also work with subcontractors from other corporations. The collaborative efforts include communicating with companies in other cities and countries through teleconferences, faxes, and e-mail messages.

Why Teamwork Is Important

Diversity of Opinion. Who has all the answers? When we look at problems individually, we tend to see issues from limited perspectives—*ours*. In contrast, teams offer many viewpoints. This is especially true if a team is cross-functional, drawing its team members from many disciplines. For instance, if a team has members from accounting, public relations, customer service, engineering, and information technology, then that diverse group can offer diverse opinions.

Checks and Balances. Diversity of opinion also provides the added benefit of checks and balances. Rarely should one individual or one department determine outcomes. When a team consists of members from different disciplines, those members can say, "Wait a minute. Your idea will negatively impact my department. Let's stop and reconsider."

Broad-Based Understanding. If decisions are made by a small group of like-minded individuals, then those decisions might surprise others in the company. Surprises are rarely good. You always want buy-in from the majority of your stakeholders. An excellent way to achieve this is through team projects. When multiple points of view are shared, a company benefits from broad-based knowledge.

Empowerment. Collaboration gives many people from many disciplines an opportunity to provide their input. When groups are involved in the decision making process, they then have a stake in the project.

Strategies for Successful Collaboration

To collaborate successfully on workplace communication projects, follow these strategies:

Develop Your Team. Who will work together? You have two choices. Either select team members by discipline (engineers with engineers, accountants with accountants, marketing experts with other people in marketing, computer programmers with computer programmers, and so on), or choose diverse team members with different skills (teams composed of an engineer, an accountant, a computer programmer, a graphic artist, a marketing specialist, etc.). The first team profile draws strength from individuals with similar skills; the second team benefits from multiple perspectives.

Choose a Team Leader. Either your teacher or your boss selects a team leader, or your group might choose the person best suited to this role. Once this person is chosen or emerges, the delegation of authority begins. Your team leader might do the following:

- Assign duties. Who will research the material, who will write the various parts of the text, who will interview people, who will proofread and revise, and who will give the oral presentation?
- Create schedules (project milestones). When will the group meet, when will research and writing be due, when are revisions required, and when is the finished project due?
- Encourage group participation. How will conflicts among group members be resolved, and how will equal participation be achieved?

Identify the Problem. Groups often are formed to solve a workplace communication challenge. Perhaps corporate profits are sagging, and a sales letter needs to bring in new clients. Maybe people are uncomfortable, dissatisfied, or overworked, and a recommendation report will suggest solutions to these problems. To write the document, the team first must identify the problems that necessitate the written correspondence.

Determine Potential Improvements. Now is the time to propose the solutions to the problems. Focus on how these potential improvements will be achieved (strategies, timetables, milestones, personnel involved, costs incurred, facilities impacted) and how the company will benefit (improved sales, better employee morale, lower cost of business, satisfied customer base).

Choose the Communication Channel. Part of solving the problem is choosing the correct communication channel. What kind of document should you create? As a team, you need to decide whether to write a report, create an intranet form, build a Web site, write an e-mail message, or present your findings orally by using PowerPoint. Some communication channels are interchangeable; others are more exclusive. Select the best communication channel for your audience and purpose.

Breach the "Gaps." For your team to succeed, challenges must be faced. These could include problems caused by verbal or nonverbal communication, poor listening skills, and conflicts within the group. Other issues that create gaps between what your team hopes to achieve and its ability to do so include the following:

- Ergonomics. A room that is too cold or too hot; badly arranged meeting space (too small, too large, not enough seats, and so on).

- **Teammates with personal problems.** Drugs, health, or family conflicts.
- **Unskilled teammates.** Coworkers who are not aware of the team's rules, the team's goals, or teammates who do not have the correct computer skills or any other technology prerequisites.
- **Insufficient resources.** This could include finances, insufficient numbers of teammates, or lack of support from management.

With the help of your boss or your team leader or through collaboration with the group, you must breach the gaps to achieve successful communication.

Complete the Team Project. When the above challenges are overcome, it is possible for the team to accomplish its goal. This might entail writing, editing, revising, and printing the finished copy (of a proposal or Web site), developing the PowerPoint slides, or practicing the presentation.

How Can You Communicate Effectively in the Workplace?

Workplace communication is a major part of your daily work experience. It takes time to construct the correspondence, and your writing has an impact on those around you. A well-written memo, letter, report, or e-mail message gets the job done and makes you look good. Poorly written correspondence wastes time and creates a negative image.

Recognizing the importance of workplace communication does not ensure that your correspondence will be well written, however. How do you effectively write the memo, letter, or report? How do you successfully produce the finished product?

To produce successful workplace communication, approach writing as a process.

The Writing Process: An Overview

Effective writing follows a process of prewriting, writing, and rewriting. Each of these steps is sequential and yet continuous. The writing process is dynamic, with the three steps frequently overlapping.

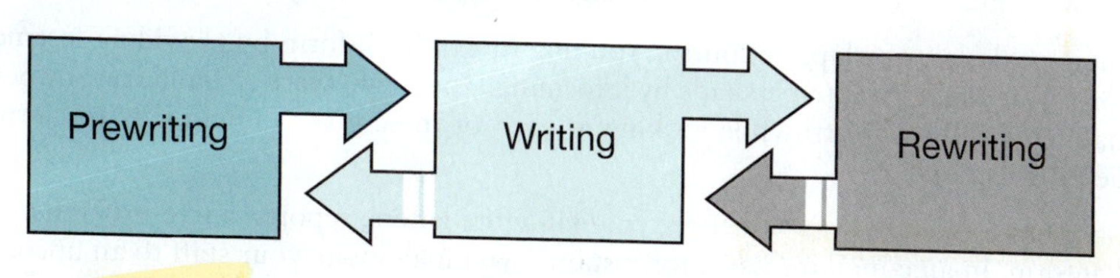

1. **Prewrite.** Prewriting encourages you to spend quality time, *prior* to writing the correspondence, generating information.
2. **Write.** Once you have gathered your data and determined your objectives, the next step is to state them. You need to draft your document. To do so, you should (a) *organize* the draft according to some logical sequence that your readers can follow easily and (b) *format* the content to allow for ease of access.
3. **Rewrite.** The final step, and one that is essential to successful writing, is to edit and rewrite your draft. This step requires that you revise the rough draft. Revision allows you to perfect your memo, letter, or report so you can be proud of your final product.

Prewriting

Prewriting, the first stage of the process, allows you to plan your communication. Through prewriting, you accomplish the following objectives.

Examine Your Purposes. Before you write the document, you need to know why you are communicating. Are you planning to write because you have chosen to do so of your own accord or because you have been asked to do so by someone else? In other words, is your motivation external or internal?

- **External Motivation.** If someone else has requested the correspondence, then your motivation is external. Your boss, for example, expects you to write a monthly status report, a performance appraisal of your subordinate, or a memo suggesting solutions to a current problem. Perhaps a vendor has requested that you write a letter documenting due dates, or a customer asks that you respond to a letter of complaint. In all of these instances, someone else has asked you to communicate.
- **Internal Motivation.** If you have decided to write on your own accord, then your motivation is internal. You need information to perform your job more effectively, so you write a letter of inquiry. You need to meet with colleagues to plan a job, so you write an e-mail message calling a meeting and setting an agenda. Perhaps you recognize a problem in your work environment, so you create a questionnaire and transmit it via the company intranet. Then, analyzing your findings, you call a meeting to report your findings. In these scenarios, you initiate the communication.

Determine Your Goals. Once you have examined why you are planning to communicate, the next step is to determine your goals in the correspondence or presentation. You might be communicating to

- inform an audience of facts, concerns, or questions you might have.
- instruct an audience by directing actions.
- persuade an audience to accept your point of view.
- build trust and rapport by managing work relationships.

These goals can overlap, of course. You might want to inform by providing instructions. You might want to persuade by informing. You might want to build trust by persuading. Still, it is worthwhile looking at each of these goals individually to clarify their distinctions.

Communicating to Inform. Often, you will write letters, reports, and e-mails merely to inform. In an e-mail message, for instance, you may invite your staff to an upcoming meeting. A trip report will inform your supervisor what conference presentations you attended or what your prospective client's needs are. A letter of inquiry will inform a vendor about questions you might have regarding her services. In these situations, your goal is not to instruct or persuade. Instead, you will share information objectively.

Communicating to Instruct. Instructions play a large role in workplace communication. You might need to tell employees under your supervision what to do. You might write an e-mail providing instructions for correctly following procedures. Your instructions could include steps for filling out employee forms, researching documents in your company's intranet data bank, using new software, or writing reports according to the company's new standards.

Communicating to Persuade. If your goal in writing is to change others' opinions or a company's policies, you need to be persuasive. For example, you might want to write a proposal, a brochure, or a flier to sell a product or a service. Maybe you will write your annual progress report to justify a raise or a promotion. Your goal in these cases is to persuade an audience to accept your point of view.

Communicating to Build Trust. Building rapport (empathy, understanding, connection, and confidence) is a very important component of your communication challenge. To maintain a successful work environment, you want to achieve the correct, positive tone in your writing.

Recognizing the goals for your correspondence makes a difference. Determining your goals allows you to provide the appropriate tone and scope of detail in your communication. In contrast, failure to assess your goals can cause communication breakdowns.

Consider Your Audience. What you say and how you say it is greatly determined by your audience. Are you writing up to management, down to subordinates, or laterally to coworkers? Are you speaking to a specialist (an expert in your field), a semispecialist (an audience with some knowledge about your field), or a lay audience (people outside your work environment)? You also must consider issues of diversity when you communicate. Each of these points is discussed in detail in Chapter 2.

Gather Your Data. Once you know why you are writing and who your audience is, the next step is deciding what to say. You can gather data in the following ways:

Answering the reporter's questions	Mind mapping
Brainstorming or listing	Outlining
Storyboarding	Creating organization charts
Flowcharting	Researching (see Ch. 7)

Each of these prewriting techniques is discussed in greater detail in Table 1.3.

Writing

Once you have gathered your data, determined your objectives, and recognized your audience, the next step is writing the document. Writing the draft lets you *organize* your thoughts in some logical, easy-to-follow sequence.

Organization. To avoid leading your readers astray, you need to organize your thoughts. As with prewriting, you have many organizational options, including the following:

- **Space (spatial organization).** You might use spatial organization when writing a technical description (see Ch. 8). A spatial organization describes an object from top to bottom, bottom to top, inside to outside, and so on.
- **Chronology.** Use chronology in instructions to show the sequence of steps. You also could use chronology in a trip or incident report to show the sequence of events.
- **Importance.** To emphasize that one point is more important than another, organize your content from most to least important or least to most important. This is useful in all written and oral workplace communication.
- **Comparison/contrast.** If you want to emphasize the similarities and differences of vendors, products, costs, locations, or personnel options, use comparison/contrast.

TABLE 1.3 PREWRITING TECHNIQUES	
Reporter's Questions By answering *who, what, when, where, why*, and *how*, you create the content of your correspondence.	***Who*** Joe Kingsberry, Sales Rep ***What*** Need to know • what our discount is if we buy in quantities • what the guarantees are • if service is provided onsite • if the installers are certified and bonded • if Acme provides 24-hour shipping ***When*** Need the information by July 9 to meet our proposal deadline ***Where*** Acme Radiators 11245 Armour Blvd. Oklahoma City, Oklahoma 45233 Jkings@acmerad.com ***Why*** As requested by my boss John, to help us provide more information to prospective customers ***How*** Either communicate with a letter or an e-mail. I can write an e-mail inquiry to save time, but I must tell Joe to respond in a letter with his signature to verify the information he provides.
Mind Mapping Envision a wheel. At the center is your topic. Radiating from this center, like spokes of the wheel, are different ideas about the topic. Mind mapping allows you to look at your topic from multiple perspectives and then cluster the similar ideas. This is a wonderful tool for visual learners.	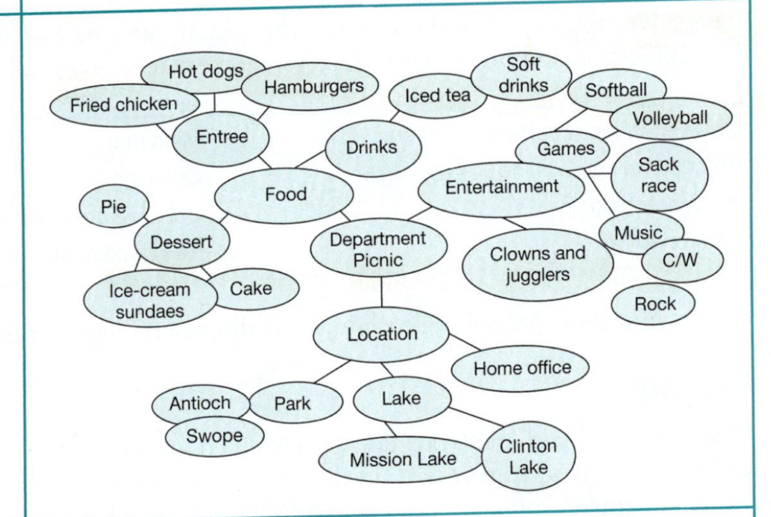
Brainstorming/ Listing Performing either individually or with a group, you can randomly suggest ideas (brainstorming) and then make a list of these suggestions. This method, which works for almost all kinds of communication, is especially valuable for team projects.	**Improving Employee Morale** • Before meetings, ask employees for agenda items (that way, they can feel empowered) • Consider flextime • Review employee benefits packages • Offer yearly awards for best attendance, highest performance, most cold calls, lowest customer complaints, etc. • Allow employee sharing for unused personal days/sick leave days • Let employees roll over personal days to the next calendar year • Include employees in decision making process • Add more personal days (as a trade-off for anticipated lower employee raises) *(continued)*

Table 1.3 Continued

Outlining This traditional method of gathering and organizing information allows you to break a topic into major and minor components. This is a commonly used, all-purpose prewriting tool.	**Topic Outline** 1.0 The Writing Process 1.1 Prewriting • Planning Techniques 1.2 Writing • All-Purpose Organizational Template • Organizational Techniques 1.3 Rewriting 2.0 Criteria for Effective Technical Writing 2.1 Clarity 2.2 Conciseness 2.3 Document Design 2.4 Audience Recognition 2.5 Accuracy
Storyboarding Storyboarding is a visual planning technique that lets you graphically sketch each page or screen of your text. This allows you to see what your document might look like.	**Brochure Storyboard** 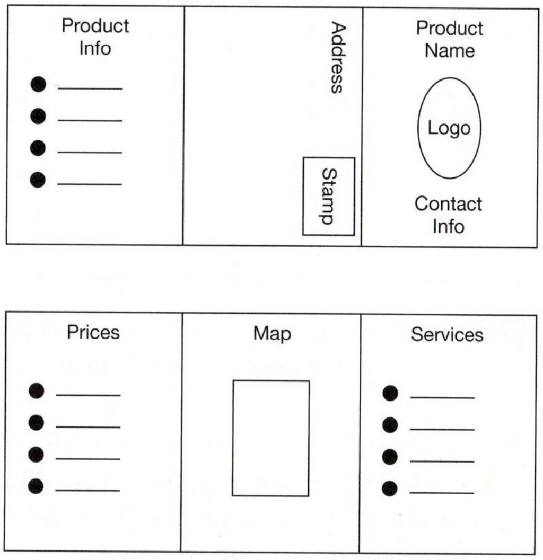 Product Info Address Stamp Product Name Logo Contact Info Prices Map Services
Organization Charts This graphic allows you to see the overall organization of a document as well as the topics to be discussed.	**Organization Chart for Web Site** 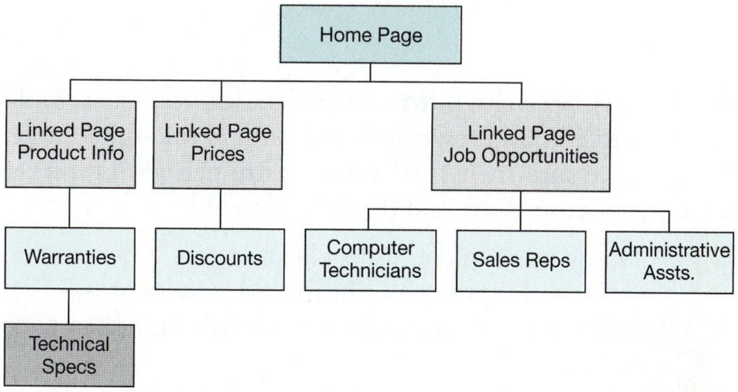 Home Page Linked Page Product Info Linked Page Prices Linked Page Job Opportunities Warranties Discounts Computer Technicians Sales Reps Administrative Assts. Technical Specs

(*continued*)

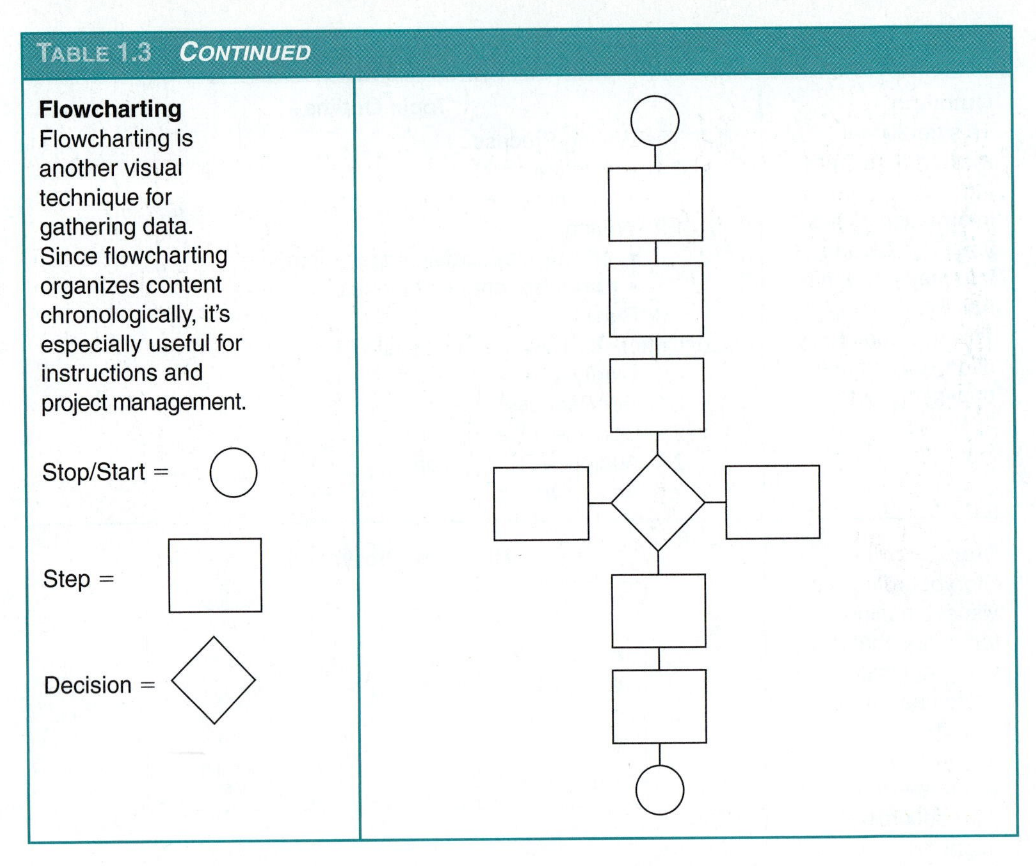

TABLE 1.3 CONTINUED
Flowcharting
Flowcharting is another visual technique for gathering data. Since flowcharting organizes content chronologically, it's especially useful for instructions and project management.
Stop/Start =
Step =
Decision =

- **Problem/solution.** Many types of workplace communication deal with business-related problems. To analyze these problems and suggest ways to resolve issues, use a problem/solution method of organization. First, highlight the problems. Next, show how these problems might be solved.

Formatting. You also must *format* your text to allow for ease of access. In addition to organizing your ideas, you need to consider how the text looks on the page. To invite your readers into the document, to make them want to read the memo, letter, or report, you need to highlight key points and break up monotonous-looking text. See Chapter 3 for details on visual communication.

Rewriting

After you have prewritten (to gather data for your text) and written your draft, your final step is to edit and rewrite. Good writers fine-tune, hone, sculpt, and polish their drafts to make sure their final versions are perfect. Editing and revision require that you look over your draft and do the following:

1. *Add* any missing detail for clarity.
2. *Delete* dead words and phrases for conciseness.
3. *Simplify* unnecessarily complex words and phrases to allow for easier understanding.
4. *Move* around information (cut and paste) to ensure that your most important ideas are emphasized.
5. *Reformat* (using highlighting techniques) to ensure reader-friendly ease of access.
6. *Enhance* the tone and style of the text.
7. *Correct* any errors to ensure accurate spelling, grammar, and content.

TABLE 1.4 CHARACTERISTICS OF GOOD WRITING

	Important	Extremely Important
Accuracy	12.2%	87.8%
Solid Spelling, Grammar, Punctuation	28.6%	71.4%

The National Commission on Writing 2005 survey emphasizes the importance of "accuracy" and "solid fundamentals," stating that good "[grammar is required and necessary] for effective writing" ("Writing: A Powerful Message" 19–20). Table 1.4 demonstrates what respondents to the survey define as characteristics of good writing.

When it comes to correct grammar, "Nearly 100 percent of respondents agree that accuracy, solid spelling, grammar and punctuation" are "extremely important" characteristics of good writing ("Writing: A Powerful Message" 19).

The Writing Process at Work

Following is a letter produced using the process approach to writing. The document, written by Buddy Ramos at ComputeToday, was a problem/solution follow-up letter to a sales representative.

Prewriting

Buddy received a phone call from an unhappy sales representative. The sales representative had not received a shipment of software and hardware on time, and the shipment was incomplete when it did arrive. While talking to the sales representative, Buddy took notes using the listing method of prewriting shown in Figure 1.2.

In addition to listing, Buddy used another prewriting technique—reporter's questions. The note tells us *who* the sales representative is (Beth); *what* her Social Security number, phone number, and sales area are; *what* her problem is (late and missing software and hardware); *where* the shipment originated (Denver); *how* much was ordered ($15,700); and *when* the shipment was due (two weeks ago). By writing down this list, Buddy is gathering data.

365-4559

Beth Fox
449-09-3441
Milwaukee, WI

1. two weeks ago
2. June $300 short
3. $700 software ordered
4. $15,000 hardware ordered
5. Denver
6. Split order

Figure 1.2 Listing

Figure 1.3 Listing Objectives

Send letter to sales rep

Send copy to manager

In letter
- Discuss problems encountered
- Show alternative method of shipment for better control

Call manager for further help if needed

After concluding his discussion with Beth, Buddy contacted his manager to decide what to do next. This time, Buddy wrote down a list of objectives, as determined by his manager, as shown in Figure 1.3.

The list again answers the reporter's questions: *what* to do (write a letter), *who* gets a copy (manager), *what* to focus on in the letter (we understand your problem; here is an alternative), and *why* to pursue the alternative (better control of shipment).

With data gathered and objectives determined, Buddy was ready to write.

Writing

First, Buddy wrote a rough draft, as shown in Figure 1.4.

Figure 1.4 Rough Draft

Dear Beth,

I appreciate your notifying me of the delay in delivery of your order. Your orders are coded to ship via Allied Shipping. Allied's stated shipping level for Milwaukee is next day. As you noticed, because of the way Allied ships merchandise, it is possible for a multiple carton order to become split. To avoid this problem, an alternative delivery is possible through a delivery service we use in your area. To do this, we need another delivery address. If you have any questions or need additional information, please contact me.

Writer's Insight

Buddy says, "When I wrote this first rough draft, I just wanted to get the information down while it was fresh in my mind. I didn't care too much about grammar or depth of detail. It was important for me to draft the letter quickly. I knew that my phone would be ringing all day, that I had meetings scheduled in the afternoon, and that other issues would arise. If I didn't get these thoughts on paper, I'd potentially forget something. Writing a perfect first draft wasn't my main goal. After all, I knew I'd edit and revise the letter later."

Rewriting

No writing is ever perfect. Every memo, letter, or report can be improved. Buddy was unsatisfied with his rough draft, so he edited and rewrote the letter (see Figure 1.5).

Based on his assessment, Buddy edited and revised the text one more time (see Figure 1.6) and mailed the letter to Beth.

Dear Beth,

Thank you for letting me know about the split deliveries of your Campaign 19 order. Our discussion last week gave me an opportunity not only to explain the situation but also to offer help.

Here's the way Allied works. Allied will deliver to your home. However, Allied sorts packages individually rather than as a group. That is, even though we send your packages to Allied as a unit, all under your name, Allied loads its trucks not by complete order but just as individual cartons. Because of this, occasionally one carton ends up on one truck while other cartons are shipped separately. You received such a split order. This is an inherent flaw in Allied's system.

Because we understand this problem, we have an alternative delivery service for you. Here is your option. Free of charge, you can have your order delivered by our delivery agent who does not split orders. Our agent, however, will deliver only within a designated area. All we need from you is an alternative address of a friend or relative in the designated delivery area.

I realize that neither of these options is perfect. Still, I wanted to share them with you. Your district manager now can help you decide which option is best for you.

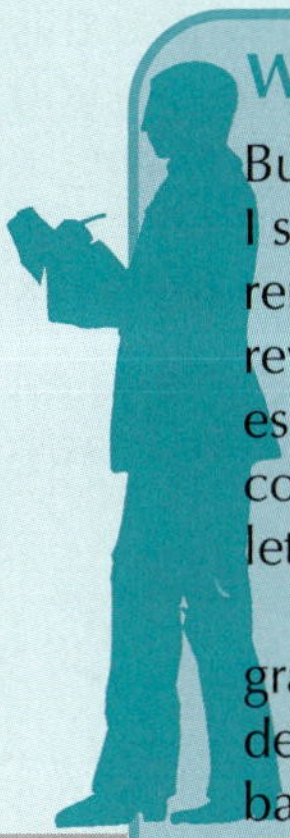

Figure 1.5 Revised Rough Draft

Writer's Insight

Buddy says, "After writing this rough draft, I set it aside for an hour or so. Then, I reread the letter and decided I needed to revise it again. In addition to adding letter essentials, such as a dateline, address, complimentary close, etc., I knew that the letter's tone had to change.

For example, in the second paragraph, I placed too much blame on our delivery service's 'inherent flaw.' That was bad business behavior. Then, in the last paragraph, I tried to shirk responsibility by telling Beth to contact our district manager for further help. Because I was Beth's primary contact, I needed to solve the problem for her, not dump it on the district manager."

Figure 1.6 Finished Letter

1818 Ram Drive
Castlehill, NJ 10023

September 21, 2006

Beth Fox
6078 Browntree
Milwaukee, WI 53131

Dear Beth:

Thank you for letting me know about the split deliveries of your Campaign 19 order. Our talk last week gives me an opportunity not only to explain the situation but also to offer help.

Allied will deliver to your home. However, Allied sorts packages individually rather than as a group. That is, even though we send your packages as a unit, all under your name, Allied loads its trucks not by complete order but by individual cartons. Because of this, occasionally one carton ends up on one truck while other cartons are shipped separately. You received such a split order.

Because we understand this problem, we have an alternative delivery service for you. Free of charge, you can have your order delivered by our delivery agent, whose system does not split orders. Our agent, however, delivers only within a designated area. If you would like this service, we need an alternate delivery address in Milwaukee.

Should you be unable to establish a different delivery address, we will work with Allied to ensure that you receive home delivery of your complete orders.

Sincerely,

Buddy Ramos
Client Executive

Writing Process Checklist

Each company you work for over the course of your career will have its own unique approach to writing business correspondence. Your employers will want you to do it their way. Company requirements vary. Different jobs and fields of employment require different types of correspondence. However, you will succeed in tackling any writing task if you have a consistent approach to writing. A process approach—prewriting, writing, and rewriting—will allow you to write any correspondence effectively. The Writing Process Checklist will give you the opportunity for self-assessment and peer evaluation of your writing.

Writing Process Checklist

__ 1. Have you used prewriting to gather data, determine objectives, and recognize your audience?

__ 2. Have you written a rough draft of your text, organizing your thoughts in some logical, easy-to-follow sequence, and formatting the text for ease of access?

__ 3. Have you edited and rewritten your rough draft using the following techniques:

- adding any missing detail for clarity?
- deleting dead words and phrases for conciseness?
- simplifying words for conciseness and easier understanding?
- moving information for emphasis?
- reformatting your text for ease of access?
- enhancing tone to build rapport?
- proofreading to correct errors?

CASE STUDIES

1. ComputeToday has decided to hire a manager of corporate communication. This person's job will be to

   - act as a liaison between ComputeToday's client executives, customers, and vendors.
   - prepare proposals (gather information through meetings and write the reports).
   - make oral presentations to potential clients.

   You have been asked to write the job description for this position, which will be advertised in the local newspaper, posted in ComputeToday's employment office, and posted online in the company's Web site: computetoday.com.

Assignment

Research the position of manager of corporate communication (either online, in the *Occupational Outlook Handbook*, or at a local company's work site). Find out the salary range for this job and the educational requirements. More importantly, based on what you have learned in this chapter about business communication, what additional job responsibilities will this position entail? What skills should this manager of corporate communication have? Based on your decisions, write the job description.

2. Amir Aksarben works in corporate communication for Prismatic Consulting Engineering. He has had the job for two months. He needs to respond to a request for proposal (RFP), posted by Oceanview City Council. The RFP is asking for bids to improve the City of Oceanview's flood control.

   Before Amir can respond to the RFP, he needs to gather more information, as follows:

   - Amir needs to communicate with Oceanview, inquiring about specific problems the city is encountering and their timeframe for making a decision. Amir's contact at Oceanview is Sally Howser, who went to college with Amir and worked with him on several school-related team projects.
   - He needs to communicate with his boss, Randy Towner, the owner and founder of Prismatic, determining what other projects Prismatic is engaged in, which engineers and architects Randy would like to assign to the construction crew for Oceanview's project, and which other Prismatic employees Randy would like to have on the proposal writing team.
   - Amir must ask his colleagues in other Prismatic departments (engineering, accounting, and drafting) for input regarding the proposal's content (timeframe to complete the job, costs, and schematics). All of these colleagues have worked on similar projects in the past.
   - He needs to tell his newly hired summer intern to start making phone calls to the city and county permit regulators to ensure that Prismatic has the appropriate work permits on file.

Based on the above information, answer the following questions, explaining your decisions:

a. Which communication channel should Amir use in each of the above instances and why?
b. Based on the different audiences, what are Amir's communication goals (inform, instruct, persuade, or build rapport)?
c. What is Amir's motivation for each of his communication activities (internal or external)?

3. Electronic City is a retailer of DVDs, televisions, CDs, computer systems, cameras, telephones, fax machines, and printers. Electronic City needs to create a Web site to market its products and services. The content for this Web site should include the following:

Prices	Store hours	Warranties	Service agreements
Job opportunities	Installation fees	Extended holiday hours	Discounts
Technical support	Product information	Special holiday sales	Delivery fees

Assignment

Review the list of Web site topics above. Using an organizational chart, decide how to group these topics. Which will be major links on the Web site's navigation bar? Which will be topics of discussion within each of the major links?

Once you have organized the links, sketch the Web site by creating a storyboard.

4. You are the special events planner in the marketing department at ***Thrill-a-Minute Entertainment Theme Park.*** You and your project team need to plan the grand opening of the theme park's newest sensation ride—*The Horror*—a wooden roller coaster that boasts a 10° gravitational drop.

   What activities should your team plan to market and introduce this special event?

Assignment

Using at least three of the prewriting techniques discussed in this chapter, gather ideas for a day-long event to introduce *The Horror.* Report your findings as follows:

a. In a brief memo or e-mail message to your marketing department boss, explain why you are writing, give options for the event, clarify which techniques you used to gather ideas, and sum up by recommending the best marketing approaches.
b. Write an e-mail to your teacher providing options for the event and explaining which techniques you used to gather data.

INDIVIDUAL AND TEAM PROJECTS

Complete the activities by doing the following:

- **Oral Presentations.** Give a three- to five-minute briefing to share with your classmates the results of your findings.
- **Written.** Write a brief (1/2 page) memo or e-mail to your teacher highlighting your findings.

1. **Visit employees from your college or university's business services.** These could include payroll, administrative computing, human resources, accounts payable, benefits, budget, catering, the bookstore, printing services, records, admission, loans, and so forth. After interviewing these employees, respond to the following:
    a. How much time do the employees of your college or university spend writing and communicating orally to students, coworkers, subordinates, and management?
    b. Ask employees from your college or university what they have written at work this week.
        - What types of communication channels did they use (e-mail, letters, memos, reports, Web pages, and so on)? Create a table showing the amount of usage for each communication channel.
        - What were their goals in writing (inform, instruct, persuade, build rapport)?
        - What were their motivations for writing (internal or external)?
2. **Visits to professionals.** Visit professionals from your area of interest or degree program. Interview these employees to learn the following:
    - What they have written at work
    - Which communication channels they use
    - The purposes of their communication activities
3. **Self-evaluation.**
    a. Calculate how much time you spend writing and communicating orally with customers, coworkers, subordinates, management, or colleagues, either on the job or in a school-related organization (club, fraternity, sorority, team, etc.). What percentage of your job or organizational activity involves communication? Make a table or pie chart to show your time spent communicating.
    b. Make a list of what you have written at work or for your school-related organization this week. What types of communication did you write, who were your audiences, and what were the goals of your communication? What percentage of your job or organizational activity involves written communication?
    c. What communication channels have you used for workplace communication or a school-related club this week? Did you use voice mail, faxes, e-mail, letters, reports, Web searches, teleconferencing, etc.?
4. **Your career choice.** Research your career choice or other potential career opportunities. To do so, visit your college or university's career placement service, look at newspaper ads, visit with human resource managers, network with employees already working in the field, or go online to search Internet job sources. Then, determine what types of communication skills are required for your potential job.

Problem Solving Think Pieces

1. In an interview, a company benefits manager said that she spent over 50 percent of her workday on communication issues. These included the following:

   - Consulting with staff, answering their questions about retirement, health insurance, and payroll deductions
   - Meeting weekly with human resources (HR) colleagues
   - Collaborating with project team members
   - Preparing and writing quarterly reports to HR supervisors
   - Teleconferencing with third-party insurance vendors regarding new services or costs
   - E-mailing supervisors and staff in response to questions
   - Calling and responding to telephone calls
   - Faxing information as requested
   - Writing letters to vendors and staff to document services

   Though she had to use various methods of both written and oral communication, the communication channels each have benefits and drawbacks. E-mailing, for example, has pluses and minuses (convenience over depth of discussion, perhaps). Think about each of the communication options above. Using the following matrix, list the benefits of each particular type of communication versus the drawbacks.

Communication Channels	Benefits	Drawbacks	Possible Solutions
One-on-one discussions			
Group meetings			
Collaborative projects			
Written reports			
Teleconferences			
E-mail			
Phone calls			
Faxes			
Letters			

2. Practicing Prewriting Techniques

   Take one of the following topics. Then, using the suggested prewriting techniques, gather data.

   - **Reporter's questions.** To gather data for your resume, list answers to the reporter's questions for two recent jobs you have held and for your past and present educational experiences.
   - **Mindmapping/clustering.** Create a mindmap for your options for obtaining college financial aid.
   - **Brainstorming/listing.** List five reasons why you have selected your degree program, why you have chosen the college or university you are

attending, why you have pledged your fraternity or sorority, or joined a college organization.

- **Outlining.** Outline your reasons for liking or disliking a current or previous job.
- **Storyboarding.** If you have a personal Web site, use storyboarding to depict graphically the various screens. If you don't have such a site, use storyboarding to depict graphically what your site's screens would include.
- **Creating an Organization Chart.** Use an Organization Chart to show the hierarchical structure of your place of business.

3. Determining Your Goals for Communicating

   When you communicate, you will do so to *persuade*, *instruct*, *inform*, or *build rapport*. From the following list, determine the goals for the communication.

   a. You write a sales letter about your home-owned painting and wallpapering company.
   b. As a supervisor, you write an e-mail to motivate a team member or subordinate who is having trouble accomplishing a task.
   c. You are creating a Web site for your company. It will provide job listings, corporate contacts, details about the company's products and services, and warranty information.
   d. As a mid-level manager, you write an e-mail to an employee, sharing good news you have just received from your supervisor who agreed to give the employee a raise.
   e. Your department needs to learn new techniques for performing a task. You will write an e-mail to your supervisor asking for funding and permission to attend a conference where these new techniques will be explained.

4. Determining How to Provide the Content

   When you communicate in the workplace (either in writing or orally), you have many communication channels to choose from. Should you write an e-mail, a memo, or a letter? Will you write a formal report? Should you create a brochure or a Web site? Should you meet with an individual, in a larger group, or via a videoconference? Is a fax appropriate?

   From the following topics, determine which communication channel you should use and explain your choice.

   - You need to ask your boss a question about a potential contract with a vendor, and you need the answer now. Your boss is in New York, and you are in Sacramento.
   - You need your boss's signature on a vendor contract, and you need it now. Your boss is in New York, and you are in Sacramento.
   - As boss, you need to invite your 15 department employees to a meeting for next week. You want to give them the date, time, location, and agenda. In addition, you want them to review a competitor's Web site before they meet with you.
   - Your company just received a flawed shipment of goods from a vendor. You need to lodge a formal complaint about this shipment since the flawed goods kept you from meeting a deadline and could lead to a lawsuit.

- Your department might need to make significant changes in employees, job responsibilities, and work procedures. All employees need to know about these potential changes as well as provide their own suggestions for improvement.

Web Workshop

1. How important is communication in the workplace? Go online to research this topic. Find five Web sites that discuss the importance of communication in the workplace, and report your discoveries to your teacher and/or class. To do so, write a brief report, memo, or e-mail message. You could also report your information orally.
2. Access American College Testing (ACT) information about workplace communication assessments at http://www.act.org/workkeys/assess/bus_writ/index.html. This site provides information about varying skill levels of workplace communication, writing examples, and a scoring guide to help you assess writing in the workplace. Read the information in this Web site and discuss your findings in small groups. Then, report to your teacher, via an e-mail message, what you have learned from this Web source.

Essential Goals of Successful Workplace Communication

OBJECTIVES

When you complete this chapter, you will be able to do the following:

1. Achieve clarity in workplace communication.
2. Write concisely, limiting the length of words, sentences, and paragraphs.
3. Proofread your correspondence to ensure accuracy.
4. Recognize different types of workplace audiences, including levels of knowledge and issues of diversity.
5. Involve the audience in your correspondence by emphasizing "you usage."
6. Achieve ethical workplace communication.
7. Test your knowledge of essential goals of successful workplace communication through end-of-chapter activities:
   - Case Studies
   - Individual and Team Projects
   - Problem Solving Think Pieces
   - Web Workshop

COMMUNICATION AT WORK

In the following scenario, Sharon Myers, Grunge Groovies' chief executive officer, asks her Web master Peter Tsui to correct his rough draft, addressing the needs of its intended audience and making the text clear, concise, accurate, and ethical.

Grunge Groovies is going online. This new Web-based shopping outlet will sell vintage clothing from the 50s through the 90s. The intended audience will focus primarily on high school students, college freshmen and sophomores, and clothing collectors. Market research has been completed, capital venture has been secured, the product is warehoused, and a Web development team has been formed. Now, it is time to create the Web site.

Peter Tsui, from Grunge Groovies' corporate communication department, has been assigned to write much of the text for the Web site. He wrote the following rough draft for the Web site's home page. His goal was to give an overview of the company.

Grunge Groovies—Who We Are and What We Offer!

Welcome to our site. Grunge Groovies offers you a plethora of antique accessories and accoutrements. If you want it, we got it! Take a gander at our gargantuan selection of various eras' clothing and appurtenances, including shirts, blouses, pants, shorts, suits, jackets, shoes, boots, as well as other paraphernalia like scarves, necklaces, bracelets, earrings, rings, hats, and you name it. We have clothing and other things from the 50s to the 90s. Our 50s' goodies covers the "beats" to the "blues brothers" clothing. We have black suits, black skinny ties, as well as black berets and turtlenecks. Our 60s' hippie stuff is way cool, such as wide, hallucinogenic ties, Nehru jackets, short-short mini-dresses, long-long granny dresses, way-out-there boots, peace symbols, fringe belts and vests, etc, etc, etc. Our 70s' clothing and items include all kinds of disco clothing including leisure suits and everything Ban-lon plus skin tight pants, shirts, white belts and boots, we even have disco dance hall balls and posters for your own disco dance floor. Our 80s' clothing focuses on what we call "business-nerd" including pocket protectors, wing-tip shoes, horned rim glasses (with the nose piece already pre-taped), we also cover the "Flash Dance" clothes like leg warmers and off-the-shoulder sweatshirts. Finally, our 90s' clothing includes everything from early 90s' grunge (raggedy cut-off shorts and flannel plaid shirts) to late 90s' hip-hop clothes and add-ons (oversized "bling bling" jewelry, sports jerseys, hooded sweatshirts, and fancy club duds). Return with us now to those thrilling days of yesteryear by purchasing accessories from *Grunge Groovies* By the way, we offer much better quality than our competitors. And for you investors, look at how our sales have gone through the roof:

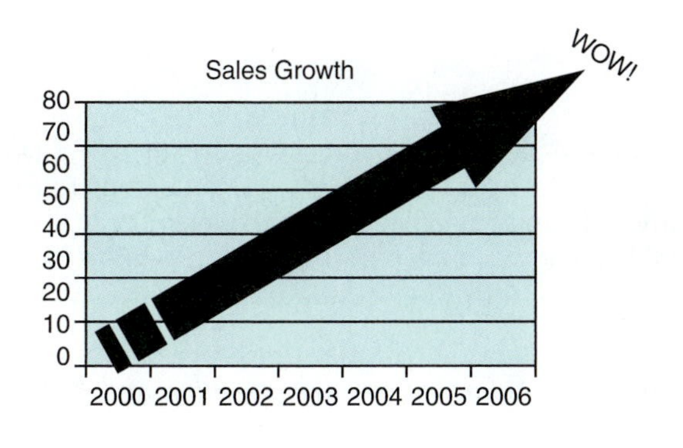

Writer's Insight

Sharon says, "Some of this word usage is too extravagant for our intended audience—high school and college students. In contrast, other words are too remedial. I'm also concerned about the length of the sentences.

Most importantly, I'm worried about the ethical correctness of Peter's assertions and the validity of the graphic. Yes, our sales have gone up, but not quite as dramatically as this graphic suggests. Yes, our product quality is good, but can we say it's 'better . . . than our competitors' without proof?

This text needs revision. I'm going to ask Peter and his team to work on clarity, conciseness, grammar, and attention to audience. I'll ask our legal staff about the ethics of some of Peter's claims."

ADDING INFORMATION TO ACHIEVE CLARITY

Sharon Myers, CEO of Grunge Groovies, says, "The ultimate goal of effective workplace communication is clarity." If you write a memo, letter, or report that is unclear to your readers, what have you accomplished? You have wasted time. If your readers must write you a follow-up inquiry to determine your needs, this wastes *their* time. Once you receive the inquiry, you must rewrite your correspondence, trying to clarify your initial intentions. You have now written twice to accomplish the same goal. This wastes *your* time.

To avoid these time-consuming endeavors, write for clarity. To accomplish this goal, answer the *reporter's questions*, and focus on *specificity.*

Answering the Reporter's Questions

To write clearly, answer the *reporter's questions* (*who, what, when, why, where,* and *how.*) Following is an example of vague writing:

before

Vague Sentence

Recently, we attended a conference to learn about HIPAA accountability.

This sentence is not informative. When was "recently"? Who is "we"? Which "conference" did you attend? What exactly is "HIPAA"? To clarify, you need to add the missing details.

A revision reads as follows:

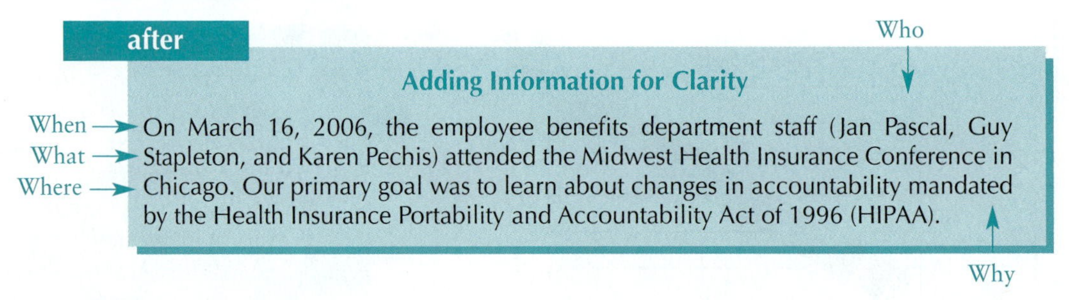

after
Who
Adding Information for Clarity
When
What
Where
On March 16, 2006, the employee benefits department staff (Jan Pascal, Guy Stapleton, and Karen Pechis) attended the Midwest Health Insurance Conference in Chicago. Our primary goal was to learn about changes in accountability mandated by the Health Insurance Portability and Accountability Act of 1996 (HIPAA).
Why

Answering the reporter's questions, *who, what, when, where, why,* and *how*, solves the problem.

Providing Specificity

A second way to achieve clarity is to add specific details to your communication. In your draft, you might have used words such as *substantial*, *several*, *near*, *many*, *few*, *recently*, *some*, *often*, or *soon*. These connotative words convey an impression. However, they do not provide quantifiable facts. How "substantial" is "substantial"? How "often" is "often"?

Look at the following example of vague writing caused by unclear adjectives.

before

Activity Report Draft

Our latest attempt at manufacturing metal backpack clasps has led to some positive results. We spent several hours in Dept. 15 trying different machine settings and techniques. Several good parts were molded using two different sheet thicknesses. Here is a summary of the findings.

First, we tried the thick sheet material. At high heats, this thickness worked well.

Next, we tried the thinner sheet material. The thinner material is less forgiving, but after adjustments we were making good parts. Still, the thin material caused the most handling problems.

Vague, connotative words are underlined in this activity report.

The engineer who wrote this report realized that it was unclear. To solve the problem, she rewrote the report, this time replacing the vague words and adding more specific information.

after

Activity Report Revision

During the week of 10/4/06, we spent approximately 12 hours in Dept. 15 trying different machine settings, techniques, and thicknesses to manufacture metal backpack clasps. Here is a report on our findings.

.030″ Thick Sheet

At 240°F, this thickness worked well.

.015″ Thick Sheet

This material is not flexible, but after decreasing the heat to 200°F, we could produce 15 percent more malleable parts. Still, material at .015″ causes handling problems.

Notice how the report's vague words have been replaced by specific information:

before: Vague Words	after: Specific Information
latest	the week of 10/4/06
some	15 percent more
several	approximately 12 hours
thick	.030″
high heats	240°F
thinner	.015″

By answering reporter's questions and by specifying, you clarify your workplace communication.

Deleting Dead Words and Phrases to Achieve Conciseness

A second essential goal of workplace communication is conciseness: brevity achieved through simple words, short sentences, and short paragraphs. A lack of conciseness can cause problems for your audience. For example, read the following sentence:

before

On two different occasions, I have made an investigation of your neighbor's adjacent property and have come to the conclusion that your fence has been incorrectly installed, extending beyond the aforementioned property line, and that the said fence must be modified by the end of this month to ensure that you avoid the possibility of legal action.

The sentence is clear; it answers reporter's questions and provides specific detail. However, it is bad writing. Though the writer has provided valuable information, the writing is wordy. This same sentence could be written concisely as follows:

after

Twice, I investigated your neighbor's property. Your fence incorrectly extends beyond the property line. To avoid a lawsuit, please move the fence by July 31.

The original sentence is 57 words long. The revision consists of three sentences, totaling 25 words. Each sentence in the revision averages about eight words. Though you need to be clear in your communication, you also must be concise for at least two reasons:

- **Space.** The length of your communication makes a difference. If your writing is too long, it might not "fit in the box." Most people, for example, want you to limit your resume to one page. That's a box. A PowerPoint screen limits you to only six or seven lines of readable text. An e-mail screen limits you to only 20 or so lines before the reader must scroll. E-mail messages or Web sites accessed on a cell phone will fit in an even smaller box, a monitor that perhaps measures only 1 inch by 1 inch.

  To ensure that your workplace communication is the appropriate length for your communication channel, delete unnecessary words and phrases.
- **Readability.** Readability is the reading level of a document. It defines whether you are writing at a 5th grade, 9th grade, or 12th grade level, for example.

  Readability is important for the following reasons. Not everyone graduates from college. Only about 33 percent of Americans graduate from college. Therefore, if you are writing at a college level, you might create readability challenges for approximately 67 percent of your audience (Mollison).

  College graduates *do not* read at a college level. College graduates read at approximately a 12th grade level, since "adults and children read at least one or two grade levels below their last school grade completed" (Hilts and Krilyk).

The *average* reading level in the United States is 8th to 9th grade. Twenty percent of the population reads at the 5th grade level and below.

Given the facts about readability, many businesses ask their employees to *write between 6th to 8th grade levels*. Regardless of your work environment, you will need to communicate to the general public. You can help most readers understand your content by writing shorter sentences.

A popular tool for determining readability is Robert Gunning's Fog Index. Gunning bases readability findings on the length of sentences and the length of words. (For more information on how to measure readability, see Figure 2.1.)

Limiting Sentence Length

To accomplish conciseness and improved readability, limit the length of your sentences. The *GNOME Documentation Style Guide* (GNOME—a Unix and Linux desktop suite and development platform) provides the following information about readability and sentence length:

Readability	Length of Sentence
Very easy to read	Average sentence length is 12 words or less.
Plain English	Average sentence length is 15 to 20 words.
Extremely difficult to read	Average sentence length exceeds 20 words.

Sentences that are approximately 10 to 12 words long are easy for your audience to read and understand. In contrast, sentences longer than 20 words can create readability challenges for your audience.

To achieve conciseness, use the following techniques for deleting dead words and phrases.

Deleting "Be" Verbs

"Be" verbs include conjugations of the verb "be": *is, are, was, were, would, will, been,* and *am.* Often, these verbs create wordy sentences. For example, look at the following sentences and the revisions:

before

Wordiness Caused by "Be" Verbs

Bill _is_ of the opinion that stock prices will decrease.

Jorge and Stacy _are_ planning to fax new invoices tomorrow.

Joan _had been_ hoping to move into her new office complex today.

I wonder if you _would be_ so kind as to provide discount information about your sales.

The senior staff accountant _is_ responsible for assuring that all liabilities _are_ reported on a yearly basis.

after

Deleting for Conciseness

Bill thinks stock prices will decrease.

(**Note:** the sentence deletes one "be" verb—*is*—but keeps another—*will.*)

Jorge and Stacy plan to fax new invoices tomorrow.

Joan hoped to move into her new office complex today.

Please provide discount information about your sales.

The senior staff accountant reports liabilities annually.

Figure 2.1
Gunning's Fog Index

How to Measure Readability

Longer words and sentences require a higher grade level of reading ability. If your words and sentences are excessively long, you might create what readability experts call "fog"—an impenetrable wall of haze caused by too many words.

A popular readability measure is Robert Gunning's **Fog Index**. Gunning's Fog Index, based on sentence length and word length, is a mathematical way of calculating reading grade levels. He suggests the following:

1. Count the number of words in successive sentences. Once you reach approximately 100 words, divide these words by the number of sentences. This will give you an average number of words per sentence.
2. Next, count the number of long words within the sentences you have just reviewed. Long words are those with *three* or *more syllables.* Do not count proper names, like Beyonce or Christopher Columbus; long words created by combining shorter words, such as *chairperson*; or three-syllable verbs created by –ed or –es endings, such as *arranges*.
3. Finally, to determine the fog index, add the number of words per sentence and the number of long words. Then multiply your total by 0.4.

To test this system, read the following paragraph (we have underlined the multisyllabic words):

> In order to facilitate an efficient report and fuel thought processes prior to the June 25 orientation, I want you to provide a brief overview of discussions recently carried out at the director and manager levels within the process. To accomplish this goal, please prepare to supply a readout of your findings and recommendations to the officer of the Southwest Group at the completion of your study period. As we discussed, the undertaking of this project implies no currently known incidences of impropriety in the Southwest Group, nor is it designed specifically to find any. Rather, it is to assure ourselves of sufficient caution, control, and impartiality when dealing with an area laden with such potential vulnerability. I am confident that we will be better served as a company as a result of this effort.

The paragraph is composed of 135 words in 5 sentences. Thus, the average number of words per sentence is 27. The paragraph contains 26 multisyllabic words. The Fog Index is as follows:

27 (average words per sentence)
+26 (multisyllabic words)
53 (total)

Figure 2.1
continued

53 (total)
× .04 (fog factor)
21.2 (Fog Index)

What does a Fog Index of 21.2 mean? Look at the following table. It gives Fog Index scores, grade levels, and readability levels. A senior in college would equal a Fog Index of 16. In contrast, 21.2 is a doctoral level of writing, far above the danger line.

Gunning Fog Index and Reading Level

	Fog Index	Grade Level	Readability Level
	17+	Graduate School	No popular magazines have readability levels this high
	16	College Senior	
	15	College Junior	
	14	College Sophomore	
	13	College Freshman	
Danger Line →	12	High School Senior	*Atlantic Monthly*
	11	High School Junior	*Time* and *Newsweek*
	10	High School Sophomore	*Reader's Digest*
	9	High School Freshman	*Good Housekeeping*
	8	Eighth Grade	*Ladies Home Journal*
	7	Seventh Grade	Modern romance novels
	6	Sixth Grade	Comic strips

Source: "The Fog Index"

Using Active Voice Versus Passive Voice

Sometimes, "be" verbs create passive voice sentences, as follows:

> It _has been_ decided that Joan Smith _will_ head our sales department.

The preceding sentence is written in the passive voice (the primary focus of the sentence, Joan Smith, is acted on rather than initiating the action). Passive voice can cause two problems.

1. Passive constructions are often unclear. Who decided that Joan Smith will head the department? To solve this problem and to achieve clarity, replace the vague indefinite pronoun "it" with a precise noun: "Larry named Joan Smith head of the sales department."

2. Passive constructions are often wordy. Passive sentences require helping verbs, such as *has been*. Revise the sentence to read "Larry named Joan Smith head of the sales department." This omits the helping verb *has been* and the "be" verb *will*.

The revision ("Larry named Joan Smith head of the sales department") is written in the active voice. When you use the active voice, your subject (Larry) initiates the action.

before

Wordiness Caused by Passive Voice

Overtime *is* favored by hourly workers.

The monthly report *must be* complete by June 19.

Your presence *is* requested at a meeting planned for Friday in Conference Room C at 10:30 a.m.

Surveys *have been* taken by the accountant to ensure accuracy.

The CPA exams *were passed* with a score of 92 by two-thirds of the 2005 graduating class.

after

Using Active Voice for Conciseness

Hourly workers favor overtime.

Complete the monthly report by June 19.

Please attend the 10:30 a.m. Friday meeting in Conference Room C.

The accountant's survey ensures accuracy.

Two-thirds of the 2005 graduating class scored 92 on the CPA exams.

When to Use Passive Voice. Occasionally, you can and should use passive voice. Passive voice is correctly used when the individual performing the action is unimportant or unknown. The following example correctly uses passive voice: "The new software was oversold by the salesperson who guaranteed ease of use." In this sentence, the individual, *salesperson*, is not as important as the inanimate object, *software*.

Deleting the Expletive Pattern

When you begin sentences with "there" or "it," you create the "expletive pattern" of sentence structure (*There* is, are, was, were, will be; *It* is, was). This is another way that passive voice occurs.

Notice how the expletive pattern again uses "be" verbs. The expletives ("there" and "it") create wordy sentences. Consider the following:

before

There are three people who will work for Acme.

It has been decided that ten engineers will be hired.

after

Three people will work for Acme.

Ten engineers will be hired.

Omitting Redundancy

When you are redundant, you say the same thing twice. Notice how the following examples show redundancies and a revised, less wordy version.

before	after
The year of 2006	2006
The month of December	December
The sum of $1,000	$1,000
Results so far achieved	Results
Regular monthly status reports	Monthly status reports
Collaborated together	Collaborated
New innovation	Innovation
Consensus of opinion	Consensus

Avoiding *Shun* Words

Another way to write more concisely is to avoid words ending in *-tion* or *-sion*—*shun* sounds. Look at the following *shun* words and their concise versions.

before	after
I would like you to make prepara*tions* for attending the conference.	Please prepare to attend the conference.
Came to the conclu*sion*	Concluded (or decided, ended, stopped)
With the excep*tion* of	Except for (or but)
Make revi*sions*	Revise
Investiga*tion* of the	Investigate (or look at, review, assess)
Consider implementa*tion*	Implement (or use)
Utiliza*tion* of	Use

Avoiding Camouflaged Words

Camouflaged words are similar to *shun* words. In both instances, a keyword is buried in the middle of surrounding words (usually helper verbs or unneeded prepositions). For example, in the phrase *make an amendment to*, the key word *amend* is camouflaged behind unnecessary words. Once you prune away these unneeded words, the key word *amend* (or *change, revise, fix*) is left, making the sentence less wordy.

Camouflaged words are common. Spotting and deleting them will make your sentences more concise. Here are some examples and their concise versions.

before	after
Make an *adjust*ment of	Adjust (or revise, alter, change, edit, fix)
Have a *meet*ing	Meet
*Thank*ing *you* in advance	Thank you
For the purpose of *discuss*ing	Discuss
Arrive at an *agree*ment	Agree
At a *later* moment	Later

Limiting Prepositional Phrases

Prepositions can be important words in your communication. They help you convey information about time and place. Common prepositions include the following:

about	below	in	through
above	beneath	inside	throughout
across	beside	into	to
after	between	like	toward
against	by	near	under
along	due	of	until
around	during	on	up
at	except	outside	upon
before	for	over	with
behind	from	since	without

Occasionally, *prepositional phrases* create wordy sentences. A prepositional phrase includes a *preposition* and a *noun* or *pronoun* that serves as the object of the preposition. For example, "at a later moment" is a prepositional phrase. It includes the preposition *at* and the noun *moment*. This prepositional phrase is wordy and can be revised to read "later." Following are other examples of wordy prepositional phrases:

before	after
in order to purchase the computer . . .	to buy the computer . . .
He spoke *at a rapid rate*.	He spoke rapidly.
She wrote *with regard to* the meeting.	She wrote about the meeting.
in the first place, . . .	First
I will call *in the near future*.	I will call soon.
I am *in receipt of* your check.	I received your check.
On two different occasions, we met.	We met twice.
The manager *of personnel* was hired.	The personnel manager was hired.
She is *in agreement*.	She agrees.
The company is *in the process of* cutting costs.	The company is cutting costs.
We are laying off employees *due to the fact that* the economy is bad.	We are laying off employees because the economy is bad.

Simplifying Words for Conciseness and Easier Understanding

To achieve clarity *and* conciseness, you also should simplify your word usage. Use easy-to-understand words. Sometimes you can accomplish this by limiting the length of your words. Long, multisyllabic words tend to cause listeners and readers problems.

A good example of a multisyllabic word is the word "multisyllabic." It consists of five syllables: *mul-ti-syl-lab-ic*.

The *GNOME Documentation Style Guide* provides the following information about readability and word length:

Readability	Word Length
Very easy to read.	No words of more than two syllables.
Plain English.	Average word has two syllables.
Extremely difficult to read.	Words with more than two syllables.

You cannot nor should you avoid words such as "accountant," "engineer," "telecommunications," "computer," or "nuclear." Though these words have more than two syllables, they are not difficult to understand. When you need longer words, use them. However, try to avoid old-fashioned, legalistic words, like "pursuant," "accordance" and "aforementioned." Too often, writers and speakers use these words to *impress* their audience. In contrast, communicators should *express* their content clearly and simply.

before

Wordy Sentence

I would like you to take into consideration the following points, which I know will assist you in better applying new OSHA rules and regulations currently burdened by the need to execute all data manually and on paper rather than through standardized, electronic transmissions.

after

Simplifying for Conciseness

Please consider the following points. This will help you apply new Occupational Safety and Health Administration (OSHA) rules by submitting data online instead of having to type all text on separate forms.

The "before" sentence is 44 words long. As noted earlier in this chapter, most people would prefer to read sentences of about 10 to 12 words. Also, the sentence contains 10 words with more than three syllables—*multisyllabic* words: "consideration," "following," "applying," "regulations," "currently," "execute," "manually," "standardized," "electronic," and "transmissions." None of these words is challenging individually. Still, the mass of syllables makes the sentence hard to understand.

In the "after" revision, the 44 words have been reduced to 32 words. Also, the original long sentence has been cut into two smaller sentences. Finally, the remaining two sentences contain only four multisyllabic words ("consider," "following," "submitting," and "separate"). The conciseness saves you and your reader time and makes the information easier to understand.

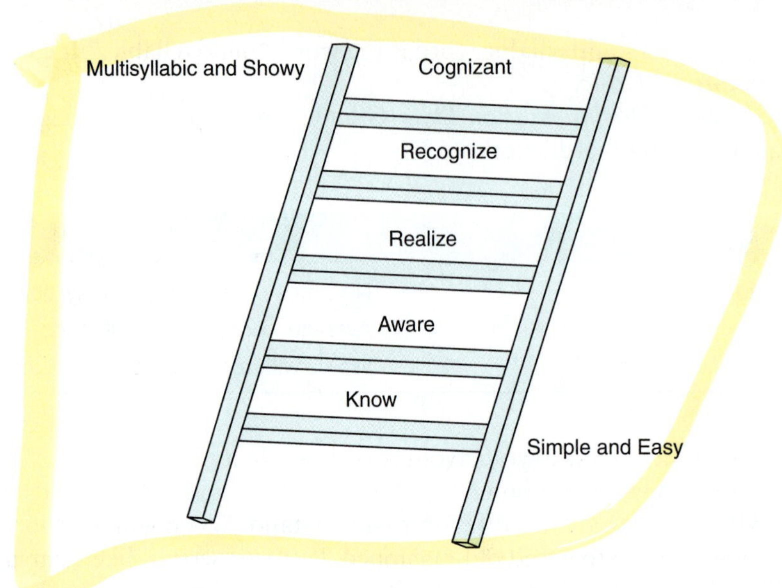

Figure 2.2 Ladder of Abstraction

Following is a list of long words that can be simplified for conciseness and easier understanding:

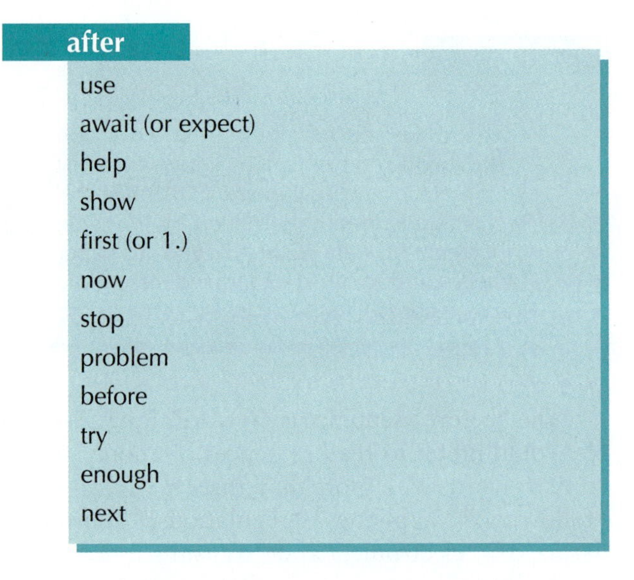

before	after
utilize	use
anticipate	await (or expect)
cooperate	help
indicate	show
initially	first (or 1.)
presently	now
prohibit	stop
inconvenience	problem
pursuant	before
endeavor	try
sufficient	enough
subsequent	next

Another way to look at simplifying word usage is to think in terms of a "ladder of abstraction," shown in Figure 2.2. Envision words along the rungs of a ladder. The most multisyllabic and pompous word sits at the top of the ladder. The simple, easy-to-understand synonym is at the bottom of the ladder. Words in between these two range in syllable length.

Achieving Accuracy in Workplace Communication

Clarity and conciseness are primary objectives of effective workplace communication. However, if your writing is clear and concise but incorrect—grammatically or textually—

then you have wasted your time and destroyed your credibility. To be effective, your workplace communication must be *accurate*. Accuracy requires that you *proofread* your text.

Proofreading Your Workplace Communication

To ensure accurate writing, use the following proofreading tips:

1. *Let someone else read it.* We miss errors in our own writing for two reasons. First, we make the error because we don't know any better. Second, we read what we think we wrote, not what we actually wrote. Another reader might help you catch errors.
2. *Read it later.* Let your correspondence sit for a while. Then, when you read it later, you'll be more objective.
3. *Read backwards.* You can't do this for content. You should read backwards only to slow yourself down and to focus on one word at a time to catch typographical errors.
4. *Read one line at a time.* Use a ruler or scroll down your PC screen to isolate one line of text. Again, this slows you down for proofing.
5. *Read long words syllable by syllable.* How is the word *responsiblity* misspelled? You can catch this error if you read it one syllable at a time (re-spon-si-bl-i-ty).
6. *Use technology.* Computer spell-checks are useful for catching most errors. They might miss proper names, homonyms (*their*, *they're*, or *there*), or incorrectly used words, such as *device* to mean *devise*.
7. *Check figures, scientific and technical equations, and abbreviations.* If you mean $400,000, don't write $40,000. Double-check any number or calculation. If you mean to say *HCl* (hydrochloric acid), don't write *HC* (a hydrocarbon).
8. *Read it out loud.* Sometimes we can hear errors that we cannot see. For example, we know that *a outline* is incorrect. It just sounds wrong. *An outline* sounds better and is correct.
9. *Try scattershot proofing.* Let your eyes roam around the page at random. Sometimes errors look wrong at a glance. If you wander around the page randomly reading, you often can isolate an error just by stumbling on it.
10. *Use a dictionary.* If you are uncertain, look it up.

If you make errors in your workplace communication, your readers will think one of two things about you: (a) they will conclude that you are uneducated, or (b) they will think that you are lazy. In either situation, you lose. Errors create a negative impression at best; at worst, a typographical error relaying false figures, calculations, amounts, equations, or scientific or medical data can be disastrous.

Achieving Audience Recognition and Involvement

To communicate successfully, you must recognize your audience's level of understanding, which could have an impact on your communication success.

Knowledge of the Subject Matter

Depending upon their knowledge of your topic, the audience will be specialists, semi-specialists, lay people, or combinations of the three.

Specialists. Specialists have the following characteristics:

- Specialists are experts in the field you are writing about.
- Specialists understand the terminology of their field.
- Specialists require minimal detail regarding standard procedures or scientific, mathematical, or technical theories.
- Specialists read to discover new knowledge or for updates regarding the status of a project.
- Specialists need little background information regarding a project's history or objectives unless the specific subject matter of the correspondence is new to them.

Semi-Specialists. Semi-specialists share the following characteristics.

- Semi-specialists are familiar with the product or service you are writing about but have job responsibilities peripheral to the subject matter.
- Because semi-specialists are familiar with the subject matter, they understand some abbreviations, jargon, and technical concepts. To ensure that readers and listeners understand your content, define terms. An abbreviation such as VLSI cannot stand alone. Define it parenthetically: very large scale integration (VLSI).

Lay People. These people share the following characteristics:

- Lay people are unfamiliar with your subject matter. Therefore, you should explain your topic clearly. Achieve clarity through precise word usage, depth of detail, and simple graphics.
- Since lay people do not understand your work environment, you must define unfamiliar terms, including acronyms and abbreviations.
- Lay people might need background information. When explaining to lay people, provide sufficient causes, results, or rationale.

Combination Audiences. Combination audiences share the following characteristics:

- Your intended audience will not necessarily be your only readers or listeners. Others might receive copies of your correspondence or hear your speech.
- Some people in a combination audience will not be familiar with the subject matter. You will have to provide background data (objectives, overviews) to clarify the information. In a short letter, memo, report, or e-mail message, this background information could be conveyed in a reference line suggesting where the readers can find out more about the subject matter if they wish—"Reference: Operations Procedure 321 dated 9/21/00." In longer reports, background data will appear in the summary or abstract as well as in the report's introduction.
- Summaries, abstracts, introductions, and references are especially valuable because reports, memos, letters, and e-mail messages are kept on file. Months (or years) later when the report is retrieved from the files, future readers will need background information to remind them of the document's context.
- Correspondence for combination audiences must have a matter-of-fact, businesslike tone. You should not be too authoritative since upper-level management

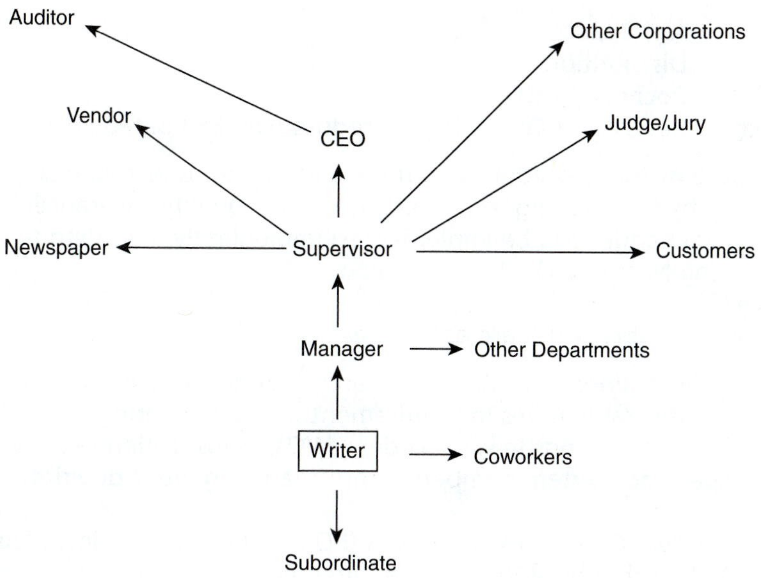

Figure 2.3 Combination Audiences

might read the memo, letter, or report. You should not be patronizing since lower-level subordinates might also read the correspondence.

- Combination audiences often need terms defined.

You might correspond with many different audiences. Figure 2.3 shows the possible diversity of such an audience.

Defining Terms for Different Audiences

To communicate successfully with combination audiences:

- Define words or phrases either parenthetically or in a glossary. For example, you could define ATM parenthetically: asynchronous transfer mode (ATM). Then readers will not misconstrue ATM as the more commonly understood *automatic teller machine*. You could also use a glossary, as follows:

HTTPS	Hypertext Transfer Protocol, Secure
TDD	telecommunication device for the deaf
TTY	teletypewriter

- Use a sentence to define terms, either in a glossary or following the unfamiliar terms.

> HyperText Transfer Protocol is a computer access code that provides secure communications on the Internet, an intranet, or an extranet.

The memo in Figure 2.4 is written to a combination audience.

Figure 2.4 Effective Memo for Combination Audiences

Designation for combination audiences

Parenthetical abbreviation

Abbreviation after prior parenthetical definition

Objective tone (passive voice) vs. a directive (active voice) that could offend management

Date: July 7, 2006
To: Distribution
From: Rochelle Kroft
Subject: Revision of Operating Procedure (OP) 354 dated 5/31/06

The reissue of this procedure was the result of extensive changes requested by Engineering, Manufacturing, and Quality Assurance. These procedural changes will be implemented immediately according to Engineering Notice (EN) 185.

Some important changes are as follows:

1. *Substitutions*: An asterisk (*) can no longer be used to identify substitutable items in requirement lists. Engineering will authorize substitutions in work orders (WOs). Substitutions also will be specified by item numbers rather than by generic description names.
2. *Product Quality Requirements* (PQRs): These will be included in WOs either by stating the requirements or by referring to previous PQRs.
3. *Oral Instructions*: When oral instructions for process adjustments are given, the engineer must be present in the department.

All managers will review OP 354 and EN 185. Next, the managers will review the changes with their supervisors to make sure that each supervisor is aware of his or her responsibilities. These reviews will occur immediately.

Distribution:

Rob Harken
Julie Burrton
Hal Lang
Sharon Myers
Manuel Ramos
Jeannie Kort
Jan Hunt
Earl Eddings

Achieving Successful Written Communication in a Multicultural and Cross-Cultural Environment

Your company might market its products or services worldwide. International business requires multicultural communication, the sharing of written and oral information between businesspeople from many different countries. Multiculturalism will not just affect you when you communicate globally. You will be confronted with multicultural communication challenges in your own city and state. Another term for this challenge is *cross-cultural communication*, writing and speaking between businesspeople of two or more different cultures within the same country (Nethery).

To write effectively for a multicultural and cross-cultural audience, consider the following guidelines:

Define Acronyms and Abbreviations

Acronyms and abbreviations cause problems for most readers. What does "CPR" mean? Is it "cardiopulmonary resuscitation" or "continuing property records"? What's "POS"? In marketing, it could mean "public opinion strategies." In criminology, however, "POS" means "proof of summons," while e-commerce commonly uses "POS" to mean "point of sale." Although you and your immediate colleagues might understand such terminology, many readers will not. This is especially true when your audience is not native to the United States.

To avoid communication problems, define your acronyms and abbreviations parenthetically the first time you use them.

Distinguish Nouns from Verbs

Many words in English act as both nouns and verbs. This is especially true with computer terms, such as "file," "scroll," "paste," "code," and "help." If your text will be translated, make sure that your reader can tell whether you are using the word as a noun or a verb (Rains 12).

Watch for Cultural Biases and Expectations

Your text will include words and graphics. As a writer, you need to realize that many colors and images that connote one thing in the United States will have different meanings elsewhere.

For example, take the idioms "in the red" and "in the black." The colors black and red have different meanings in different cultures. *Red* in the United States connotes danger. Therefore, "in the red" suggests a financial problem. In China, however, the word *red* has a positive connotation, which would skew your intended meaning. The word *black* often implies death and danger, yet "in the black" suggests financial stability. Such contradictions could confuse readers in various countries.

Avoid Humor and Puns

Humor is not universal. The United States has regional humor. If a joke is good in the South but not in the North, how could that same joke be effective overseas? Microsoft's software package Excel is promoted by a logo that looks like an X superimposed over an L. This visual pun works in the United States where the letters X and L are pronounced like the name of the software package. If readers are not familiar with English, however, they might miss this clever sound-alike image (Horton 686).

Avoid Figurative Language

Many Americans use sports images to figuratively illustrate points. Americans "tackle" a chore; in business, a "good defense is the best offense"; people "huddle" to make decisions; if a sale isn't made, you might have "booted the job"; if a sale is made, you "hit a home run." If something is easy, it's a "slam dunk." Each of these sports images might mean something to native speakers, but they do not necessarily communicate worldwide (Weiss 14).

Be Careful with Numbers, Measurements, Dates, and Times

Numbers and Measurements. The United States uses inches, feet, and yards. However, most of the world measures in metric units. If you write 18 high × 20 wide × 30 deep, what are the measurements? There is a huge difference between 18 inches, 20 inches, and 30 inches, and 18 millimeters, 20 millimeters, and 30 millimeters.

Dates. People in the United States abbreviate dates as MM/DD/YY: 05/03/06. In many other countries, however, dates are abbreviated DD/MM/YY. Thus, 05/03/06 will be perceived as March 5, 2006, instead of May 3, 2006. See Table 2.1 for additional examples.

Time. Time is another challenge. Look at Table 2.2 to see how different countries indicate time.

In addition to different ways of writing time, you must also remember that even within the United States, 1:00 p.m. does not mean the same thing to everyone. Is that Central Time, Pacific Time, Mountain Time, or Eastern Time? Add to this the problems with world time zones, and the challenge increases. You must define time zones clearly.

To solve these problems, determine your audience and make changes accordingly. That might mean

- writing the date out completely (January 12, 2006).
- telling the reader what standard of measurement you will use ("This document provides all measurements in metric units.").
- telling the reader what scheme of time presentation you will use ("This document relates time using a 24-hour clock rather than a 12-hour clock.").
- using multiple formats ("Let's meet at 2:30 p.m./14:30.").
- avoiding vague words such as "today," "tomorrow," or "yesterday."

TABLE 2.1 DIFFERENT WAYS OF UNDERSTANDING AND WRITING THE DATE 05/03/06

Country	Dates
United States	May 3, 2006
United Kingdom	March 5, 2006
France	5 mars 2006
Germany	5. Marz 2006
Sweden	06-05-03
Italy	5.3.06

TABLE 2.2 DIFFERENT WAYS OF WRITING THE SAME TIME

Country	Time
United States	5:15 p.m.
France	17:15
Germany	17.15
Quebec, Canada	17 h 15

before

To: Jose Guerrero, Mexico City, Mexico Office; Yong Kim, Hong Kong Office; Hans Rittmaster, Berlin Office
From: Leonard Liss, New York Office
Subject: Agenda for Teleconference

Time to wrap up that deal. If we don't finish the project soon, we're all behind the eight ball. So, here's what I'm planning for the 03/07/06, 12:00 discussion:

- Restructured design—rather than build the part at 8 x 10 x 23, let's consider a smaller design.
- Shipping method—let's use a new carrier/vendor. We've not had good luck with Flyrite Overnight. I'm open to your suggestions. Let's think outside the box.

Brainstorm before our teleconference so we can tackle this topic pronto. My boss needs our suggestions ASAP, so I need it even sooner. I know you'll come through with flying colors.

Figure 2.5 Poorly Written E-mail Message for a Multicultural Audience

Slang words will not communicate internationally.

Dates must be written out to ensure clarity. For time, be sure to include a.m. or p.m. and the time zone.

For dimensions, state whether you are using inches or meters.

Words like "pronto" and "ASAP" are regionalisms that will not communicate internationally.

Figure 2.5 is an example of poor e-mail communication for a multicultural audience. Figure 2.6 corrects the multicultural communication problems shown in Figure 2.5.

Avoiding Biased Language

In addition to recognizing your audience's level of knowledge and cultural diversity, you also must consider your audience's age, physical limitations, and gender.

Ageist Language

A word such as "elderly" could imply feebleness. The words "old folks" create a negative image. Similarly, the words "girls" and "boys" in reference to employees diminish their position and authority.

before

Professor Jones, an elderly teacher at State University, is publishing a textbook despite his age.

You girls need to respond within one hour to all insurance claims.

after

Professor Jones, a State University teacher, is publishing a textbook.

Please respond within one hour to all insurance claims.

Figure 2.6
Improved E-mail Message for a Multicultural Audience

after

To: Jose Guerrero, Mexico City, Mexico Office; Yong Kim, Hong Kong Office; Hans Rittmaster, Berlin Office
From: Leonard Liss, New York Office
Subject: Agenda for Teleconference

We need to complete our team project. Doing so will allow our company to meet our client's deadline. A teleconference is our best way to communicate, given everyone's diverse locations. I have made all the technical arrangements.

Writing out the date and including the time zone ensures clarity. →

The teleconference is scheduled for March 7, 2006, at 12:00 noon, Pacific Standard Time. During the teleconference, we will discuss the following:

The audience clearly will know that the dimensions are given in inches. →

- Restructured design—rather than build the part at 8" x 10" x 23" as planned (all dimensions in inches) we should consider a smaller design.
- Shipping method—a new carrier or vendor might save us money and time. Flyrite Overnight has increased their shipping fees by 25 percent. What do you think? I am open to your creative ideas.

Discuss this issue with your coworkers and supervisors before our teleconference. This will allow us to use time effectively. My manager needs our suggestions by March 8, 2006, 4:00 p.m. Central Standard Time. With your help, I know our company will make the correct decisions.

Biased Language about People with Disabilities

The word "handicap" creates a negative image. "Disability" is generally preferred. However, any euphemism can be offensive. You should avoid reference to a person's disabilities. If you need to refer to a physical problem, do so without negative characterizations.

before

Debbie Brown, a blind market researcher, won "Employee of the Month."

The AIDS *victim* changed insurance carriers.

John *suffers* from diabetes.

Sheila is *confined* to a wheelchair.

after

Debbie Brown, a market researcher, won "Employee of the Month."

The AIDS patient changed insurance carriers.

John is diabetic.

Sheila uses a wheelchair.

Sexist Language

Women constitute close to half of the workforce in the United States. Thus, you must avoid gender-biased language in your communication. Gender-biased language occurs through

- Omission
- Unequal treatment
- Stereotyping
- Word choice (pronouns and nouns)

Omission. When your writing ignores women or refers to them as secondary, you are being biased. The following are examples of biased comments and their nonbiased alternatives.

before	after
The computer information specialists and their wives and children attended the company picnic.	The computer information specialists and their families attended the company picnic.
The congressional legislation on foreign trade agreements was proposed by a woman, Claire McGowan.	The congressional legislation on foreign trade agreements was proposed by Claire Mc-Gowan.
When conducting his quarterly review, the auditor must always check for errors.	When conducting a quarterly review, the au-ditor must always check for errors.
As we acquired scientific knowledge, men began to examine long-held ideas more critically.	As we acquired scientific knowledge, people began to examine long-held ideas more critically.

Unequal Treatment. Modifiers that describe women in physical terms not applied to men are inappropriate.

before	after
The poor women and exhausted men could no longer work on the report.	The exhausted women and men could no longer work on the report.
Mrs. Acton, a very pretty young woman, is Joe Granger's assistant.	Jan Acton is Joe Granger's assistant.

Stereotyping. If your writing implies that only men do one kind of job and only women do another kind of job, you are stereotyping. For example, if you suggest that men hold all management jobs and women hold all subordinate positions, this is stereotyping.

before	after
Current tax regulations allow a head of household to deduct for support of his children.	Current tax regulations allow a head of household to deduct for child support.
The manager is responsible for the productivity of his department; the foreman is responsible for the work of his linemen.	Management is responsible for departmental productivity. Supervisors are responsible for their personnel.
The secretary brought her boss his coffee.	The secretary brought the boss's coffee.
The teacher must be sure her lesson plans are filed.	The teacher must file all lesson plans.

Pronouns. Biased language disappears when you use pronouns and nouns that treat all people equally. Pronouns such as *he*, *him*, or *his* are masculine. Sometimes you read disclaimers by manufacturers stating that "although these masculine pronouns are used, they are not intended to be biased. They are only used for convenience." This is an unacceptable statement. Use of *he*, *him*, and *his* exclusively creates a masculine image.

To avoid this bias, avoid masculine pronouns. Instead, use the plural, generic *they* or *their*. You also can use *he or she* and *his or her*. Sometimes you can solve the problem by omitting all pronouns.

before	after
Sometimes the doctor calls on his patients in their homes.	Sometimes the doctor calls on patients in their homes.
The typical child does his homework after school.	Most children do their homework after school.
A good lawyer will make sure that his clients are aware of their rights.	A good lawyer will make sure that his or her clients are aware of their rights.

Nouns. Use nouns that are nonbiased. To achieve this, avoid nouns that exclude women and denote that only men *or* women are involved. These are called "gender-tagged" words. Most of them have an actual gender attached to them. Usually, this involves the word "man." However, the suffix "ess" is a feminine ending, also creating a gender tag, such as "princess."

before	after
mankind	people
manpower	workers, personnel
the common man	the average citizen
wise men	leaders
businessmen	businesspeople
policemen	police officers
firemen	firefighters
foreman	supervisor
chairman	chairperson *or* chair
stewardess	flight attendant
waitress	server

Figure 2.7 Biased Writing

DiBono and Sons Landscaping
1349 Elm
Oakland Hills, IA 20981

April 2, 2006

Owner
TurboCharge, Inc.
2691 Sommers Rd.
Iron Horse, IA 20992

Dear Sir:

Biased and outdated salutation, implying only a man will read the letter

Interested in beautifying your company property, with no care or worries? We have been in business for over 25 years helping men just like you. Here's what we can offer:

1. Shrub Care. Your workmen will no longer need to water shrubs. We'll take care of that with a sprinkler system geared toward your business's unique needs.
2. Seasonal System Checks. Don't worry about your foreman having to ask his workers to turn on the sprinkler system in the spring or turn it off in the winter. Our repairmen take care of that for you as part of our contract.
3. New Annual Plantings. We plant rose bushes and daffodils that are so pretty, your secretary will want to leave her desk and pick a bunch for her office. Of course, you might want to do that for her yourself ☺.
4. Trees. We don't just plant annuals. We can add shade to your entire property. Just think how beautiful your parking lot will look with elegant elms, oaks, and maples. Then, when you have that company picnic, your employees and their wives will be surrounded by beauty.

Biased "gender-tagged" word

Stereotyping

Call today for prices. Let our skilled craftsmen work for you!

Sincerely,

Richard DiBono

Figure 2.7 is a letter using biased words. Figure 2.8 revises this letter to avoid biased writing.

Achieving Audience Involvement

Recognizing your audience entails knowing their levels of knowledge, their language of origin and culture, and diversity issues, such as age, physical limitations, and gender. By understanding these audience concerns, you can communicate more effectively.

Figure 2.8
Unbiased Revision

Unbiased words are highlighted for emphasis.

DiBono and Sons Landscaping
1349 Elm
Oakland Hills, IA 20981

April 2, 2006

Owner
TurboCharge, Inc.
2691 Sommers Rd.
Iron Horse, IA 20992

Subject: Sales Information

Interested in beautifying your company property, with no care or worries? We have been in business for over 25 years helping business owners just like you. Here's what we can offer:

- Shrub Care. Your employees will no longer need to water shrubs. We'll take care of that with a sprinkler system geared toward your business's unique needs.
- Seasonal System Checks. Don't worry about your supervisor having to ask his or her workers to turn on the sprinkler system in the spring or turn it off in the winter. Our staff takes care of that for you as part of our contract.
- New Annual Plantings. We plant rose bushes and daffodils that are so pretty, all of your employees will want a vase of flowers on their desks.
- Trees. We don't just plant annuals. We can add shade to your entire property. Just think how beautiful your parking lot will look with elegant elms, oaks, and maples. Then, when you have that company picnic, your employees and their families and friends will be surrounded by beauty.

Call today for prices. Let our skilled experts work for you!

Sincerely,

Richard DiBono

An additional goal is to achieve audience involvement. Effective workplace communication must build rapport to achieve audience buy-in.

Involve Your Audience with "You Usage"

Limit your use of first-person pronouns, such as "I," "me," and "my." Your audience wants to know what is in it for them, not how you will benefit. To achieve a different focus on the audience or the team, emphasize "you usage." Use second-person pronouns, such as "you" and "your." Pronouns such as "we," "us," and "our" also are effective to create a sense of team between the writer and audience. By using pronouns, you personalize the communication, creating a person-to-person tone.

Pronoun	Focus
You Your	→ The reader
We Us Our	→ The team
I Me My	→ The ego

Look at the following poorly written, impersonal examples that fail to build rapport. The revisions include personal pronouns, positive words, and a motivational tone.

before

Impersonal

Dear Sir:
With regard to lost policy 123, enclosed is a lost policy form. Complete this form and return it ASAP. Upon receipt, the company will issue a replacement policy.

Claims Procedure

To obtain service under the Emissions Performance Warranty, take the vehicle to the company dealer as soon as possible after it fails an I/M test along with documentation showing that the vehicle failed an EPA emissions test.

after

"You Usage"

Dear Mr. Salinas:
Your lost policy 123 can be replaced easily. I've enclosed a form to help us replace it for you. All you need to do is fill it out. As soon as we get the form, we'll send you your replacement policy.

Claims Procedure

How do you get service under the Emissions Performance Warranty? Take your car to the dealer as soon as possible after it has failed an Environmental Protection Agency (EPA) test. Be sure to bring along the document that shows your car failed the test.

Audience Checklist

The Audience Checklist will give you the opportunity for self-assessment and peer evaluation of your writing.

Audience Checklist

__ 1. What is the audience's level of understanding regarding the subject matter?
- Specialist
- Semi-specialist
- Lay
- Combination

__ 2. Have you defined acronyms, abbreviations, and jargon?

__ 3. Have you supplied enough background data?

__ 4. Have you considered diversity?

__ 5. Have you avoided biased language that could offend various age groups, people of different sexual orientations, people with disabilities, or people of different cultures and religions?

__ 6. Have you considered that people from different countries and people for whom English is a second language might be your audience?

__ 7. Have you avoided sexist language?

Achieving Ethical Communication in the Workplace

You are a biomedical equipment salesperson. Your company has created a new piece of equipment to be marketed worldwide. Part of the sales literature that your boss tells you to share with potential customers contains the following sentence:

> **Note:** Our product has been tested for defects and safety by trained technicians.

When read literally, this sentence is accurate. The product has been tested, and the technicians are trained. However, you know that the product has been tested for only 24 hours by technicians trained on site without knowledge of international regulations.

As a loyal employee, are you required to do as your boss requests? Even though the statement is not completely true, can you legally include it in your sales literature?

The answer to both questions is no. You have an ethical responsibility to write the truth. Your customers expect it, and it is in the best interests of your company. Of equal importance, including the sentence in your sales literature is misleading. Though the sentence is essentially true, it implies something that is false. Readers will assume that the product has been thoroughly tested by technicians who have been correctly trained. Thus, the sentence deceives the readers.

Why Business Ethics Are Important

Business schools and management experts stress the importance of ethics. Many problems can occur when businesses fail to maintain ethical standards. These problems can include dissatisfied customers, large legal judgments, prison terms, antitrust litigation, loss of goodwill, lost sales, fines, and bankruptcies.

Knowing this, however, does not make communicating in the workplace easy. Ethical dilemmas exist in corporations. What should you do when confronted with such problems?

Strategies for Ethical Workplace Communication

The International Association of Business Communication (IABC) (see Figure 2.9) and the Society for Technical Communication (STC) (see Figure 2.10) provide two sources for ethical standards in the workplace.

When writing or speaking, apply the following strategies to your workplace communication.

Use Language and Visuals with Precision. You are writing a sales brochure for a Florida resort hotel. You state that the hotel is "in easy walking distance to the Gulf." The hotel actually is located a half mile from the beach. The road is unpaved and uphill. Though the original sales line implies ease of access, for many people this

International Association of Business Communication
Code of Ethics for Professional Communicators

Preface

Because hundreds of thousands of business communicators worldwide engage in activities that affect the lives of millions of people, and because this power carries with it significant social responsibilities, the International Association of Business Communication developed the Code of Ethics for Professional Communicators.

The Code is based on three different yet interrelated principles of professional communication that apply throughout the world.

These principles assume that just societies are governed by a profound respect for human rights and the rule of law; that ethics, the criteria for determining what is right and wrong, can be agreed upon by members of an organization; and, that understanding matters of taste requires sensitivity to cultural norms.

These principles are essential:

- Professional communication is legal.
- Professional communication is ethical.
- Professional communication is in good taste.

Recognizing these principles, members of IABC will:

- engage in communication that is not only legal but also ethical and sensitive to cultural values and beliefs;
- engage in truthful, accurate and fair communication that facilitates respect and mutual understanding; and,
- adhere to the . . . articles of the IABC Code of Ethics for Professional Communicators.

Because conditions in the world are constantly changing, members of IABC will work to improve their individual competence and to increase the body of knowledge in the field with research and education.

Figure 2.9 The IABC Code of Ethics
Source: "Code of Ethics"

uphill walk could prove to be difficult. In this case, you would not be using words precisely.

Precision also is required when you use visuals to convey information. Look at Figures 2.11 and 2.12. Both figures show that XYZ's sales have risen. However, Figure 2.11 suggests that sales have risen consistently and dramatically. The weight of the line even exclaims success. Figure 2.12 is more accurate and honest. It shows sales growth, but it also depicts peaks and valleys.

The precise use of language and visuals also involves intellectual property laws as they relate to the Internet. If you and your company "borrow" from an existing Internet site, thus infringing upon that site's copyright, you can be assessed actual or statutory damages.

Figure 2.10 STC Code for Communicators
Source: "Code for Communicators"

STC Code for Communicators

As a technical communicator, I am the bridge between those who create ideas and those who use them. Because I recognize that the quality of my services directly affects how well ideas are understood, I am committed to excellence in performance and the highest standards of ethical behavior.

I value the worth of the ideas I am transmitting and the cost of developing and communicating those ideas. I also value the time and effort spent by those who read or see or hear my communication.

I therefore recognize my responsibility to communicate technical information truthfully, clearly, and economically.

My commitment to professional excellence and ethical behavior means that I will

- Use language and visuals with precision.
- Prefer simple, direct expression of ideas.
- Satisfy the audience's need for information, not my own need for self-expression.
- Hold myself responsible for how well my audience understands my message.

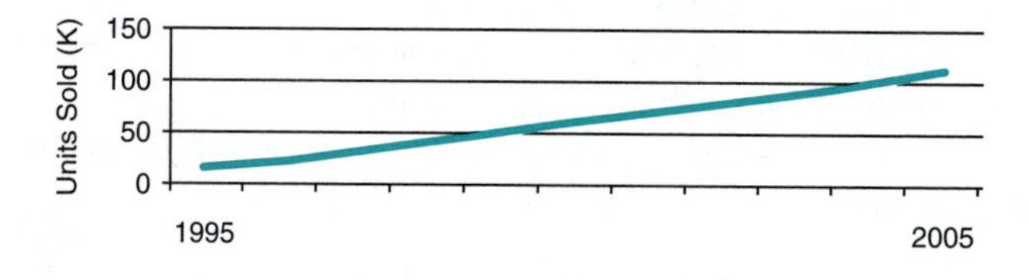

Figure 2.11 Imprecise Depiction of Sales Growth

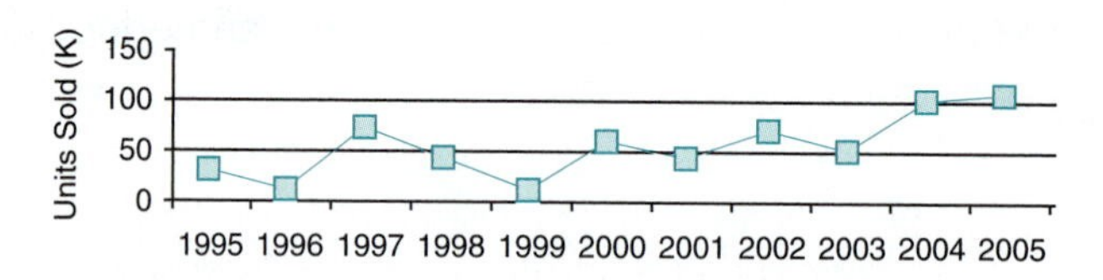

Figure 2.12 Precise Depiction of Sales Growth

In addition to financial damages, your company, if violating intellectual property laws, could lose customers, damage its reputation, and lose future capital investments. To solve these problems and to protect your own property rights, you should

- assume that any information on the Internet is covered under copyright protection laws unless proven otherwise.
- obtain permission for use from the original creator of graphics or text.
- cite the source of your information.
- create your own graphics and text.
- copyright any information you create.
- place a copyright notice at the bottom of your Web site (Johnson, 17).

Prefer Simple, Direct Expression of Ideas. While writing a proposal, you might be asked to use legal language to help your company avoid litigation (Bowman and Walzer). Following is a typical warranty, written using long sentences and difficult words.

Company Warranty Information

Acme's liability for damages from any cause whatsoever, including fundamental breach, arising out of this Statement of Limited Warranty, or for any other claim related to this product, shall be limited to the greater of $10,000 or the amount paid for this product at the time of the original purchase, and shall not apply to claims for personal injury or damages to personal property caused by Acme's negligence, and in no event shall Acme be liable for any damages caused by your failure to perform your responsibilities under this Statement of Limited Warranty, or for loss of profits, lost savings, or other consequential damages, or for any third-party claims.

Will your audience be able to understand this warranty's long sentence structure and complex words? Readers will become frustrated and fail to recognize what is covered under the warranty. Although you are following your boss's directions, you are not communicating clearly.

Writing can be legally binding and easy to understand. The first requirement does not negate the second. As a successful writer, you should help your audience understand the message by using simple words and the direct expression of ideas whenever possible.

Satisfy the Audience's Need for Information, Not Your Own Need for Self-Expression. The previously mentioned warranty might be considered sophisticated in its word usage and, thus, "more professional." The warranty, however, does not communicate what the audience needs. Such elaborate and convoluted writing will only satisfy one's need for self-expression; that is not the goal of effective workplace communication.

Hold Yourself Responsible for How Well the Audience Understands the Message. Ethical standards for successful communication require the writer to always remember the readers—the people who read corporate annual reports to understand how their stocks are doing, who read brochures before making travel arrangements, or who read proposals to determine whether to purchase a service. An ethical writer remembers that these people will be confused by inaccessible stock reports and misled by inaccurate proposals or brochures.

Take the time to check your facts. Present your information precisely, and communicate clearly so that your audience is safe and satisfied. You are responsible for your message.

CASE STUDIES

1. Sharon Myers is CEO of Grunge Groovies. This Web-based shopping outlet sells vintage clothing (see the **Communication at Work** scenario that begins this chapter). Her employee, Peter Tsui, has written text for the company's home page.
   However, Sharon has concluded that his writing is not clear, concise, or appropriate for the intended audience, and it is grammatically flawed. In addition, Sharon thinks that his graphic is unethical. The company has experienced sales growth, but the growth has not been as spectacular as the graphic depicts. Their sales have been as follows: $10,000 in 2000; $30,000 in 2001; $20,000 in 2002; $60,000 in 2003; $30,000 in 2004; $60,000 in 2005; and $70,000 in 2006.
   Sharon now is asking Peter to make his text more readable and accurate. To accomplish Sharon's goals, Peter has to delete dead words and phrases for conciseness, simplify his excessive word usage, correct his grammatical errors, and ensure ethical communication.

Assignment

Read Peter's text below. Then, revise it, using the techniques you learned in this chapter.

Grunge Groovies*—Who We Are and What We Offer!*

Welcome to our site. Grunge Groovies offers you a plethora of antique accessories and accoutrements. If you want it, we got it! Take a gander at our gargantuan selection of various eras' clothing and appurtenances, including shirts, blouses, pants, shorts, suits, jackets, shoes, boots, as well as other paraphernalia like scarves, necklaces, bracelets, earrings, rings, hats, and you name it. We have clothing and other things from the 50s to the 90s. Our 50s' goodies covers the "beats" to the "blues brothers" clothing. We have black suits, black skinny ties, as well as black berets and turtlenecks. Our 60s' hippie stuff is way cool, such as wide, hallucinogenic ties, Nehru jackets, short-short mini-dresses, long-long granny dresses, way-out-there boots, peace symbols, fringe belts and vests, etc., etc., etc. Our 70s' clothing and items include all kinds of disco clothing including leisure suits and everything Banlon plus skin tight pants, shirts, white belts and boots, we even have disco dance hall balls and posters for your own disco dance floor. Our 80s' clothing focuses on what we call "business-nerd" including pocket protectors, wing-tip shoes, horned rim glasses (with the nose piece already pre-taped), we also cover the "Flash Dance" clothes like leg warmers and off-the-shoulder sweatshirts. Finally, our 90s' clothing includes everything from early 90s' grunge (raggedy cut-off shorts and flannel plaid shirts) to late 90s' hip-hop clothes and add-ons (oversized jewelry, sports jerseys, hooded sweatshirts, and fancy club duds). Return with us now to those thrilling days of yesteryear by purchasing accessories from Grunge Groovies By the way, we offer much better quality than our competitors. And for you investors, look at how our sales have gone through the roof:

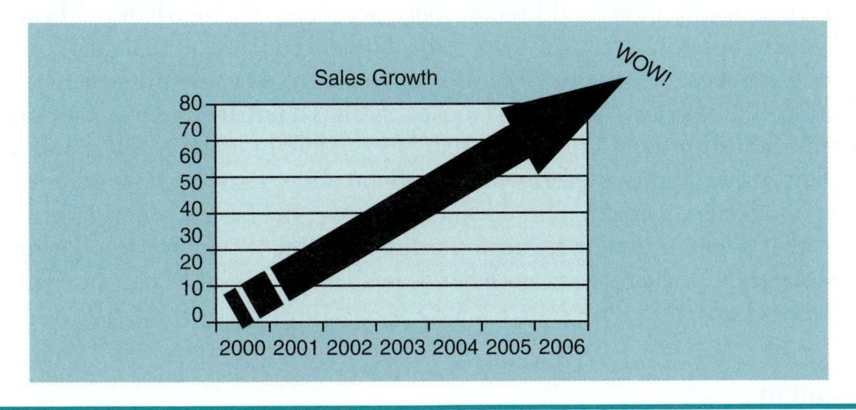

2. CompToday computer hardware company must abide by the Sarbanes-Oxley Act, passed by Congress in response to accounting scandals. This act specifically mandates the following related to documentation standards:

   Section 103: Auditing, Quality Control, and Independence Standards and Rules.

   Companies must "prepare, and maintain for a period of not less than 7 years, audit work papers, and other information related to any audit report, in sufficient detail to support the conclusions reached in such report."

   Section 401(a): Disclosures in Periodic Reports; Disclosures Required.

   "Each annual and quarterly financial report . . . must be presented so as not to contain an untrue statement or omit to state a material fact necessary in order to make the pro forma financial information not misleading."

   Beverly Warden, technical documentation specialist at CompToday, is responsible for managing the Sarbanes-Oxley reports. She is being confronted by some ethical issues.

   To help Beverly prepare the first annual report, her chief financial officer (CFO) has given her six months of audits (January through June). These prove that the company is meeting its accounting responsibilities. However, Beverly's report covers the entire year, including July through December. Section 103 states that the report must provide "sufficient detail to support the conclusions reached in [the] report."

   Are the company's first six months' audits sufficient? If Beverly writes a report stating that her company is in compliance, is she abiding by her communicator's ethical principles, which state that a writer's work is "consistent with laws and regulations"?

Assignment

What are her ethical and legal responsibilities?

1. Share your findings in an oral presentation.
2. Write a letter, memo, report, or e-mail to your teacher stating your opinion regarding this issue.

Beverly's CFO also has told her that during the year, the company fired an outside accounting firm and hired a new one to audit the company books. The first firm expressed concerns about several bookkeeping practices. The newly hired firm, providing a second opinion after reviewing the books, concluded that all bookkeeping practices were acceptable. The CFO sees no reason to mention the first firm.

Section 401(a) states that reports must contain no untrue statements or omit a material fact. Ethical principles also say that her writing must be truthful and accurate, to the best of her ability. Beverly can report factually that the new accounting firm finds no bookkeeping errors. Should she also report the first accounting firm's assessment? Is that a material fact? If she omits any mention of the first accounting firm, as her boss suggests, is she meeting both her writer's responsibilities and the requirements of Sarbanes-Oxley?

Assignment

What are her ethical and legal responsibilities?

1. Share your findings in an oral presentation.
2. Write a letter, memo, report, or e-mail to your teacher stating your opinion regarding this issue.

INDIVIDUAL AND TEAM PROJECTS

1. Adding Information for Clarity

a. *Reporter's Questions*

The following sentences are vague. Different readers will interpret them differently. Revise these sentences by answering *reporter's questions* (who, what, when, where, why, and how).

1. We need this information for the meeting.
2. The machinery will replace the broken equipment in our department.
3. If we fail to meet their request, we will lose the account.
4. Weather problems in the area resulted in damage to the computer systems.
5. If we cannot solve this problem, we will not meet the customer's deadline.

b. *Specificity*

The following sentences are vague. Different readers will interpret them differently. Revise these sentences by using precise words.

1. We need reports as soon as possible.
2. Failure to meet the deadline could have a negative impact.
3. Insufficient personnel caused the most recent occurrences.
4. Fire in the office led to substantial losses.
5. By not completing the recent deal, we lost a large percentage of our business.

2. Deleting Wordiness for Conciseness

a. *Readability—Fog Index*

1. Read the following paragraph. Use Robert Gunning's Fog Index to determine the paragraph's readability. Figure out the average number of words

per sentence, count the number of multisyllabic words, and then determine the grade level for the paragraph. Revise the paragraph to achieve clarity and conciseness.

> Ramifications of yesterday's revised implementation schedule are significant because doing as requested by management could lead to missed deadlines as well as the potential for production malfunctions. I respectively request that management reconsider these suggestions, taking into consideration the short-term longevity of our employees, many of whom are newly hired. Instead, I am of the opinion that any inconveniences our company might experience due to revising the schedule will be offset by the inestimable values we will derive. I am cognizant that changes are challenging to make, but management might consider doing so at this point in time to benefit employee morale.

b. *Passive Voice*

Passive voice often leads to vague, wordy sentences. Revise the following sentences by rewriting them in the active voice.

1. Installation of the new network-wide software was carried out by the information technology department.
2. Benefits were derived when George attended the conference.
3. The information was demonstrated and explained in great detail by the training supervisor.
4. Discussions were held with representatives from Allied, who supplied analytical equipment for automatic upgrades.
5. The symposium on departmental rules and regulations was attended by the nursing staff.

c. *Sentence Length*

The following sentences are unnecessarily wordy. They contain expletives, "be"-verb constructions, redundancies, "shun" words, camouflaged words, and wordy prepositional phrases. Revise the sentences to make them more concise.

1. In regard to the progress reports, they should be absolutely complete by the fifteenth of each month.
2. I wonder if you would be so kind as to answer a few questions about your proposal.
3. I am in receipt of your memo requesting an increase in pay and am of the opinion that it is not merited at this time due to the fact that you have worked here for only one month.
4. In this meeting, our intention is to acquire a familiarization with this equipment so that we might standardize the replacement of obsolete machinery throughout our entire work environment.
5. There is the possibility that we will implement these suggestions early next month.

3. Simplifying for Easier Understanding

Multisyllabic words can create long sentences and confusion. To limit sentence length, limit word length. Find shorter, easier-to-understand words to replace the following words.

1. aforementioned
2. adjacent
3. approximately
4. ascertain
5. attached herewith
6. consequence
7. elucidate
8. demonstrate
9. frequently
10. identify
11. numerous
12. occasion
13. residence
14. subsequent
15. terminate

4. Correcting Errors for Accuracy

The following memo contains grammatical errors. The errors include problems with spelling and punctuation, as well as agreement, capitalization, and number usage. Correct the errors to ensure professionalism.

Date Febuary 12, 2006
To: Martha Collins
From: Richard Davis

Subject: 2006 Digital Carriers

Attached is the supplemental 2006 Digital Carriers reports that is required to support this years growth patterns. As we have discussed in previous phone conversations the January numbers show a decrease in traffic but forecasts still suggest increased traffic therefore we are issuing plans for this contingency.

If the January forecasts prove to be accurate the carriers being placed in the network via these plans will support our future growth accept for areas where growth cannot be predicted. Some areas for example are to densely populated for forecasting. Because the company did not hire enough survey personal to do a thorough job.

Following is an update of our suggested revisions;

Digital Carriers Needed	**Capitol Costs**
52,304	$3,590,625

If your going to hire anyone to provide follow-up forecasts, they should have sufficient lead time. The survey teams, if you want a successful forecast, needs at least three months. 25 team members should be sufficient.

If we can provide farther information please let us know.

5. Recognizing Audience

a. List 10 terms (jargon, acronyms, and/or abbreviations) unique to your degree program. Then, envisioning a lay audience, parenthetically define and briefly explain these terms.

To test the success of your communication abilities, orally share these highly technical terms with other students who have different majors. First, state the term to see if they understand it. If they do not, provide the parenthetical definition. How much does this help? Do they understand now? If not, add the third step—the brief explanation. How much information do the readers need before they understand your highly technical terms?

b. Rewrite the following sentences for multicultural, cross-cultural audiences.

1. Let's meet at 8:30 p.m.
2. The best size for this new component is 16 × 23 × 41.
3. To keep us out of the red, we need to round up employees who can put their pedal to the metal and get us out of this hole.
4. After you correctly code the code, file the changes in the appropriate file.
5. The meeting is planned for 07/09/06.

c. Rewrite the following flawed correspondence. Be sure to consider your multicultural/cross-cultural audience's needs.

To:	Andre Castro, Barcelona; Sunyun Wang, Singapore; Nachman Sumani, Tel Aviv
From:	Ron Shields, New York
Subject:	Brainstorming

I need to pick your brains, fellows. We've got a big one coming up, a killer deal with a major European player. Before I can make the pitch, however, let's brainstorm solutions. The client needs a proposal by 12/11/06, so I need your input before that date. Give me your ideas about the following:

1. What should we charge for our product if the client buys in bulk?
2. What's our turnaround time for production?
3. What plans should we suggest for international rollout?
4. What kinds of training should we implement for the new systems?

E-mail me your feedback by tomorrow, 1:00 p.m. my time, at the latest. Trust me, guys. If we boot this one, everyone's bonuses will be lost ;)

PROBLEM SOLVING THINK PIECES

In teams or individually, consider the following issues. Then, provide your answers either by writing a memo, e-mail message, or report, or by making an oral presentation.

1. This chapter emphasizes the importance of "you usage"—pronouns to build rapport and to create a personalized tone. However, in some instances, personalization might not be appropriate in workplace communication. Decide when correspondence could or should avoid "you usage." Find examples from the workplace or online to prove your point.

2. According to ethical standards, visuals should be precise. Look at the following bar chart. It shows that during one quarter, the company lost over $50,000. However, the visual could be considered misleading, since the bar showing a loss is not drawn to the same scale as the other bars.

   Based on what you have read in this chapter regarding ethics, decide whether this visual is acceptable or not. Share your findings either orally or in a written report. If you conclude that the graphic is ethically misleading, revise it for accuracy.

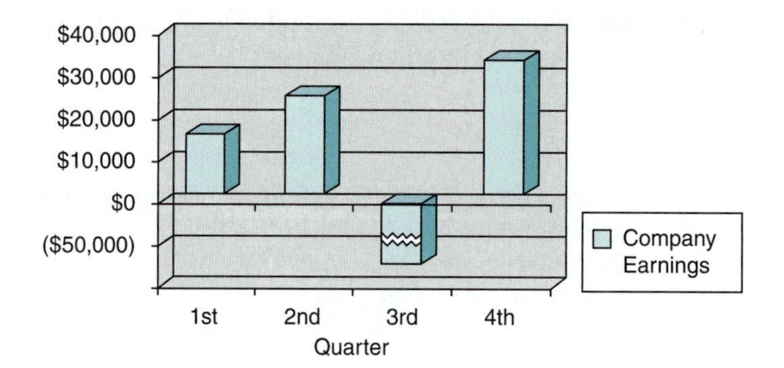

3. Many businesses ask their employees to *write between 6th to 8th grade levels* to make correspondence easier to read. However, there might be instances where correspondence (memos, letters, e-mail messages, reports, etc.) could be or should be written at a higher grade level. When do you believe that it is acceptable for workplace communication to be written at a "higher" level? Explain your decision, either in writing (memo, e-mail message, or report), or present your findings in an oral presentation. Find examples of workplace communication that, in your opinion, have been appropriately written at a "college" level. Explain your decisions.

4. Home Mortgage Bank (HMB) has suffered several lawsuits recently. A former employee sued the company, contending that it practiced "ageism" by promoting a younger employee over him. In an unrelated case, another employee contended that she was denied a raise due to her ethnicity. To combat these concerns, HMB has instituted new human resources practices. Their goal is to ensure that the company meets all governmental personnel regulations, in-

cluding those required by the Equal Employment Opportunity Commission (EEOC), the Family Medical Leave Act (FMLA), and the Americans with Disabilities Act (ADA). HMB is committed to achieving diversity in its workplace.

HMB now needs to hire a new office manager for one of its branch operations. It has two outstanding candidates. Following are their credentials:

- Carlos Gutierrez is a 27-year-old recent recipient of a Master's in Business Administration (MBA) from an acclaimed business school. At this university, he learned many modern business applications, including capital budgeting, human resource management, organizational behavior, diversity management, accounting and marketing management, and team management strategies. He has been out of graduate school for only two years, but his work during that time has been outstanding. As an employee at one of HMB's branches, he has already impressed his bosses by increasing the branch's market share by 28 percent through innovative marketing strategies he learned in college. In addition, his colleagues enjoy working with him and praise his team-building skills. Carlos has never managed a staff, but he is filled with promise.
- Rose Massin is 47 years old and has been out of the workforce for 12 years. During that absence, she raised a family of three children. Now, the youngest child is in school, and Rose wants to reenter the job market. She has a Bachelor's degree in business and was the former office manager of this HMB branch. Thus, she has management experience, as well as mortgage experience.

  While she was branch manager, she did an outstanding job. This included increasing business, working well with employees and clients, and maintaining excellent relationships with lending banks and realtors. She was a highly respected employee and was involved in civic activities and community volunteerism. In fact, she is immediate past president of the local Rotary club. Though she has been out of the workforce for years, she has kept active in the city and has maintained excellent business contacts. Still, she's a bit rusty on modern business practices.

Assignment

Who will you hire from these candidates? Based on the information provided, make your hiring decision. This is a judgment call. Be sure to substantiate your decision with as much proof as possible. Then share this finding as follows:

1. Give a three- to five-minute presentation to share with your classmates the results of your decision.
2. Write a memo, e-mail message, or report about your findings.

WEB WORKSHOP

1. You are the director of human resources for your company. Your job is to write a nondiscrimination policy. To do so, access an online search engine to find other companies' nondiscrimination policies. Compare and contrast what you find. Then, based on your research, write your company's policy.

2. You are head of international relations at your corporation. Your company is preparing to go global. To ensure that your company is sensitive to multicultural concerns, research the cultural traits and business practices in five countries of your choice. To do so, access an online search engine and type in "multicultural business practices in __________" (specify the country's name). Report your findings either orally or in a memo, report, or e-mail message.

Visual Communication
Page Layout and Graphics

OBJECTIVES

When you complete this chapter, you will be able to do the following:

1. Understand the importance of visual communication.
2. Achieve effective page layout.
3. Use graphics to achieve conciseness and clarity.
4. Apply the ten effective traits of visual aids.
5. Create effective tables and figures.
6. Test your knowledge of visual communication through end-of-chapter activities:
   - Case Studies
   - Individual and Team Projects
   - Problem Solving Think Pieces
   - Web Workshop

COMMUNICATION AT WORK

To help his clients understand complex figures, Bert Lang includes visual aids in his proposals.

Bert Lang is an investment banker at Country Commercial Bank. He is writing a proposal to a potential client, Sylvia Light, a retired public health nurse. Sylvia is 68 years old and worked for the Texas Public Health Department for 36 years. She has earned her State of Texas retirement and Social Security benefits. She now has $315,500 allocated as follows: $78,000 in an individual retirement account (IRA), $234,000 in a low-earning certificate of deposit (CD), and $3,500 in her checking account.

Sylvia contacted Bert, asking him to help her organize her portfolio for a comfortable retirement. Bert has studied Sylvia's various accounts and considered her lifestyle and expenditures. Now, he is ready to write the proposal.

In this proposal, Bert wants to show Sylvia how to reallocate her funds. She should keep some ready money available and invest a portion of capital for long-term returns. Currently, too much of her money is tied up in a CD earning 1.1 percent. Bert plans to propose that Sylvia could reallocate her funds as follows:

- $110,000 in an annuity
- $78,000 in an individual retirement account (IRA)
- $45,000 in municipal bonds
- $37,000 in a stock fund
- $42,000 in certificates of deposit (CD)
- $3,500 in a checking account

Like most people, Sylvia is unfamiliar with financial planning. Though she was an expert in her health field, tuberculosis treatment, money matters confuse her. Numbers alone will not explain Bert's vision for her money management.

To make this proposal visually appealing and more readily understandable to Sylvia, Bert will use visual aids. He will provide Sylvia a pie chart to show how he wants to invest her money. Bert will use a line graph to predict how much more money she can earn by reallocating her assets. Finally, he will create a table to clarify the types of investments, the amount in each investment, the interest to be earned, and the fees.

Although Sylvia has always been fiscally conservative, Bert hopes that he can explain the need for growth of capital even in retirement. The graphic aids will visually enhance his written explanation.

The Importance of Visual Communication

Bert Lang, investment banker at Country Commercial Bank, realizes that communicating complex numbers, sums of money, and percentages is challenging. For clients, words are not the only tool in workplace communication. Mr. Lang knows that what he writes is important, but how the text looks on the page and visual aids are equally important. Pages full of wall-to-wall words will not be effective workplace communication. Page layout is not a costly frill. Effective document design is good for your company's business.

The Workplace Communication Context

People read workplace communication for information about a product or service, such as deadlines, costs, personnel, warranties, schedules, delivery dates, and so forth. They read this correspondence while they talk on the telephone, while they commute to work, or while they walk to meetings.

Given these contexts, your readers want you to provide them information quickly, information they can understand at a glance. Reading word after word, paragraph after paragraph takes time and effort, which most readers cannot spare. Good workplace communication allows readers rapid access to information.

Clarity of Content

Visual aids can clarify complex information. Long paragraphs of detailed numbers or other information are hard to read and hard to understand. Graphics, including figures and tables, help readers see the following:

- Trends
- Comparisons
- Percentages
- Facts and figures

Damages and Dangers

If your intended readers fail to read your text because it is visually inaccessible, imagine the possible repercussions. They could damage equipment by not recognizing important information that you have buried in dense blocks of text. The readers could give up on the text and call your company's toll-free hotline for assistance. This wastes your readers' and your coworkers' time and energy. Worse, your readers might hurt themselves and sue your company for failing to highlight potential dangers.

Corporate Identity

Your document—whether it is a memo, letter, report, proposal, Web site, or brochure—is a visual representation of your company, graphically expressing your company's identity. It might be the only way you meet your clients. If your text is unappealing, that is the corporate image your company conveys to the customer. Visually unappealing and inaccessible correspondence can negatively affect your company's sales and reputation. In today's competitive workplace, any leverage you can provide your company is a plus. Document design and visual aids are two ways to appeal to a client.

ACHIEVING EFFECTIVE PAGE LAYOUT

To clarify the importance of page layout, look at the inaccessible meeting minutes in Figure 3.1.

To make these minutes more inviting and achieve more effective page layout, provide your readers visual appeal through *headings, chunking, order, access,* and *variety*.

Headings and Talking Headings

To improve your page layout and make content accessible, use headings and talking headings. Headings—words or phrases such as "Introduction," "Discussion," "Conclusion," "Problems with Employees," or "Background Information"—highlight the content in a particular section of a document. When you begin a new section, you should use a new heading. In addition, use subheadings if you have a long section under one heading. This will help you break up a topic into smaller, more readable units of text.

Talking headings, in contrast, are more informative than headings. A heading helps your readers navigate the text by guiding them to key parts of a document. However, headings such as "Introduction," "Discussion," and "Conclusion" do not tell the readers what content is included in the section. Talking headings, such as "Human Resources Committee Reviews 2006 Benefits Packages," informatively clarify the content that follows.

Figure 3.1
Inaccessible Meeting Minutes

before

MINUTES

The meeting at the Carriage Club was attended by members and guests. After the dinner, Roger Traver introduced the guest speaker, George Smith, university chancellor, and noted his accomplishments and experiences prior to education—U.S. Navy commander, Oak Ridge Laboratory researcher, and politician. Dr. Smith's talk, "Industry and Education Collaboration," was very interesting and included a history of special projects enjoyed by both academics and corporate heads. Dr. Smith suggested that we engineers could work with education to (1) provide training seminars, (2) help in urban development, and (3) provide intern opportunities. Recent industry-education collaborations include training seminars in computers, fiber optics, and human resource options. The chancellor's primary thrust was a request for $100,000 in financial aid for urban development. He said money had already been donated from three sources: a large realty firm, Capital Homes, had given $20,000; a philanthropic group, We Care, had donated a matching $20,000; Dr. Smith's university gave a matching $20,000. The remaining $40,000, Dr. Smith hoped, would come from industry donations. Finally, the chancellor noted that industry could help itself, as well as the community, by providing internships for university undergraduate majors. These internships could either be semester- or year-long arrangements, whereby students would work for minimum wage to learn more about the day-to-day aspects of their chosen fields. The chancellor said that these internships would not only increase the students' theoretical knowledge of engineering by giving them hands-on experience but also make them better future employees for the host engineering companies. Everyone would benefit. Dr. Smith noted that the students would receive a grade and credit for their work. After the speech, our VP introduced new business, calling for nominations for next year's officers; gave us the agenda for our next meeting; and adjourned the meeting.

These minutes are neither clear nor concise. You are given too much data in an unappealing format. Wall-to-wall words are visually unappealing and inaccessible.

One way to create a talking heading is to use a subject (someone or something performing the action), a verb (the action), and an object (something acted upon) (Christensen).

Example

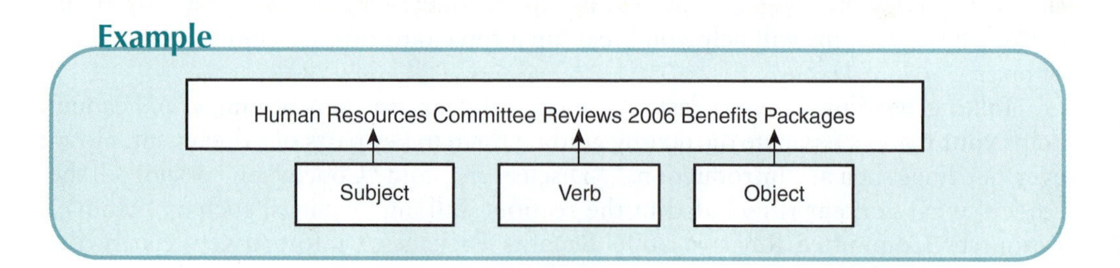

TABLE 3.1 TALKING HEADINGS
Sentences Used as Talking Headings
Rude Customer Service Leads to Sales Losses
Accounting Department Requests Feedback on Benefits Package
Corporate Profit Sharing Decreases to 27 percent
Parking Lot Congestion Angers Employees
Phrases Used as Talking Headings
Budget Increases Frozen until 2007
Outsourced Workers Leading to Corporate Layoffs
Harlan Cisneros—New Departmental Supervisor
EEOC: Questions about Company Hiring Practices

Another way to create talking headings is to use informative phrases, such as "Problems Leading to Employee Dissatisfaction," "Uses of Company Cars for Personal Errands," and "Cost Analysis of Technology Options for the Accounting Department."

Table 3.1 provides examples of informative talking headings:

Chunking

An easy way to improve your document's design is to break text into smaller chunks of information, a technique called chunking. When you use chunking to separate blocks of text, you help your readers understand the overall organization of your correspondence. They can see which topics go together and which are distinct.

Chunking to organize your text is accomplished by using any of the following techniques:

- White space (horizontal spacing between paragraphs, created by double or triple spacing)
- Rules (horizontal lines typed across the page to separate units of information)
- Section dividers and tabs (used in longer reports to create smaller units)
- Headings

Order

Once a wall of unbroken words has been separated through headings and chunking to help the reader navigate the text, the next benefit a reader wants from your text is a sense of order. What is most important on the page? What is less important? What is least important?

You can help your audience prioritize information by ordering ideas. One way to accomplish this goal is by using a hierarchy of headings set apart from each other through various techniques, such as typeface, type size, capitalization, density, and position.

Typeface. Some different typefaces (or fonts) include Times New Roman, Courier, *Monotype Corsiva*, **Forte**, *Bellevue*, Arial, CASTELLAR, **Impact**, Rockwell, and Tekton.

Whichever typeface you choose, it will either be a serif or sans serif typeface. Serif type has "feet" or decorative strokes at the edges of each letter. This typeface is commonly used in text because it is easy to read, allowing the reader's eyes to glide across the page. Times New Roman is a serif font, for example.

SERIF ⟵ decorative feet

Sans serif (as seen in this parenthetical comment) is a block typeface that omits the feet or decorative lines. This typeface is often used for headings. Arial is a sans serif font, for example.

SANS SERIF ⟵ no decorative feet

Though you have many font typefaces to choose from, all are not appropriate for business documents. Times New Roman and Arial are the best to use for letters, memos, reports, resumes, and proposals, because these font types are most professional looking and are easiest to read.

Type Size. Another way of prioritizing for your readers is through the size of your type. For example, a first-level heading could be in 18-point type. The second-level heading would then be set in 16-point type, the third-level heading in 14-point type, and the fourth-level heading in 12-point type. Avoid using headings smaller than 12-point font. A smaller font size is hard to read.

Capitalization. You can order headings through capitalization. An "all cap" heading (ALL CAP) would take priority over headings typed in "title case." "Title Case" includes a mixture of upper- and lower-case characters.

Density. The weight of the type also prioritizes your text. Type density is created by boldfacing.

Position. Your headings can be centered, aligned with the left margin, indented, or outdented (hung heads). No one approach is more valuable or more correct than another. The key is consistency. If you center your first-level heading, for example, and then place subsequent heads at the left margin, this should be your model for all chapters or sections of that report.

Access

A fourth way to assist your audience is by helping them access information rapidly. You can use any of the following highlighting techniques to help the readers filter out extraneous or tangential information and focus on key ideas:

White Space. In addition to horizontal space created by double or triple spacing you also can create vertical space by indenting. This vertical white space breaks up the monotony of wall-to-wall words. White space invites your readers into the text and helps the audience focus on the indented points you want to emphasize.

Bullets. Bullets, used to emphasize items within an indented list, are created by using asterisks (*), hyphens (-), bullet characters (•), and Webdings or Wingdings (✓).

Numbering. Whereas bullets create lists of items equal in importance, numbered lists show sequence or importance. Use a numbered list for step-by-step procedures or to show that the first point is more important than subsequently numbered points.

Underlining. Use underlining cautiously. If you underline too frequently, none of your information will be emphatic. One underlined word or phrase will call attention to itself and achieve reader access. If you underline too many words or phrases, you lose the emphasis and visually distract your readers.

Italics. *Italics*, like underlining, should be used with caution. Italicize to emphasize a keyword or phrase. Do not overuse this highlighting technique.

Text Boxes. An excellent way to emphasize a key point is by creating a text box. You can add to the emphasis by using shadows as well.

Note

Be sure to hand-tighten the nuts at this point. Once you have completed the installation, go back and *securely* tighten all nuts.

A text box with shadow highlights important information. Adding "NOTE" calls further attention to the point.

Color. A word typed in red, for example, could be used to emphasize the potential for **danger. Warnings** are highlighted in orange, and **cautions** in yellow.

You can use background colors to emphasize ideas, but be careful. Background colors can be distracting, and you must choose text colors that are easy to read against the colored background.

Gradients can create an interesting background.

For readability, the best color combination is black text on a white background.

Inverse Type. To help readers access information, use inverse type—printing white on black, versus black on white.

When it comes to highlighting techniques, more is *not* better. A few highlighting techniques help your readers filter out background data and focus on key points. Too many highlighting techniques are distracting and clutter page layout.

Variety

You might want to use smaller or larger paper; vary the weight of your paper, using 10-pound, 12-pound, or heavier card stock; or even print your text on colored paper.

You can vary the document design as follows:

Choose Landscape Orientation. Rather than print your text using a vertical, portrait orientation ($8^1/_2'' \times 11''$), you could choose to print the text horizontally, using a landscape orientation ($11^1/_2'' \times 8''$).

Use More Columns. Provide your reader two to five columns of text for variety. Columns are commonly used in brochures, newsletters, fliers, and other marketing publications.

Vary Gutter Width. Columns of text are separated by vertical white space called gutters. You can vary these spaces for more visual appeal.

Portrait Orientation

($8^1/_2$ x 11)

Landscape Orientation

(11 x $8^1/_2$)

Landscape Orientation with Three Columns

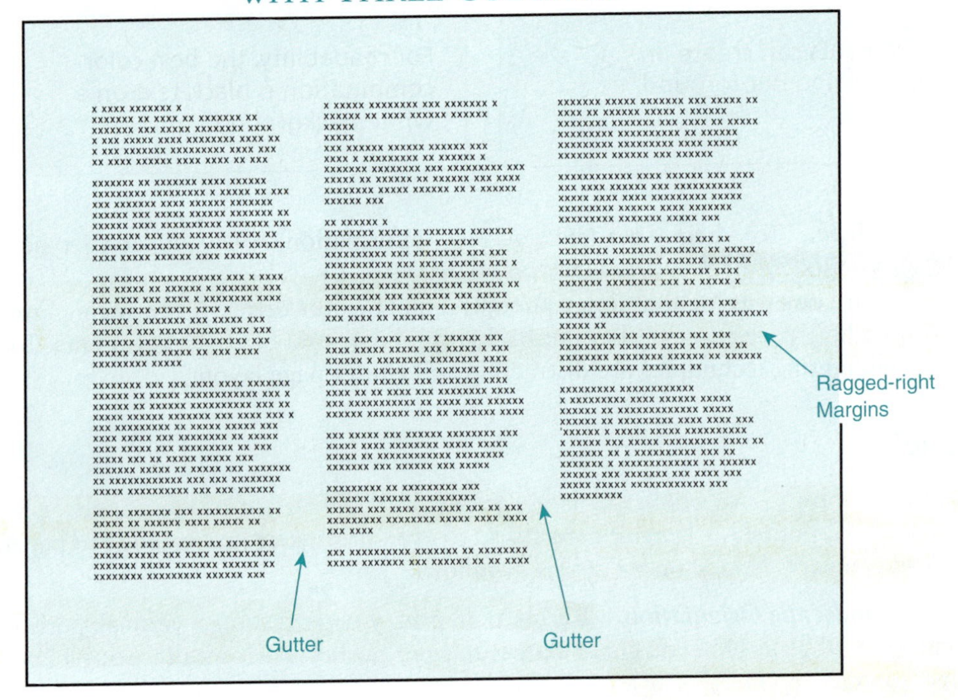

Use Ragged-Right Margins. Once, fully justified text (both right and left margins aligned) was considered professional, giving the text a clean look. Now, however, studies confirm that right-margin-aligned text is harder for the audience to read. It is too rigid. In contrast, ragged-right type (the right margin is not aligned) is easier to read and more pleasing to the eye. Ragged-right text is a visually appealing page layout preference (Everson 397).

Communicating with Graphics

Although you can vary your page layout by printing horizontally and by using multiple columns, the audience is still confronted by words, words, and more words. Many readers will not want to wade through text.

Words are not your only means of communication. You can achieve effective workplace communication by varying your method of delivery. Graphics are an excellent alternative. Many people are more comfortable grasping information visually than verbally. Although it is a cliché, a picture is often worth a thousand words.

To clarify the importance of graphics within text, look at Figure 3.2, which adds a pie chart to the meeting minutes in addition to headings, subheadings, and bullets.

The Importance of Graphics

Although your writing may have no grammatical or mechanical errors and you may present valuable information, you won't communicate effectively if your information is inaccessible.

The goal of effective workplace communication is to communicate information easily. To help readers digest data easily or clearly see comparative changes, supplement or replace text with graphics. In workplace communication, visual aids help you achieve conciseness and clarity.

Conciseness

Visual aids allow you to present large amounts of information in a small space. Words used to convey data can double, triple, or even quadruple the space needed to report information. By using graphics, you can also delete many dead words and phrases.

Clarity

Visual aids can clarify complex information. Graphics help readers more easily understand trends; compare and contrast information; clarify difficult percentages, facts, and figures; and create visual impact (as shown in Table 3.2).

Traits of Effective Visual Aids

Successful tables and figures accomplish the following:

- Are integrated with the text (i.e., the graphic complements the text; the text explains the graphic).
- Are appropriately located (preferably immediately following the text referring to the graphic and not a page or pages later).
- Add to the material explained in the text (without being redundant).
- Communicate important information that could not be conveyed easily in a paragraph or longer text.
- Do not contain details that detract from rather than enhance the information.
- Are an effective size (not too small or too large).
- Are readable.
- Are correctly labeled (with legends, headings, and titles).
- Follow the style of other figures or tables in the text.
- Are well conceived and carefully executed.

after

First level, hung head, all cap, Arial

MINUTES

The meeting at the Carriage Club was attended by 30 members and guests. After the dinner, Roger Traver introduced the guest speaker, George Smith, university chancellor, and noted his accomplishments and experiences prior to education:

- U.S. Navy commander
- Oak Ridge Laboratory researcher
- Politician

Bulleted points for easy access

Dr. Smith's talk, "Industry and Education Collaboration" was very interesting and included a history of special projects enjoyed by both academics and corporate heads. Dr. Smith suggested that we engineers could work with education to accomplish two goals.

Urban Development

Second-level heading

The chancellor's primary thrust was a request for $100,000 in financial aid for urban development. He said money had already been donated from three sources, but business and industry can still help. The following pie chart clarifies what money has been pledged and how industry donations are still needed.

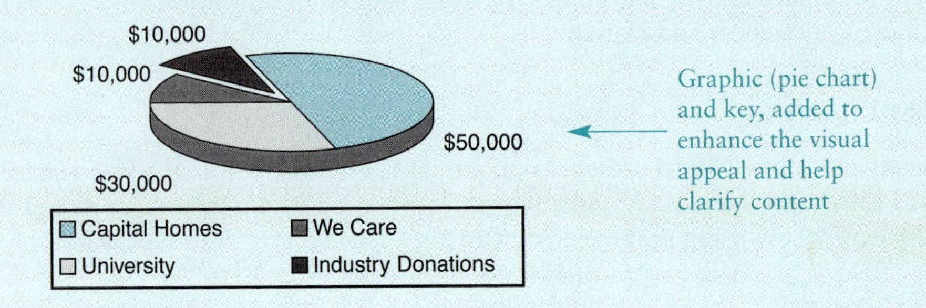

Internships

The chancellor noted that industry could help itself, as well as the community, by providing internships for university undergraduate majors:

- Semester-long internships
- Year-long internships

Students would work for minimum wage to learn more about the day-to-day aspects of their chosen field. The chancellor said that these internships would not only increase the students' theoretical knowledge of engineering by giving them hands-on experience but also make them better future employees for the host engineering companies. Everyone would benefit. Dr. Smith noted that the students would receive a grade and credit for their work.

Training Seminars

Third-level heading

Recent industry-education collaborations include training seminars in computers, fiber optics, and human resource options.

Conclusion

After the speech, our VP introduced new business, calling for nominations for next year's officers; gave us the agenda for our next meeting; and adjourned the meeting.

Figure 3.2 Improved Document Design of Meeting Minutes

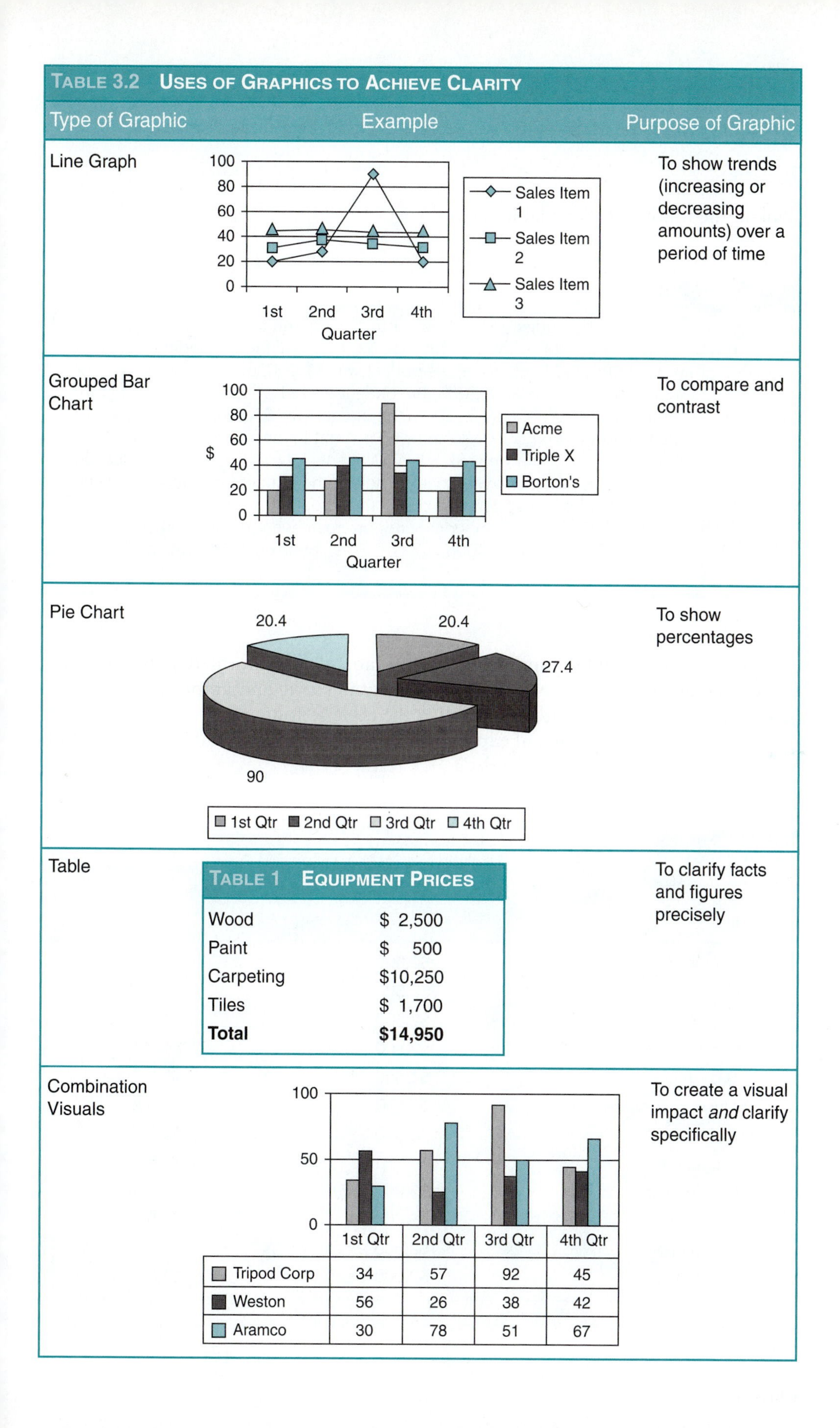

TABLE 3.2 USES OF GRAPHICS TO ACHIEVE CLARITY

Type of Graphic	Example	Purpose of Graphic
Line Graph		To show trends (increasing or decreasing amounts) over a period of time
Grouped Bar Chart		To compare and contrast
Pie Chart		To show percentages
Table		To clarify facts and figures precisely
Combination Visuals		To create a visual impact *and* clarify specifically

TABLE 1 EQUIPMENT PRICES

Wood	$ 2,500
Paint	$ 500
Carpeting	$10,250
Tiles	$ 1,700
Total	**$14,950**

	1st Qtr	2nd Qtr	3rd Qtr	4th Qtr
Tripod Corp	34	57	92	45
Weston	56	26	38	42
Aramco	30	78	51	67

Types of Graphics

No matter what type of graphic you use, it will either be a table or a figure.

Tables

Look at the following unreadable text.

before

In January 2006, the budget was $1,500, but the expenditure for that month was $2,000. In February 2006, the budget was $1,500, but the expenditure for that month was $2,500. In March 2006, the budget was $1,000, but the expenditure for that month was $2,500. In April 2006, the budget was $1,000, but the expenditure for that month was $2,500. In May 2006, the budget was $5,000, but the expenditure for that month was $1,500. No money was spent in June 2006. The budget had been $5,000. In July 2006, only $2,500 was spent. July's budget was $5,000. In August 2006, again no money was spent, whereas the budget for August had been $2,500. In September and October 2006, the expenditure of $5,000 matched the budget. Similarly, the November expenditure matched the budgeted amount of $2,500. Finally, in December 2006, $2,000 was spent compared to the budget of $1,500.

Notice how this text is improved when you tabulate the information in a table. Because effective workplace communication integrates text and graphics, you should provide an introductory sentence prefacing the table as follows:

This table reveals the actual expenditure for each month in 2006 versus the budgeted amount for those same months.

after

Expenditure vs. Budgeted Amounts for 2006

Month	Expenditure	Budgeted Amount
January	$2,000	$1,500
February	$2,500	$1,500
March	$2,500	$1,000
April	$2,500	$1,000
May	$1,500	$5,000
June	$ 0	$5,000
July	$2,500	$5,000
August	$ 0	$2,500
September	$5,000	$5,000
October	$5,000	$5,000
November	$2,500	$2,500
December	$2,000	$1,500

This table has advantages for both the writer and the reader.

- The headings eliminate needless repetition of words, thereby making the text more readable.
- The audience easily can see the comparison between the expenditures and budgeted amounts.
- The table highlights the content's significant differences.
- The table allows for easy future reference. Tables could be created for each year. Then the reader quickly could compare the changes in expenditures.
- If this information is included in a report, the writer will reference the table in the List of Illustrations. This creates easy access for the reader.

Criteria for Effective Tables

To construct tables correctly, do the following:

1. Number tables in order of presentation (i.e., Table 1, Table 2, Table 3, etc.).
2. Title every table. In your text, refer to the table by its number, not its title. Simply say, "Table 1 shows . . . ," "As seen in Table 1," or "The information in Table 1 reveals. . . . "
3. Present the table as soon as possible after you have mentioned it in your text. Preferably, place the table on the same page as the appropriate text, not on a subsequent, unrelated page or in an appendix.
4. Do not present the table until you have mentioned it.
5. Use an introductory sentence or two to lead into the table.
6. After you have presented the table, explain its significance. You might write, "The expenditures in both March and April exceeded the budgets by $3,000."
7. Write headings for each column. Choose terms that summarize the information in the columns. For example, you could write "% of Error," "Length in Ft.," or "Amount in $."
8. Since the size of columns is determined by the width of the data or headings, you may want to abbreviate terms. If you use abbreviations, however, be sure your audience understands your terminology.
9. Center tables between right and left margins. Do not crowd them on the page.
10. Separate columns with ample white space, vertical lines, or dashes.
11. Show that you have omitted information by printing two or three periods, a hyphen, or dash in an empty column.
12. Be consistent when using numbers. Use either decimals or numerators and denominators for fractions. You could write $3^1/_4$ and $3^3/_4$ or 3.25 and 3.75. If you use decimal points for some numbers but other numbers are whole, include zeroes. For example, write 9.00 for 9.
13. If you do not conclude a table on one page, on the second page write *Continued* in parentheses after the number of the table and the table's title.

Figures

Another way to enhance your workplace communication is to use figures. Whereas tables eliminate needless repetition of words, figures highlight and supplement important points in your writing. Like tables, figures help you communicate with your reader.

Types of figures include the following:

- Bar charts
- Pictographs
- Gantt charts
- Pie charts
- Line charts
- Combination charts
- Flowcharts
- Organizational charts
- Line drawings
- Renderings and virtual reality drawings
- Photographs
- Icons

All of these types of figures can be computer generated using an assortment of computer programs. The program you use depends on your preference, software, and hardware.

Criteria for Effective Figures

To construct figures correctly, do the following:

1. Number figures in order of presentation (i.e., Figure 1, Figure 2, Figure 3, etc.).
2. Title each figure. When you refer to the figure, use its number rather than its title: for example, "Figure 1 shows the relation between the average price for houses and the actual sales prices."
3. Preface each figure with an introductory sentence.
4. Do not use a figure until you have mentioned it in the text.
5. Present the figure as soon as possible after mentioning it instead of several paragraphs or pages later.
6. After you have presented the figure, explain its significance. Do not let the figure speak for itself. Remind the reader of the important facts you want to highlight.
7. Label the figure's components. For example, if you are using a bar or line chart, label the x- and y-axes clearly. If you are using line drawings, pie charts, or photographs, use clear call-outs (names or numbers that indicate particular parts) to label each component. You also can include telescopic views to highlight small details in your graphic.
8. When necessary, provide a legend or key at the bottom of the figure to explain information. For example, a legend in a bar or line chart will explain what each different colored line or bar means. In line drawings and photographs, you can use numbered call-outs in place of names. If you do so, you will need a legend at the bottom of the figure explaining what each number means.
9. If you abbreviate any labels, define these in a footnote. Place an asterisk (*) or a superscript number ([1, 2, 3]) after the term and then at the bottom of the figure where you explain your terminology.
10. If you have taken information from another source, note this at the bottom of the figure.

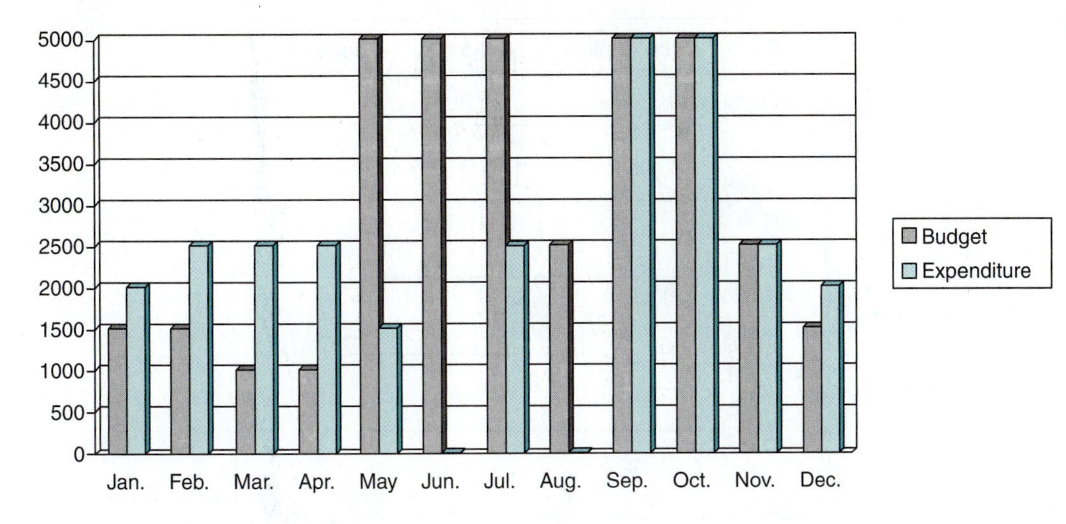

Figure 3.3 3-D Vertical Bar Chart

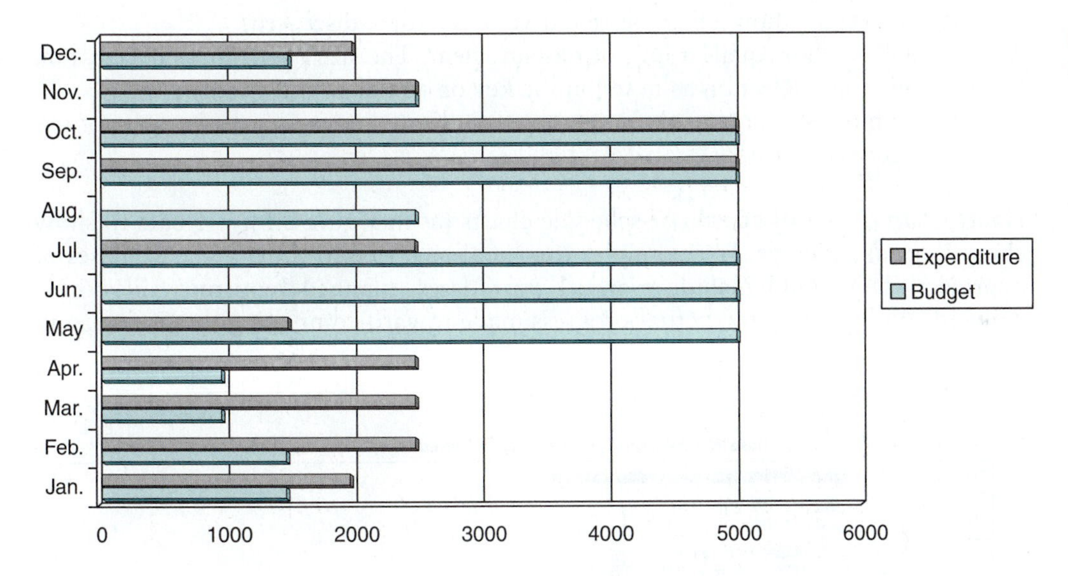

Figure 3.4 Horizontal Bar Chart

11. Frame the figure. Center it between the left and right margins or place it in a text box.
12. Size figures appropriately. Do not make them too small or too large.

Bar Charts. Bar charts show either vertical bars (as in Figure 3.3) or horizontal bars (as in Figure 3.4). These bars are scaled to reveal quantities and comparative values. You can shade, color, and crosshatch the bars to emphasize the contrasts. If you do so, include a key explaining what each symbolizes, as in Figure 3.3 and 3.4.

Pictographs. Pictographs (as in Figure 3.5) use picture symbols instead of bars to show quantities. To create effective pictographs, do the following:

Figure 3.5
Pictograph

Pictographs add interest to your text. Be sure that the picture clearly represents the topic.

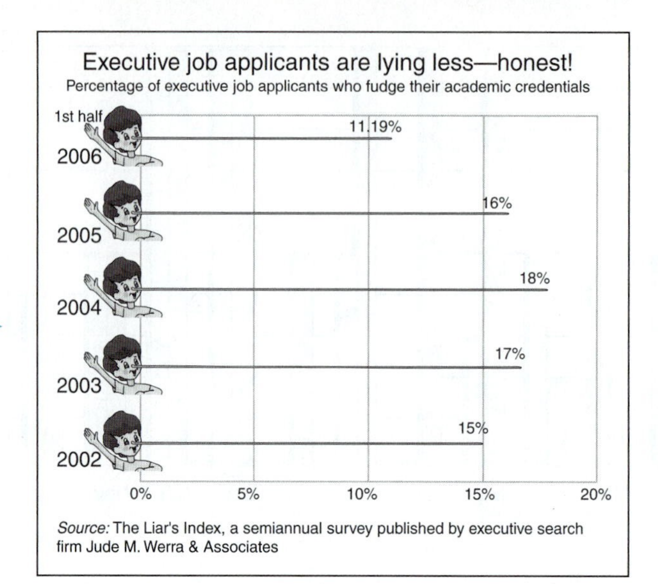

1. The picture should be representative of the topic discussed.
2. Each symbol equals a unit of measurement. The size of the units depends on your value selection as noted in the key or on the x- and y-axes.
3. Use more symbols of the same size to indicate a higher quantity; do not use larger symbols.

Gantt Charts. Gantt charts, or schedule charts (as in Figure 3.6), use bars to show chronological activities. For example, your goal might be to show a client phases of a project. This could include planned start dates, planned reporting milestones, planned completion dates, actual progress made toward completing the project, and

Figure 3.6 Gantt Chart

Gantt charts are an excellent visual aid for conveying project start dates, anticipated completion times, and current status.

The key or legend helps your audience understand the meaning of each bar.

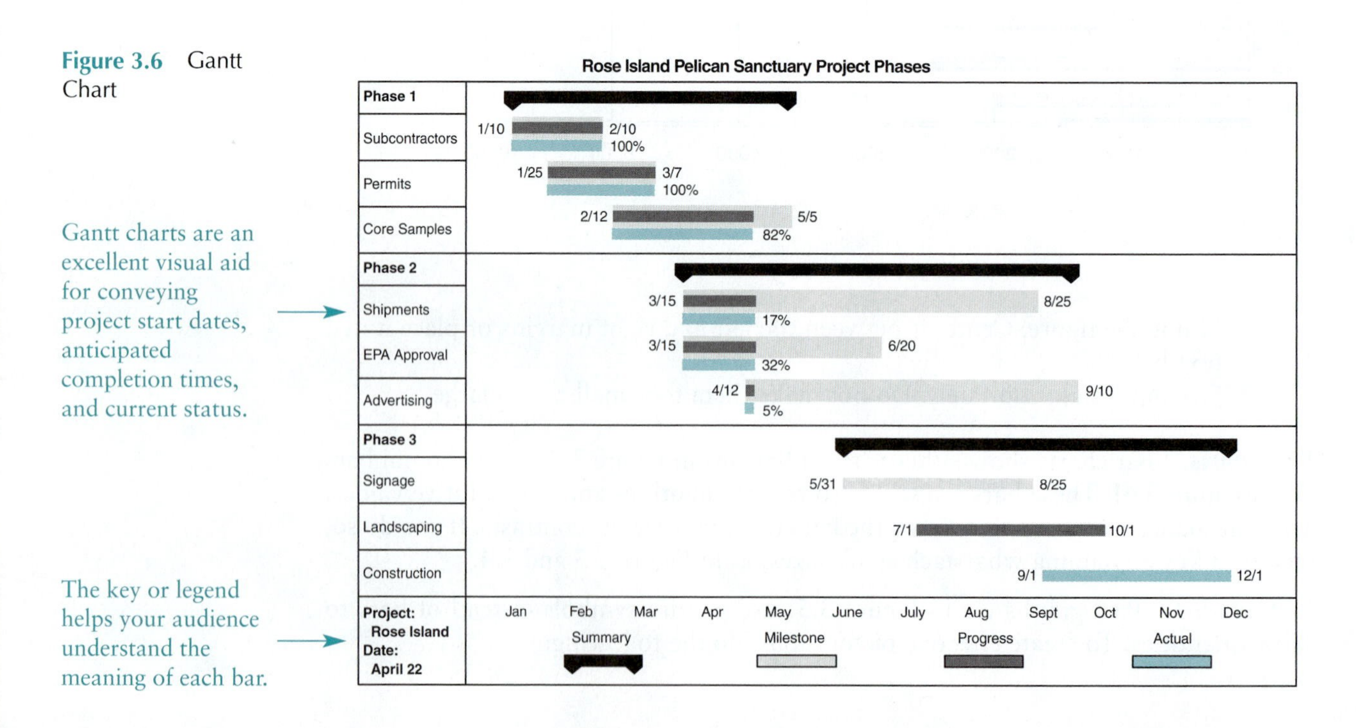

work remaining. Gantt charts are an excellent way to represent these activities visually. They are often included in proposals to project schedules or in reports to show work completed.

To create successful Gantt charts, do the following:

1. Label your x- and y-axes. For example, if the y-axis represents the various activities scheduled, then the x-axis represents time (either days, weeks, months, or years).
2. Provide grid lines (either horizontal or vertical) to help your readers pinpoint the time accurately.
3. Label your bars with exact dates for start or completion.
4. Quantify the percentages of work accomplished and work remaining.
5. Provide a legend or key to differentiate between planned activities and actual progress.

Pie Charts. Use pie charts (as in Figure 3.7) to illustrate portions of a whole. The pie chart represents information as pie-shaped parts of a circle. The entire circle equals 100 percent, or 360 degrees. The pie pieces (the wedges) show the various divisions of the whole.

To create effective pie charts, do the following:

1. Be sure that the complete circle equals 100 percent, or 360 degrees.
2. Begin spacing wedges at the 12 o'clock position.
3. Use shading, color, or crosshatching to emphasize wedge distributions.
4. Use horizontal writing to label wedges.
5. If you do not have enough room for a label within each wedge, provide a key defining what each shade, color, or crosshatching symbolizes.
6. Provide percentages within wedges; when possible.
7. Do not use too many wedges; this would crowd the chart and confuse readers.
8. Make sure that different sizes of wedges are fairly large and dramatic.

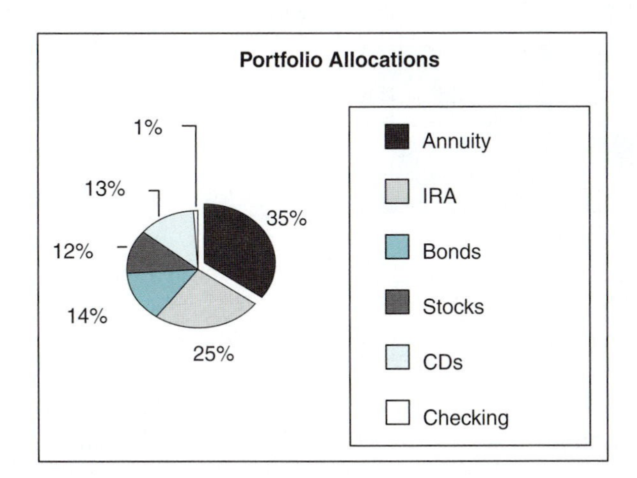

Figure 3.7 Pie Chart

Writer's Insight

Bert Lang, investment banker to Sylvia Light, says, "I knew that raw numbers would not communicate clearly to my customer. When I propose investment plans, it's just too hard for most people to visualize where their money will be located.

To solve this problem, I created a pie chart for Ms. Light. I also emphasized where I planned to invest the largest percentage of her money (annuity allocations) by pulling a wedge apart from the pie.

Finally, I made sure that I quantified the allocations by providing the percentages and using a legend to identify each wedge."

Line Charts. Line charts reveal relationships between sets of figures. To make a line chart, plot sets of numbers and connect the sets with lines. These lines create a picture showing the upward and downward movement of quantities, such as in Figure 3.8.

Figure 3.8 Line Chart

For clarity, line charts need an x- and y-axis. Be sure to label each axis and title the chart.

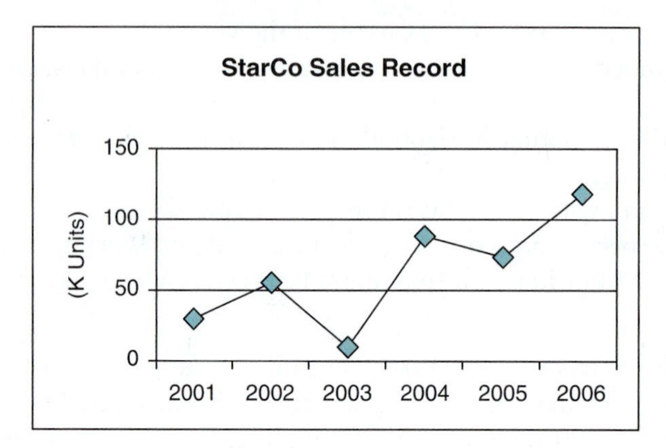

Combination Charts. A combination chart reveals relationships between two sets of figures. To do so, it uses a combination of figure styles, such as a bar chart and an area chart (as shown in Figure 3.9) or a bar chart and a line chart. The value of a combination chart is that it adds interest and distinguishes the two sets of figures by depicting them differently.

Figure 3.9 Combination Chart

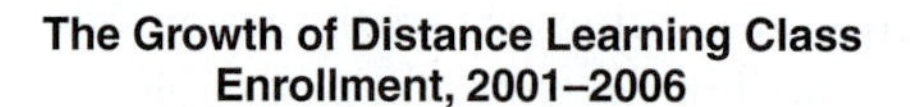

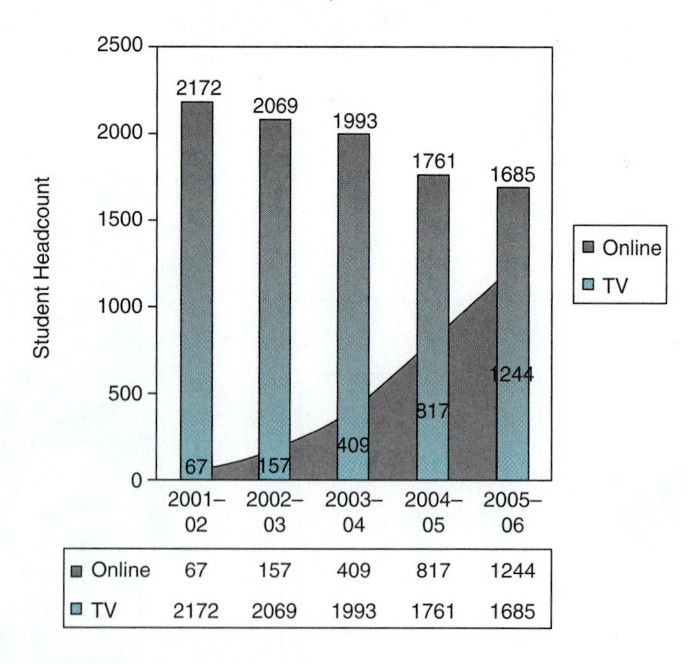

Flowcharts. You can show a chronological sequence of activities using a flowchart. When using a flowchart, remember that ovals represent starts and stops, rectangles represent steps, and diamonds equal decisions (see Figure 3.10).

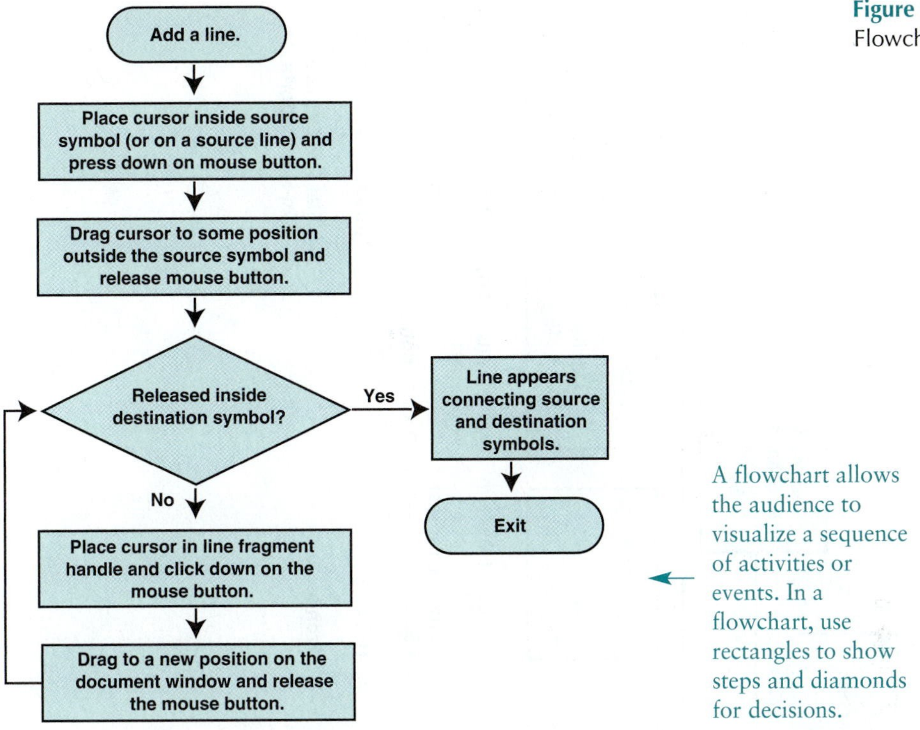

Figure 3.10
Flowchart

A flowchart allows the audience to visualize a sequence of activities or events. In a flowchart, use rectangles to show steps and diamonds for decisions.

Organizational Charts. Organizational charts (as in Figure 3.11) show the chain of command in an organization. You can use boxes around the information or use white space to distinguish among levels in the chart. An organizational chart helps your readers see where individuals work within a business and their relation to other workers.

Line Drawings. Use line drawings to show the important parts of a mechanism or to enhance your text cosmetically. To create line drawings, do the following:

1. Maintain correct proportions in relation to each part of the object drawn.
2. If a sequence of drawings illustrates steps in a process, place the drawings in left-to-right or top-to-bottom order.
3. Using call-outs to name parts, label the components of the object drawn (see Figure 3.12).
4. If there are numerous components, use a letter or number to refer to each part. Then reference this letter or number in a key (see Figure 3.13).
5. Use exploded views (Figures 3.12 and 3.13) or cutaways (Figure 3.14) to highlight a particular part of the drawing.

Renderings and Virtual Reality Drawings. Two different types of line drawings are renderings and virtual reality views. Both offer 3-D representations of buildings or sites. Often used in the architectural and engineering industry, these 3-D drawings (as shown in Figures 3.15 and 3.16) help clients get a visual idea of what services your company can provide.

Renderings and virtual reality drawings add lighting, materials, and shadow and reflection mapping to mimic the real world and allow customers to see what a building or site will look like in a photo-realistic setting.

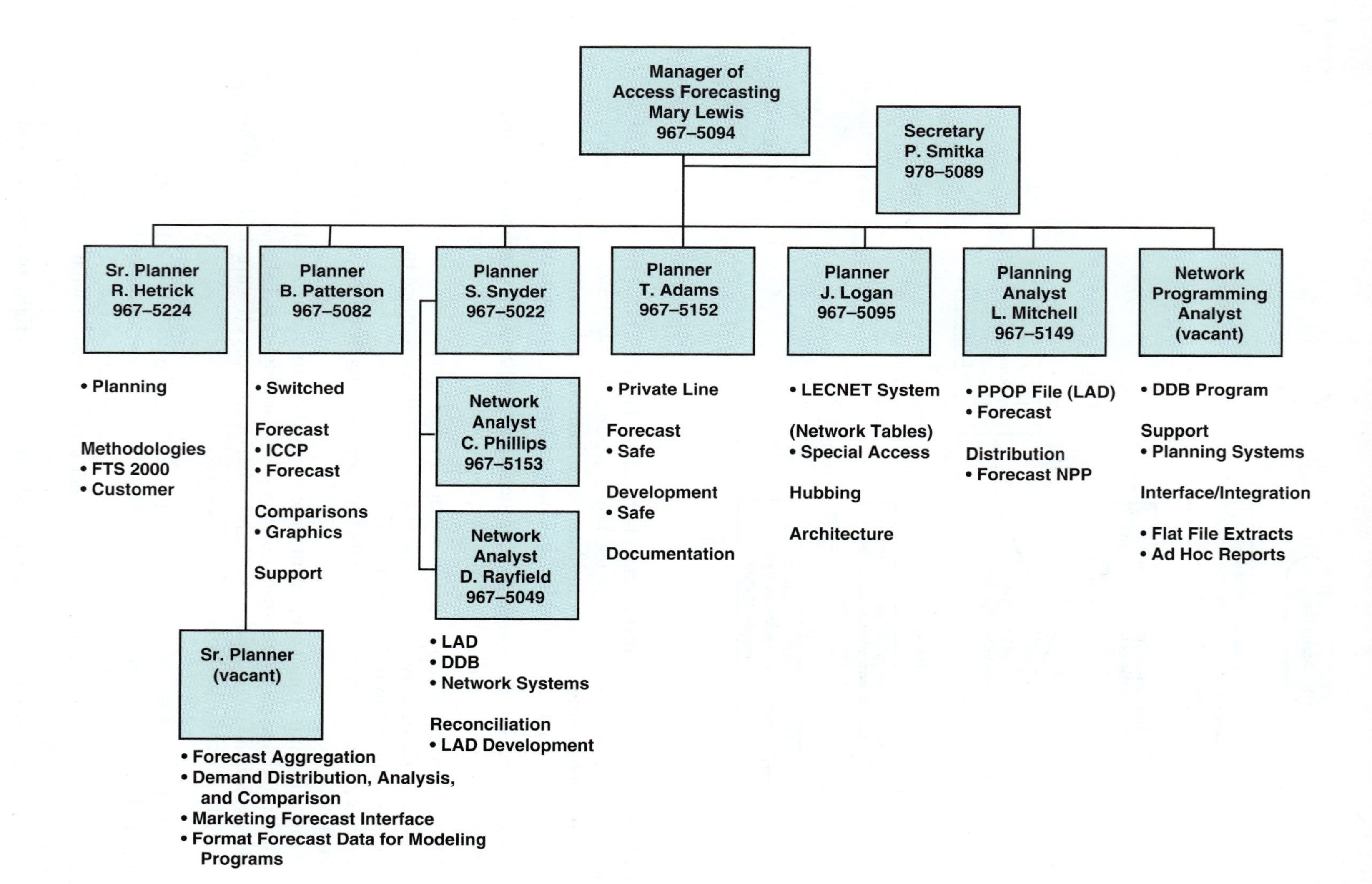

Figure 3.11 Organizational Chart

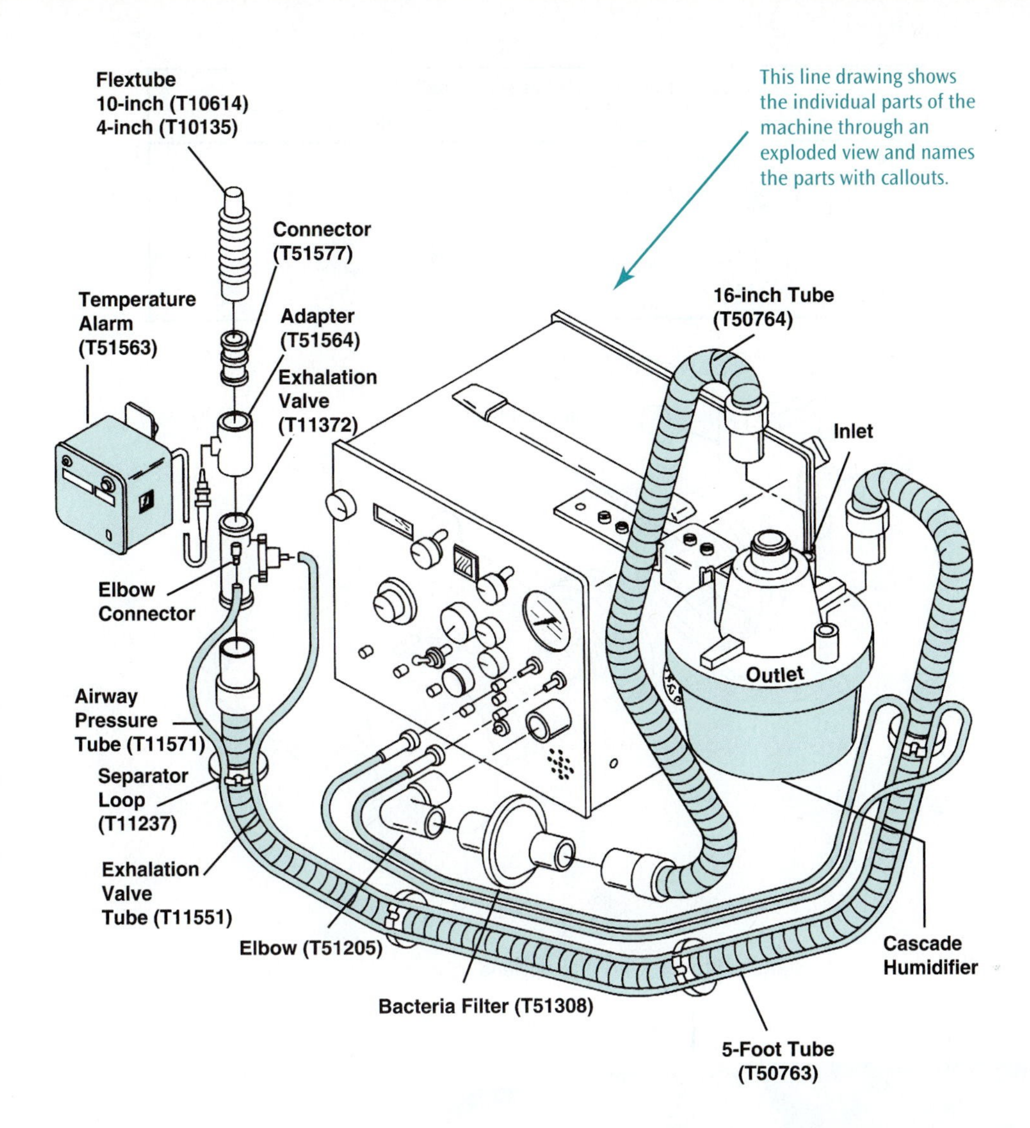

Figure 3.12 Line Drawing of Ventilator

Photographs. A photograph can illustrate your text effectively. Like a line drawing, a photograph can show the components of a mechanism. If you use a photo for this purpose, you will need to label (name), number, or letter parts and provide a key. Photographs are excellent visual aids because they emphasize all parts equally. Their primary advantage is that they show something as it truly is.

Photographs have one disadvantage, however. They are difficult to reproduce. Whereas line drawings photocopy well, photographs do not. See Figure 3.17.

Icons. "In the United States, according to *Nation's Business* magazine, an estimated 15 million adults holding jobs at the beginning of the 21st century were functionally illiterate. The American Council of Life Insurance reported that 75 percent of the Fortune 500 companies provide some level of remedial training for their workers"

Figure 3.13 Line Drawing of Exhalation Valve (exploded view with key)

Exhalation Valve Parts List		
Item	Part Number	Description
1	000723	Nut
2	003248	Cap
3	T50924	Diaphragm
4	Reference	Valve Body
5	Reference	Elbow Connector
—	T11372	Exhalation Valve

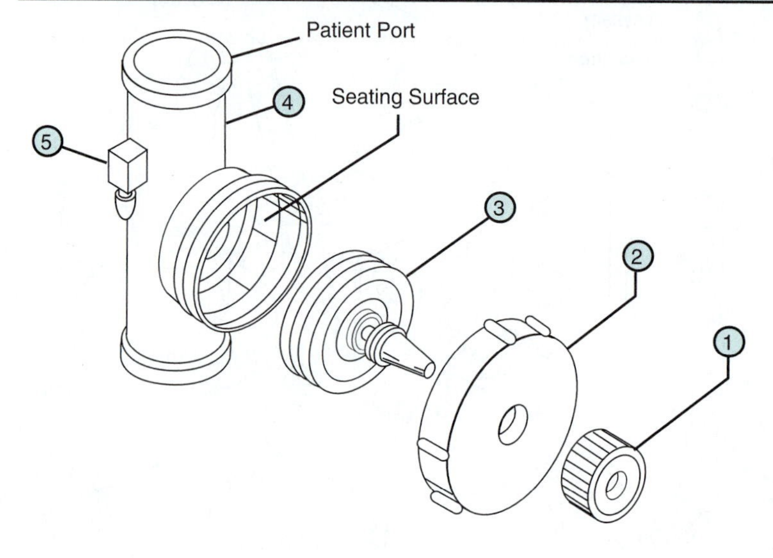

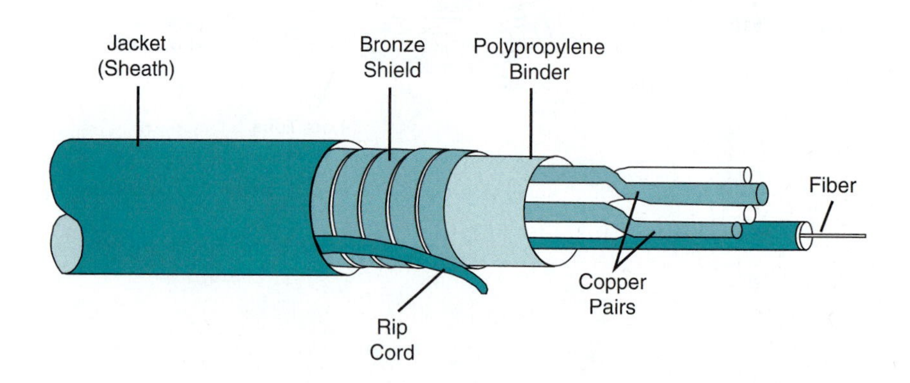

Figure 3.14 Line Drawing of Cable (cutaway view)

("Functional illiteracy"). In today's global economy, consumers speak diverse languages. Given these facts, how can business employees communicate to people who cannot read and to people who speak different languages? Icons offer one solution. Icons (as in Figure 3.18) are visual representations of a capability, a danger, a direction, an acceptable behavior, or an unacceptable behavior.

For example, the computer industry uses an icon of an open manila folder to represent a computer file. In manuals, a jagged lightning stroke iconically represents the danger of electrocution. On streets, an arrow represents the direction we should travel; on computers, the arrow shows us which direction to scroll. Universally depicted stick

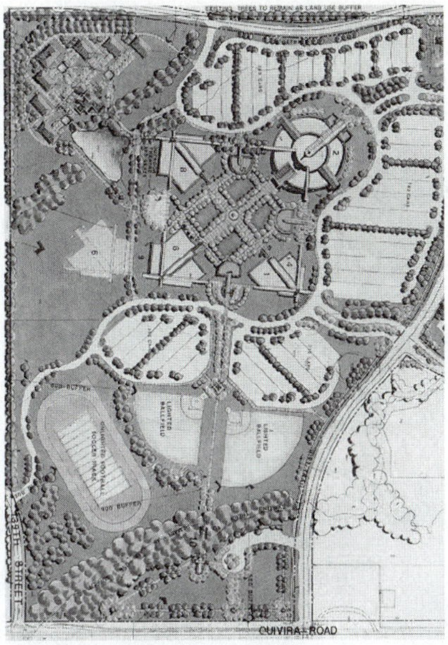

Figure 3.15 Architectural Rendering (Courtesy of George Butler Associates, Inc.)

Architects and engineers use renderings (drawings) to help clients visualize proposed constructions.

Figure 3.16 Virtual Reality Drawing (Courtesy of Johnson County Community College)

figures of men and women greet us on restroom doors to show us which rooms we can enter and which rooms we must avoid.

When used correctly, icons can save space, communicate rapidly, and help readers with language problems understand the writer's intent.

To create effective icons, follow these suggestions:

1. Keep it simple. You should try to communicate a single idea. Icons are not appropriate for long discourse.

Figure 3.17 Photograph of Mechanical Piping (Courtesy of George Butler Associates, Inc.)

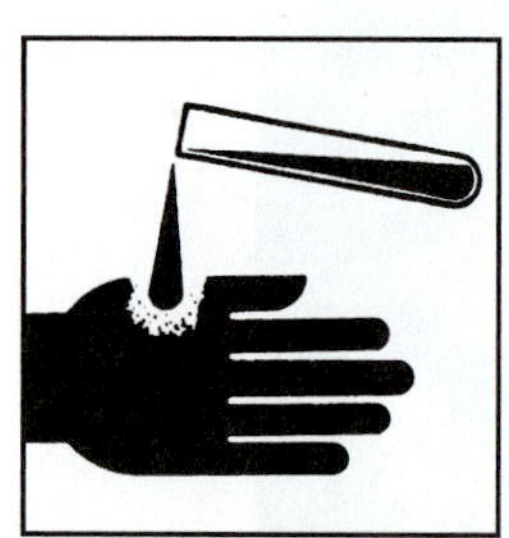

Figure 3.18 Icons of Hazards

2. Create a realistic image. This could be accomplished by representing the idea as a photograph, drawing, caricature, outline, or silhouette.
3. Make the image recognizable. A top view of a telephone or computer terminal is confusing. A side view of a playing card is completely unrecognizable. Select the view of the object that best communicates your intent.
4. Avoid cultural and gender stereotyping. For example, if you are drawing a hand, you should avoid showing any skin color, and you should stylize the hand so it is neither clearly male nor female.
5. Strive for universality. Stick figures of men and women are recognizable worldwide.

Page Layout and Graphics Checklist

The page layout and graphics checklist will give you the opportunity for self-assessment and peer evaluation of your writing.

Page Layout and Graphics Checklist

Page Layout

__ 1. Did you make the text accessible through headings and talking headings?

__ 2. Is your document organized, using chunking, white space, horizontal rules, or section dividers?

__ 3. Did you prioritize the order of information by changing typeface, type size, density, and spacing?

__ 4. Have you created accessible information with white space, bullets, numbers, boldface, italics, all caps, etc.?

__ 5. Does your page layout have variety through the use of columns, page orientation, and margins?

Graphics

__ 1. Are your graphics integrated in the text through lead-in comments or follow-up explanations?

__ 2. Are the graphics appropriately located immediately following the text that refers to the graphics?

__ 3. Does the visual add to the text without being redundant?

__ 4. Did you communicate important information not easily conveyed in words?

__ 5. Are your graphics an effective size (not too big to be distracting or too small to read)?

__ 6. Are your graphics correctly labeled with legends and titles?

__ 7. Do all graphics maintain a consistent style?

__ 8. Are all graphics consecutively numbered throughout your text or presentation?

__ 9. Have you used call-outs or keys to label your figure components?

__10. Have you chosen the appropriate type of graphic (table or figure including the correct type of figure) to convey your information?

CASE STUDIES

1. Bert Lang is an investment banker at Country Commercial Bank, as noted in this chapter's beginning scenario. He is writing a proposal to a potential client, Sylvia Light, a retired public health nurse.

   She now has $315,500 in savings, allocated as follows: $78,000 in an individual retirement account (IRA), $234,000 in a low-earning certificate of deposit (CD), and $3,500 in her checking account.

   Sylvia contacted Bert, asking him to help her organize her portfolio for a comfortable retirement. Bert has studied Sylvia's various accounts and considered her lifestyle and expenditures. Now, he is ready to write the proposal. Bert plans to propose that Sylvia could reallocate her funds as follows:

   - $110,000 in an annuity
   - $78,000 in an individual retirement account (IRA)
   - $45,000 in municipal bonds
   - $37,000 in a stock fund
   - $42,000 in certificates of deposit (CD)
   - $3,500 in a checking account

   To make this proposal visually appealing and more readily understandable to Sylvia, Bert will use visual aids.

Assignment

a. Create a table to show how he wants to invest her money.
b. Create a bar chart comparing her current allocations versus his proposed allocations.

2. Your company is submitting a proposal to a client. The proposal is about the creation of a company newsletter geared to the client's employees. Part of this proposal will include a tentative project schedule. You plan to inform the client that you will begin the project May 1 and conclude September 7.

   Project activities include the following: (1) meetings between the newsletter staff and the client's marketing department, scheduled for May 1 through May 5, to determine objectives; (2) employee interviews regarding company recreational events and employee celebrations (birthdays, births, awards, etc.), scheduled for May 8 through May 19; (3) meetings with the accounting department, scheduled for May 22 through May 31, to ascertain stock information, annual corporate earnings, and employee benefits; (4) research regarding newsletter layout options (paper size, weight, color, and number of pages suggested per edition), scheduled for June 5 through June 16; (5) cost estimates for newsletter production based on the above findings, scheduled for June 19 through June 30; (6) follow-up meetings with client management to approve the cost estimates and tentative page layout, scheduled for July 3 through July 5; (7) writing the first draft of a newsletter, scheduled for July 10 through July 21; (8) submission of the draft to client management for approval, scheduled for July 24; (9) follow-up interviews of employees and the accounting department to update information for the final newsletter draft, scheduled for August 1 through August 12; (10) newsletter production, scheduled for August 15 through August 26; and (11) project completion—newsletters delivered to each client employee—scheduled for September 7.

Assignment

Create a Gantt chart conveying the above information.

3. Reformat the following case study to improve its document design.

> The City of Waluska wants to provide its community a safe and reliable water treatment facility. The goal is to protect Waluska's environmental resources and to ensure community values.
>
> To achieve these goals, the city has issued a request for proposal to update the Loon Lake Water Treatment Plant (LLWTP). The city recognizes that meeting its community's water treatment needs requires overcoming numerous challenges. These challenges include managing changing regulations and protection standards, developing financially responsible treatment services, planning land use for community expansion, and upholding community values.
>
> For all of the above reasons, Hardtack and Sons (H & S) Engineering is your best choice. We understand the project scope and recognize your community's needs.
>
> We have worked successfully with your community for a decade, creating feasibility studies for Loon Lake toxic control, developing odor-abatement procedures for your streams and creeks, and assessing your water treatment plant's ability to meet regulatory standards.
>
> H & S personnel are not just engineering experts. We are members of your community. Our dynamic project team has a close working relationship with your community's regulatory agencies. Our Partner in Charge, Julie Schopper, has experience with similar projects worldwide, demonstrated leadership and has the ability to communicate effectively with clients.
>
> H & S offers the City of Waluska an integrated program that addresses all your community's needs.
>
> We believe that H & S is your best choice to ensure that your community receives a water treatment plant ready to meet the challenges of the twenty-first century.

4. Jim Goodwin owns an insurance agency, Goodwin and Associates Insurance (GAI). Letters are a major part of his workplace communication with vendors, clients, and his insurance company's home office. However, many of Jim's employees fail to recognize the importance of workplace communication. To prove how important their communication is to the company, Jim plans to summarize the amount of writing they do on the job. He has determined that they write the following each week:

   - **57 inquiry letters.** Insurance coverage changes constantly. To clarify these changes for customers, employees write inquiries, asking the home office questions about new insurance laws, levels of coverage, coverage options, and rate changes.

- **253 response letters.** A key to GAI's success is acquiring new customers. When potential customers call, e-mail, or write letters asking for insurance quotes, Jim and his employees write response letters.
- **43 cover (transmittal) letters.** Once customers purchase insurance policies, Jim and his coworkers mail these policies, prefaced by cover letters. The transmittal letters clarify key points within the documents.
- **12 good news letters.** Jim's clients receive discounts when they add new cars to an existing policy or include both car and home coverage. GAI writes good news letters to convey this information.
- **25 adjustment letters.** When accidents occur or when losses take place, GAI writes adjustment letters stating that the claim is covered.
- **2 order letters.** Like any business, GAI needs office supplies. To obtain these goods, GAI writes order letters to office suppliers.
- **7,000 e-mail messages.** Jim and his 20 employees send and receive a minimum of 50 e-mail messages each day.

Assignment

To show his employees the importance of workplace communication, Jim wants to create a visual aid depicting this data. Create the appropriate graphic for GAI's employees.

INDIVIDUAL AND TEAM PROJECTS

1. **Page Layout**

   a. Bring samples of workplace communication to class. These could include letters, brochures, newsletters, instructions, reports, or advertisements from magazines or newspapers.

   Assess the page layout of each sample. Determine which samples have successful document designs and which samples have poor document designs. Base your decisions on the criteria provided in this chapter.

   Either orally or in writing, share your findings with others in the class or with your instructor.

   b. Reformat the following memo to improve its document design.

DATE: November 30, 2006
TO: Jan Hunt
FROM: Tom Langford
SUBJECT: CLEANING PROCEDURES FOR MANUFACTURING WALK-IN OVENS #98731, #98732, AND #98733

The above-mentioned ovens need extensive cleaning. To do so, vacuum and wipe all doors, walls, roofs, and floors. All vents/dampers need to be removed, and a tack cloth must be used to remove all loose dust and dirt. Also, all filters need to be replaced.

I am requesting this because loose particles of dust/dirt are blown onto wet parts when placed in the air-circulating ovens to dry. This causes extensive rework. Please perform this procedure twice per week to ensure clean production.

2. Creating Figures and Tables

a. Present the following information in a pie chart, a bar chart, and a table.

> In 2005, the Interstate Telephone Company bought and installed 100,000 relays. They used these for long-range testing programs that assessed failure rates. They purchased 40,000 Nestor 221s; 20,000 VanCourt 1200s; 20,000 Macro R40s; 10,000 Camrose Series 8s; and 10,000 Hardy SP6s.

b. Using the 2005 information presented above and the 2006 data, show the comparison between 2005 and 2006 purchases through two pie charts, a grouped bar chart, and a table.

> In 2006, after assessing the success and failure of the relays, the Interstate Telephone Company made new purchases of 200,000 relays. It bought 90,000 VanCourt 1200s; 50,000 Macro R40s; 30,000 Camrose Series 8s; and 30,000 Hardy SP6s. No Nestors were purchased.

c. Create a table for the following information.

> When the voltage out is 13 V, the frequency out is 926 Hz. When the voltage out is 12.5 V, the frequency out is 1.14 K. When the voltage out is 12 V, the frequency out is 1.4 K. When the voltage out is 11 V, the frequency out is 1.8 K. When the voltage out is 10 V, the frequency out is 2.3 K. When the voltage out is 9 V, the frequency out is 2.8 K. When the voltage out is 8 V, the frequency out is 3 K. When the voltage out is 7 V, the frequency out is 0 Hz.

d. Create a line chart to show trends. To do so, select any topic you like. The subject matter, however, must include varying values. For example, present a line chart of your grades in one class, your salary increases (or decreases) at work, the week's temperature ranges, your weight gain or loss throughout the year, the miles you have run during the week or month, amounts of money you have spent on junk food during the week, and so forth.

e. Create a flowchart. To do so, select a topic about which you can write an instruction—the steps for changing oil, winterizing your house, planting trees, seeding your yard, refinishing furniture, making a cake, changing a tire, interviewing for a job, wallpapering a room, building a deck, installing a fence, writing a proposal, or giving a speech. Flowchart the sequential steps for any of these procedures.

PROBLEM SOLVING THINK PIECES

1. Angel Guerrero, Computer Information Systems technologist at ***HeartHome Insurance***, was responsible for making an inventory of his company's hardware. He learned the following: the company had 75 laptops, 159 PCs, 27 printers, 10 scanners, 59 handheld computers (PDAs), 238 cell phones with e-mail capabilities, and 46 digital cameras.

   To write his inventory report, Angel needs to chart the above data.

   Which type of visual aid should Angel use? Explain your answer based on the information provided in this chapter. Create the appropriate visual aid.

2. Minh Tran works in the marketing department at ***Thrill-a-Minute Entertainment Theme Park*** (TET). Minh and her project team need to study entry prices, ride prices, food and beverage prices, and attendance of their park versus their primary competitor, Carnival Towne (CT).

   Minh and her team have found that TET charges $16.50 admission, while CT charges $24.95. Most of TET's rides are included in the entry price, but special rides (the Horror, the Bomber, the Avenger, and Peter Pan's Train) cost $2.50 each. At CT, the entry fee covers many rides, excluding Alice's Teacup, Top-of-the-World Ferris Wheel, and the Zinger, which cost $2.00 each. Food and beverages at TET cost $1.95 for a hot dog, $2.50 for a burger, $3.95 for nachos, and $1.50 to 2.50 for drinks. At CT, food and beverages cost $1.75 for hot dogs, $2.75 for burgers, $2.50 for nachos, and $1.50 and 2.50 for drinks. Attendance at TET last year was 250,000, while attendance at CT was 272,000.

   What type of visual aid should Minh and her team use to convey this information? Explain your decision based on the criteria for graphics provided in this chapter. Create the appropriate visual aid.

3. Toby Hebert is human resource manager at ***Crab Bayou Industries*** (Crab Bayou, LA), the world's largest wholesaler of frozen Cajun food. Toby and her management are concerned about the company's hiring trends. A prospective employee complained about discriminatory hiring practices at Crab Bayou.

   To prove that the company has not practiced discriminatory hiring practices, Toby has studied the last ten years' hires by age. She found that in 1995, the average age per employee was 48; in 1996, the average age was 51; in 1997, the average age was 47; in 1998, the average age was 52; in 1999, the average age was 45; in 2000, the average age was 47; in 2001, the average age was 42; in 2002, the average age was 39; in 2003, the average age rose to 42; in 2004 and 2005, the average age fell to 29 and 30 respectively (due to a large number of early retirements).

   What type of visual aid should Toby use to convey this information? Explain your decision, based on the criteria for graphics provided in this chapter. Create the appropriate visual aid.

4. Yasser El-Akiba is a member of his college's International Students' Club. Yasser is a native of Israel. Other members of the club are from other countries: 3 from Australia, 2 from Ecuador, 8 from Mexico, 5 from Africa, 2 from England, 3 from Canada, 4 from the Dominican Republic, and 9 from China.

   What kind of visual aid could Yasser create to show his club members' homelands? Defend your decision based on criteria for graphics in this chapter. Create the appropriate visual aid.

Web Workshop

1. On the Internet, access 10 corporate Web sites. Study them and make a list of the techniques used for visual communication. Which Web sites are successful, and why? Which Web sites are unsuccessful, and why? How would you redesign the less successful Web sites to achieve better visual communication?

2. Using an Internet search engine, type in phrases such as "automobile sales+line graph," "population distribution by age+pie chart," or "California+organizational chart." Create similar phrases for bar charts, pictographs, flowcharts, tables, and so forth. Open several links from your Web search and study the examples you have found. Which examples of graphics are successful, and why? Which examples of graphics are unsuccessful, and why? Explain your reasoning based on this chapter's criteria.

Correspondence
Memos and Letters

OBJECTIVES

When you complete this chapter, you will be able to do the following:

1. Understand the differences between memos and letters.
2. Follow all-purpose templates to write memos and letters.
3. Use memo samples as guidelines for memo components, organization, writing style and tone.
4. Evaluate your memos and letters with checklists.
5. Correctly use the eight essential letter components: the writer's address, the date, an inside address for the recipient, a salutation, the body of the letter, a complimentary close, and the writer's signed and typed names.
6. Write different types of letters, including the following:
   - Inquiry
   - Cover (Transmittal)
   - Complaint
   - Adjustment
   - Sales
7. Follow the writing process—prewriting, writing, and rewriting—to create memos and letters.
8. Test your knowledge of correspondence through end-of-chapter activities:
   - Case Studies
   - Individual and Team Projects
   - Problem Solving Think Pieces
   - Web Workshop

COMMUNICATION AT WORK

In this scenario, a biotechnology company frequently corresponds through letters and memos.

CompuMed

CompuMed, a wholesale provider of biotechnology equipment, is home-based in Reno, NV. CompuMed's CEO, Jim Goodwin, plans to capitalize on emerging nanotechnology to manufacture and sell the following:

- Extremely lightweight and portable heart monitors and ventilators.
- Pacemakers and hearing aids, 1/10 the size of current products on the market.
- Microscopic bio-robotics that can be injected in the body to manage, monitor, and/or destroy blood clots, metastatic activities, arterial blockages, alveoli damage due to carcinogens or pollutants, and scar tissues creating muscular or skeletal immobility.

CompuMed is a growing company with over 5,000 employees located in 24 cities and 3 states. To manage this business, supervisors and employees write on average over 20 letters and 15 memos a day.

The letters are written to many different audiences and serve various purposes. CompuMed must write letters for employee files, to customers, to job applicants, to outside auditors, to governmental agencies involved in biotechnology regulation, to insurance companies, and more. They write

- *Letters of inquiry* to retailers seeking product information (technical specifications, pricing, warranties, guarantees, credentials of service staff, and so forth).
- *Cover letters* prefacing CompuMed's proposals.
- *Complaint letters* written to parts manufacturers if and when faulty equipment and materials are received in shipping and *adjustment letters* to compensate retailers when problems occur.
- *Sales letters* to computer and biotechnology retailers.

CompuMed's managers and employees also write memos to accomplish a variety of goals:

- document work accomplished
- call meetings and establish meeting agendas
- request equipment from purchasing
- preface internal proposals
- highlight productivity and problems

Writing memos and letters are two important types of workplace communication at CompuMed.

THE DIFFERENCES BETWEEN MEMOS AND LETTERS

This chapter focuses on traditional correspondence—memos and letters. To give you an overview of the differences and similarities between memos and letters, look at Table 4.1.

MEMOS

Reasons for Writing Memos

Memos are an important means by which employees communicate with each other. Memos, hard-copy correspondence written within your company, are important for several reasons. In 2004, The National Commission on Writing substantiated the importance of correspondence (memos and letters). Their survey of 120 major companies employing approximately 8 million workers found that 70 percent of the companies require the writing of memos and letters ("Writing: A Ticket to Work").

Next, you will write memos to a wide range of readers. This includes your supervisors, coworkers, subordinates, and multiple combinations of these audiences. Since memos usually are copied (cc: complimentary copies) to many readers, a memo sent to your boss could be read by an entire department, the boss's boss, and colleagues in other departments.

Because of their frequency and widespread audiences, memos could represent a major component of your interpersonal communication skills within your work environment.

Furthermore, memos are very flexible and can be written for many different purposes. Consider these options.

TABLE 4.1 MEMOS VS. LETTERS

Characteristics	Memos	Letters
Destination	Internal: correspondence written to colleagues within a company.	External: correspondence written to people outside the business.
Format	Identification lines include "Date," "To," "From," and "Subject." The message follows.	Includes letterhead address, date, reader's address, salutation, text, complimentary close, and signatures.
Audience	Generally specialists or semi-specialists, mostly business colleagues.	Generally semi-specialists and lay readers, such as vendors, clients, stakeholders, and stockholders.
Topic	Generally topics related to internal corporate decisions; abbreviations and acronyms often allowed.	Generally topics related to vendor, client, stakeholder, and stockholder interests; abbreviations and acronyms usually defined.
Tone	Informal due to peer audience.	More formal due to audience of vendors, clients, stakeholders, and stockholders.
Attachments or Enclosures	Hard-copy attachments can be stapled to the memo. Complimentary copies (cc) can be sent to other readers.	Additional information can be enclosed within the envelope. Complimentary copies (cc) can be sent to other readers.
Delivery Time	Determined by a company's in-house mail procedure.	Determined by the destination (within the city, state, or country). Letters could be delivered within 3 days but may take more than a week.
Security	If a company's mail delivery system is reliable, the memo will be placed in the reader's mailbox. Then, what the reader sees on the hard copy page will be exactly what the writer wrote. Security depends on the ethics of coworkers and whether the memo was sent in an envelope.	The U.S. Postal Service is very reliable. Once the reader opens the envelope, he or she sees exactly what the writer wrote. Privacy laws protect the letter's content.

- **Documentation.** Expenses, incidents, accidents, problems encountered, projected costs, study findings, hirings, firings, reallocations of staff or equipment, etc.
- **Confirmation.** A meeting agenda, date, time, and location; decisions to purchase or sell; topics for discussion at upcoming teleconferences; conclusions arrived at; fees, costs, or expenditures, etc.
- **Procedures.** How to set up accounts, research on the company intranet, operate new machinery, use new software, apply online for job opportunities through the company intranet, create a new company Web site, solve a problem, etc.
- **Recommendations.** Reasons to purchase new equipment, fire or hire personnel, contract with new providers, merge with other companies, revise current practices, renew contracts, etc.
- **Feasibility.** Studying the possibility of changes in the workplace (practices, procedures, locations, staffing, equipment, missions/visions), etc.
- **Status.** Daily, weekly, monthly, quarterly, biannually, yearly statements about where you, the department, or the company is regarding many topics (sales, staffing, travel, practices, procedures, finances, etc.)
- **Directive (delegation of responsibilities).** Informing subordinates of their designated tasks.

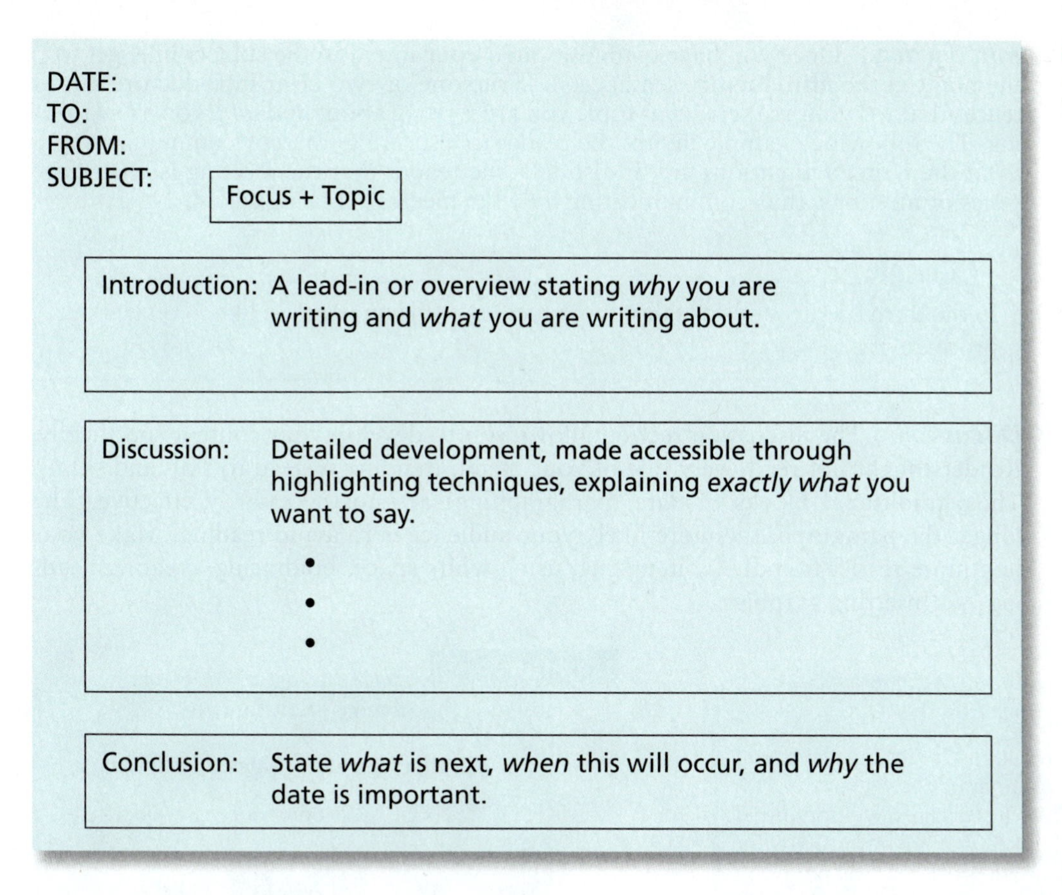

DATE:
TO:
FROM:
SUBJECT:
Focus + Topic
Introduction: A lead-in or overview stating why you are writing and what you are writing about.
Discussion: Detailed development, made accessible through highlighting techniques, explaining exactly what you want to say.
•
•
•
Conclusion: State what is next, when this will occur, and why the date is important.

Figure 4.1 All-Purpose Memo Template

- **Inquiry.** Asking questions about upcoming processes or procedures.
- **Cover.** Prefacing an internal proposal, long report, or other attachments.

Criteria for Writing Memos

Memos contain the following key components:

- Memo Identification Lines—Date, To, From, and Subject
- Introduction
- Discussion
- Conclusion
- Audience Recognition
- Appropriate Memo Style and Tone

Figure 4.1 shows an ideal, all-purpose organizational template that works well for memos.

Subject Line. The subject line summarizes the memo's content. One-word subject lines do not communicate effectively, as in the following flawed subject line. The "before" sample has a *topic* (a what) but is missing a *focus* (a what about the what).

before
Subject: Sales
after
Subject: Report on Quarterly Sales

Introduction. Once you have communicated your intent in the subject line, get to the point in the introductory sentence(s). Write one or two clear introductory sentences that tell your readers *what* topic you are writing about and *why* you are writing. The following example invites the reader to a meeting, thereby communicating *what* the writer's intentions are. It also tells the reader that the meeting is one of a series of meetings, thus communicating *why* the meeting is being called.

Example

In the third of our series of sales quota meetings this quarter, I'd like to review our productivity.

Discussion. The discussion section allows you to develop your content specifically. Readers might not read every line of your memo (tending instead to skip and skim). Thus, traditional blocks of data (paragraphing) are not necessarily effective. The longer the paragraph, the more likely your audience is to avoid reading. Make your text more reader-friendly by itemizing, using white space, boldfacing, creating headings, or inserting graphics.

before

Unfriendly Text

This year began with an increase, as we sold 4.5 million units in January compared to 3.7 for January 2005. In February we continued to improve with 4.6, compared with 3.6 for the same time in 2005. March was not quite so good, as we sold 4.3 against the March 2005 figure of 3.9. April was about the same with 4.2, compared to 3.8 for April 2005.

after

Reader-Friendly Text

Comparative Quarterly Sales (in Millions)

	2005	2006	Increase/Decrease
Jan.	3.7	4.5	0.8+
Feb.	3.6	4.6	1.0+
Mar.	3.9	4.3	0.4+
Apr.	3.8	4.2	0.4+

Conclusion. Conclude your memo with "thanks" and/or a directive action. A pleasant conclusion could motivate your readers, as in the following example. A directive close tells your readers exactly what you want them to do next or what your plans are (and provides dated action).

If our quarterly sales continue to improve at the current rate, we will double our sales expectations by 2006. Congratulations! Next Wednesday (12/22/06), please provide next quarter's sales projections and a summary of your sales team's accomplishments.

Audience Recognition. Since letters go outside your company, your audience is usually a semi-specialist or lay, demanding that you define your terms specifically. In memos your in-house audience is easy to address (usually a specialist or semi-specialist). You often can use more acronyms and internal abbreviations in memos than you can in letters.

You will write the message to "Distribution" (listing a group of readers) or send the memo to one reader but "cc" (send a "carbon copy" or "complimentary copy") to other readers. Thus, you might be writing simultaneously to your immediate supervisor (specialist), to his or her boss (semi-specialist), to your colleagues (specialists),

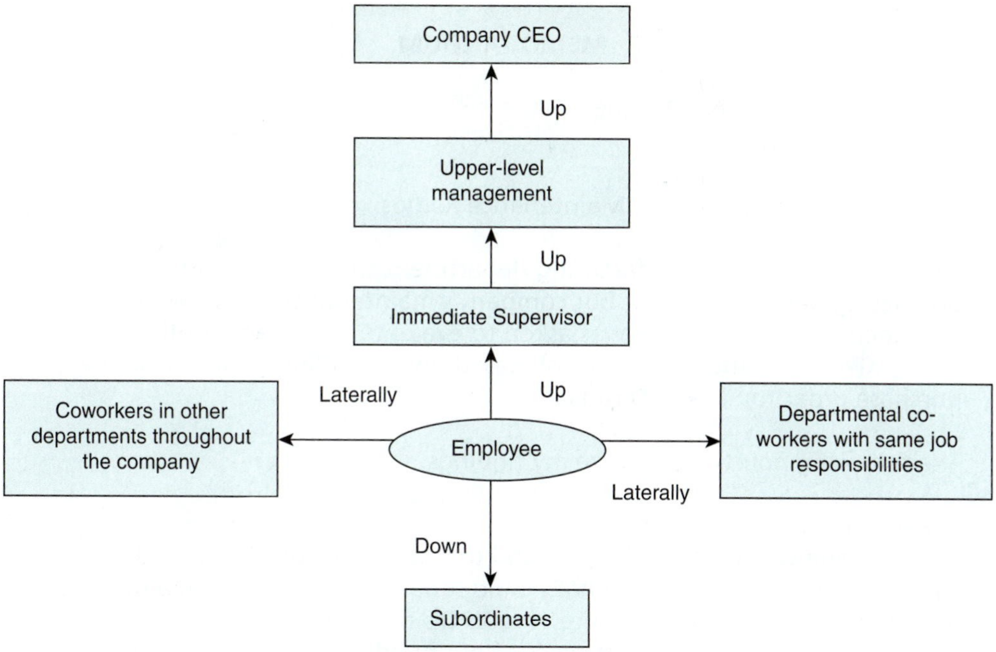

Company CEO

Up

Upper-level management

Up

Immediate Supervisor

Up

Laterally

Coworkers in other departments throughout the company

Employee

Departmental co-workers with same job responsibilities

Laterally

Down

Subordinates

Figure 4.2 Vertical and Lateral Communication within a Company

and to a CEO (semi-specialist). To accommodate multiple audiences, use parenthetical definitions, such as Cash In Advance (CIA) or Continuing Property Records (CPR). Audience recognition and involvement are discussed in Chapter 2.

Style and Tone. Because memos are usually only one page long, use simple words, short sentences, specific detail, and highlighting techniques. In addition, strive for an informal, friendly tone. Memos are part of your interpersonal communication abilities, so a friendly tone will help build rapport with colleagues.

In memos, audience determines tone. For example, you cannot write directive correspondence to supervisors mandating action on their part. It might seem obvious that you can write directives to subordinates, but you should not use a dictatorial tone. Though the subordinates are under your authority, they must still be treated with respect. You will determine the tone of your memo by deciding if you are writing vertically (up to management or down to subordinates) or laterally (to coworkers), as shown in Figure 4.2.

before

Unfriendly, Demanding Style

We will have a meeting next Tuesday, Jan. 11, 2006. Exert every effort to attend this meeting. Plan to make intelligent comments regarding the new quarter projections.

after

Friendly, Personal Style

Let's meet next Tuesday (Jan. 11, 2006). Even if you're late, I'd appreciate your attending. By doing so you can have an opportunity to make an impact on the new quarter projections. I'm looking forward to hearing your comments.

Sample Memos

See Figures 4.3 and 4.4 for sample memos.

"Distribution" indicates that the memo is being sent to a number of employees. Notice that at the bottom of the memo, a distribution list is provided.

The introduction states the purpose of this memo, providing dates, personnel, options, and intended action.

The memo's discussion analyzes the criteria used to decide which radio to purchase. Tables are used to ensure reader-friendly ease of access.

The conclusion summarizes the importance of the action. It also ends in a personalized and positive tone to ensure reader involvement and to build rapport.

MEMORANDUM

DATE: December 12, 2006
TO: Distribution
FROM: Luann Brunson
SUBJECT: Replacement of Maintenance Radios

On December 5, the manufacturing department supervisor informed the purchasing department that our company's maintenance radios were malfunctioning. Purchasing was asked to evaluate three radio options (the RPAD, XPO 1690, and MX16 radios). Based on my findings, I have issued a purchase order for 12 RPAD radios.

The following points summarize my findings.

1. *Performance*
During a one-week test period, I found that the RPAD outperformed our current XPO's reception. The RPAD could send and receive within a range of 5 miles with minimal interference. The XPO's range was limited to 2 miles, and transmissions from distant parts of our building broke up due to electrical interference.

2. *Specifications*
Both the RPAD and the MX16 were easier to carry, because of their reduced weight and size, than our current XPO 1690s.

	RPAD	XPO 1690	MX16
Weight	1 lb.	2 lbs.	1 lb.
Size	5″ × 2″	8″ × 4″	6″ × 1″

3. *Cost*
The RPAD is our most cost-effective option because of quantity cost breaks and maintenance guarantees.

	RPAD	XPO 1690	MX16
Cost per unit	$70.00	$215.00	$100.00
Cost per doz.	$750.00	$2,580.00	$1,100.00
Guarantees	1 year	6 months	1 year

Purchase of the RPAD will give us improved performance and comfort. In addition, we can buy 12 RPAD radios for approximately the cost of 4 XPOs. If I can provide you with additional information, please call. I'd be happy to meet with you at your convenience.

Distribution: M. Ellis, P. Michelson, M. Rhinehart, R. Travers, T. Schroeder, R. Xidis

Figure 4.3 Comparison/Contrast Memo Recommending Action

Memo

CompuMed

DATE: November 11, 2006
TO: CompuMed Management
FROM: Bill Baker, Human Resources Director
SUBJECT: Information about Proposed Changes to Employee Benefits Package

As of January 1, 2007, CompuMed will change insurance carriers. This will affect all 5,000 employees' benefits packages. I have attached a proposal, including the following:

1. Reasons for changing from our current carrier. page 2
2. Criteria for our selection of a new insurance company. pages 3–4
3. Monthly cost for each employee. pages 5–6
4. Overall cost to CompuMed. page 7
5. Benefits derived from the new healthcare plan. page 8

Please review the proposal, survey your employees' responses to our suggestions, and provide your feedback. We need your input by December 1, 2006. This will give the human resources department time to consider your suggestions and work with insurance companies to meet employee needs.

Enclosure: Proposal

Figure 4.4 Cover Memo Prefacing Attachments

MEMO CHECKLIST

The Memo Checklist will give you the opportunity for self-assessment and peer evaluation of your writing.

Memo Checklist

Memo Checklist

__ 1. Does the memo contain identification lines (Date, To, From, and Subject)?

__ 2. Does the subject line contain a topic and a focus?

__ 3. Does the introduction clearly state
- Why this memo has been written?
- What topic the memo is discussing?

__ 4. Does the body explain exactly what you want to say?

__ 5. Does the conclusion
- tell when you plan a follow-up or when you want a response?
- explain why this dated action is important?

__ 6. Are highlighting techniques used effectively for document design?

__ 7. Is the memo concise?

__ 8. Is the memo clear,
- achieving specificity of detail?
- answering reporter's questions?

__ 9. Does the memo recognize audience,
- defining acronyms or abbreviations where necessary for various levels of readers (specialists, semi-specialists, and lay)?

__10. Did you avoid grammatical errors? Errors will hurt your professionalism. See Appendix A for grammar rules.

LETTERS

Reasons for Writing Letters

Letters are external correspondence that you send from your company to a colleague working at another company, to a vendor, to a customer, to a prospective employee, and to stakeholders and stockholders. Letters leave your work site (as opposed to memos, which stay within the company).

Because letters are sent to readers in other locations, your letters not only reflect your communication abilities but also are a reflection of your company. This chapter provides letter components, formats, criteria, and examples to help you write the following kinds of letters: inquiry, cover (transmittal), complaint, adjustment, and sales.

Essential Components of Letters

Your letter should be typed or printed on $8^{1}/_{2}'' \times 11''$ paper. Leave 1″ to $1^{1}/_{2}''$ margins at the top and on both sides. Choose an appropriately businesslike font (size and style), such as Times New Roman or Arial (12 point). Though "designer fonts," such as Comic Sans and Shelley Volante, are interesting, they tend to be harder to read and less professional (see Chapter 3 for more information on font selection and readability).

Your letter should contain the essential components shown in Figure 4.5.

Writer's Address. This section contains either your personal address or your company's address. If the heading consists of your address, you will include your street address; the city, state, and zip code. The state may be abbreviated with the appropriate two-letter abbreviation.

If the heading consists of your company's address, you will include the company's name; street address; and city, state, and zip code.

Date. Document the month, day, and year when you write your letter. You can write your date in one of two ways: May 31, 2006, or 31 May 2006. Place the date one or two spaces below the writer's address.

Reader's Address. Place the reader's address two lines below the date.

- Reader's name (If you do not know the name of this person, begin the reader's address with a job title or the name of the department.)
- Reader's title (optional)
- Company name
- Street address
- City, state, and zip code

Salutation. The traditional salutation, placed two spaces beneath the inside address, is *Dear* and your reader's last name, followed by a colon (Dear Mr. Smith:).

You can also address your reader by his or her first name if you are on a first-name basis with this person (Dear John:). If you are writing to a woman and are unfamiliar with her marital status, address the letter Dear Ms. Jones. However, if you know the

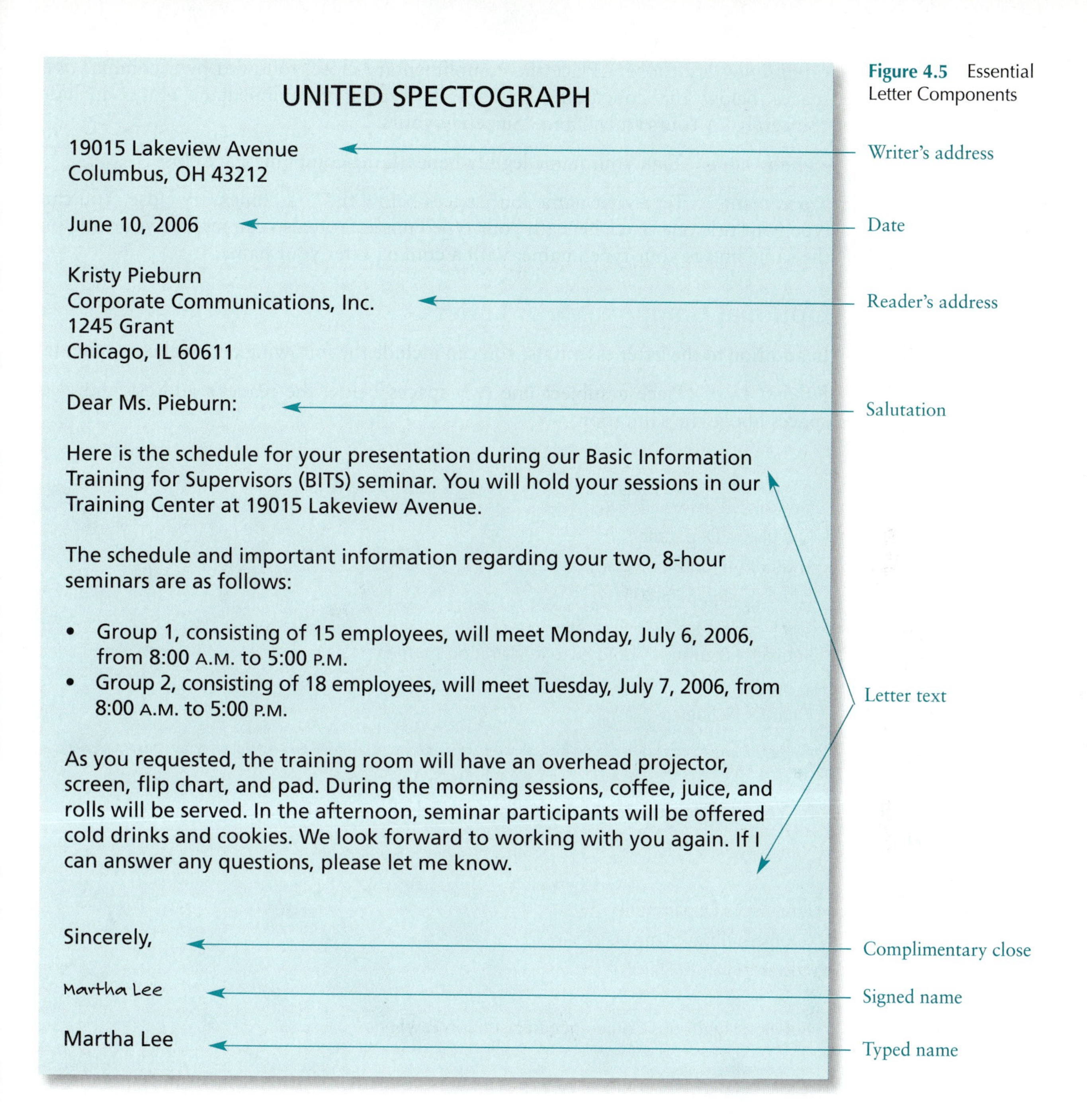
UNITED SPECTOGRAPH

19015 Lakeview Avenue
Columbus, OH 43212

June 10, 2006

Kristy Pieburn
Corporate Communications, Inc.
1245 Grant
Chicago, IL 60611

Dear Ms. Pieburn:

Here is the schedule for your presentation during our Basic Information Training for Supervisors (BITS) seminar. You will hold your sessions in our Training Center at 19015 Lakeview Avenue.

The schedule and important information regarding your two, 8-hour seminars are as follows:

- Group 1, consisting of 15 employees, will meet Monday, July 6, 2006, from 8:00 A.M. to 5:00 P.M.
- Group 2, consisting of 18 employees, will meet Tuesday, July 7, 2006, from 8:00 A.M. to 5:00 P.M.

As you requested, the training room will have an overhead projector, screen, flip chart, and pad. During the morning sessions, coffee, juice, and rolls will be served. In the afternoon, seminar participants will be offered cold drinks and cookies. We look forward to working with you again. If I can answer any questions, please let me know.

Sincerely,

Martha Lee

Martha Lee

Figure 4.5 Essential Letter Components

woman's marital status, you can address the letter accordingly: Dear Miss Jones: *or* Dear Mrs. Jones: *or* Dear Dr. Jones:

Letter Body. Begin the body of the letter two spaces below the salutation. The body includes your introductory paragraph, discussion paragraph(s), and concluding paragraph. The body should be single spaced with double spacing between paragraphs. Whether you indent the beginning of paragraphs or leave them flush with the left margin is determined by the letter format you employ.

Complimentary Close. Place the complimentary close, followed by a comma, two spaces below the concluding paragraph. Typical complimentary closes include "Sincerely," "Yours truly," and "Sincerely yours."

Signed Name. Sign your name legibly beneath the complimentary close.

Typed Name. Type your name four spaces below the complimentary close. You can type your title one space beneath your typed name. You also can include your title on the same line as your typed name, with a comma after your name.

Optional Components of Letters

In addition to the letter essentials, you can include the following optional components.

Subject Line. Place a subject line two spaces below the reader's address and two spaces above the salutation.

Dr. Ron Schaefer
Linguistics Department
Southern Illinois University
Edwardsville, IL 66205

Subject: Linguistics Conference Registration Payment

Dear Dr. Schaefer:

You also could use a subject line instead of a salutation.

Linguistics Department
Southern Illinois University
Edwardsville, IL 66205

Subject: Linguistics Conference Registration Payment

A subject line not only helps readers understand the letter's intent but also (if you are uncertain of your reader's name) helps you avoid such awkward salutations as "To Whom It May Concern," "Dear Sirs," and "Ladies and Gentlemen." In the simplified format, both the salutation and the complimentary close are omitted, and a subject line is included.

New Page Notations. If your letter is longer than one page, cite your name, the page number, and the date on all pages after page 1. Place this notation either flush with the left margin at the top of subsequent pages or across the top of subsequent pages. (You must have at least two lines of text on the next page to justify another page.)

Left margin, subsequent page notation

Mabel Tinjaca Page 2 May 31, 2006

Across top of subsequent pages

Mabel Tinjaca 2	May 31, 2006

Writer's and Typist's Initials. If the letter was typed by someone other than the writer, include both the writer's and the typist's initials two spaces below the typed signature. The writer's initials are capitalized, the typist's initials are typed in lowercase, and the two sets of initials are separated by a colon. If the typist and the writer are the same person, this notation is not necessary.

Sincerely,

W. T. Winnery
WTW:mm

Enclosure Notation. If your letter prefaces enclosed information, such as an invoice or report, mention this enclosure in the letter and then type an enclosure notation two spaces below the typed signature (or two spaces below the writer and typist initials). The enclosure notation can be abbreviated "Enc."; written out as "Enclosure"; show the number of enclosures, such as "Enclosures (2)"; or specify what has been enclosed—"Enclosure: January Invoice."

Copy Notation. If you have sent a copy of your letter to other readers, show this in a copy notation. A complimentary copy is designated by a lowercase "cc." List the other readers' names following the copy notation. Type the copy notation two spaces below the typed signature or two spaces below either the writer's and typist's initials or the enclosure notation.

Sincerely,

Brian Altman
Enclosure: August Status Report
cc: Marcia Rittmaster and Larry Rochelle

Formatting Letters

Three common types of letter formats include **full block** (Figure 4.6), **full block with subject line** (Figure 4.7), and **simplified** (Figure 4.8). Two popular and professional formats used in business are full block and full block with subject line. With both formats, you type all information at the left margin without indenting paragraphs, the

1–$1^1/_2$″ margins on all sides of the letter

State Health Department
1890 Clark Road
Jefferson City, MO 67220

Single space within the paragraphs.

Double space between the paragraphs.

June 6, 2006 double space above and below the date

Dale McGraw, Manager
Elmwood Mobile Home Park
Elmwood, MO 64003

Dear Mr. McGraw: double space above and below the salutation

On April 19, 2006, Ryan Duran and I, environmental specialists from the health department, conducted an inspection of the Elmwood Mobile Home Park Wastewater Treatment Facility. The purpose was to assess compliance with the following: the state's Clean Water Law, Clean Water Commission regulations, and your facility's plan for pollution control. The inspection also would allow the state to promote proper operation of wastewater facilities and to provide technical assistance where needed to the Elmwood Mobile Home management.

Though the Elmwood Mobile Home pollution control plan had expired in 2005, a consent judgment was issued by the state's attorney general's office. The county court stipulated a timeline for correction by connection to an available sewer system. Your mobile home park's wastewater system has continually discharged to the Little Osage River. A copy of the abatement order, which requires that monthly discharge monitoring reports (DMRs) be submitted by the 28th of the month following the reporting periods, is attached. All DMRs for the previous twelve months have been received, and reported pollution parameters are not within limits. Due to the plant's performance, the stream was placed on the 1998 303 (d) stream for impairment by the Elmwood Mobile Home.

As part of the inspection, a review of the facility's DMR was conducted. Twenty-four-hour composite samples were collected using a composite sampler. Enclosed are the results of the 24-hour composite samples collected on April 20, 2006. Every one of the problems documented is an infraction that must be addressed.

Within 30 days of receipt of this letter, please submit to the health department written documentation describing steps taken to correct each of the concerns identified in the enclosure. Also include engineering reports, and submit a timeframe to eliminate the problems. Thank you for your cooperation.

Sincerely, double space before "Sincerely"

Harvey Haddix 4 spaces between "Sincerely" and the typed signature
Environmental Manager

Enclosure

Figure 4.6 Full Block Format

State Health Department
1890 Clark Road   Jefferson City, MO 67220

June 6, 2006

Dale McGraw, Manager
Elmwood Mobile Home Park
Elmwood, MO 64003

Subject: Pollution Control Inspection

Dear Mr. McGraw:

On April 19, 2006, Ryan Duran and I, environmental specialists from the health department, conducted an inspection of the Elmwood Mobile Home Park Wastewater Treatment Facility. The purpose was to assess compliance with the following: the state's Clean Water Law, Clean Water Commission regulations, and your facility's plan for pollution control. The inspection also would allow the state to promote proper operation of wastewater facilities and to provide technical assistance where needed to the Elmwood Mobile Home management.

Though the Elmwood Mobile Home pollution control plan had expired in 2005, a consent judgment was issued by the state's attorney general's office. The county court stipulated a timeline for correction by connection to an available sewer system. Your mobile home park's wastewater system has continually discharged to the Little Osage River. A copy of the abatement order, which requires that monthly discharge monitoring reports (DMRs) be submitted by the 28th of the month following the reporting periods, is attached. All DMRs for the previous twelve months have been received, and reported pollution parameters are not within limits. Due to the plant's performance, the stream was placed on the 1998 303 (d) stream for impairment by the Elmwood Mobile Home.

As part of the inspection, a review of the facility's DMR was conducted. Twenty-four-hour composite samples were collected using a composite sampler. Enclosed are the results of the 24-hour composite samples collected on April 20, 2006. Every one of the problems documented is an infraction that must be addressed.

Within 30 days of receipt of this letter, please submit to the health department written documentation describing steps taken to correct each of the concerns identified in the enclosure. Also include engineering reports, and submit a timeframe to eliminate the problems. Thank you for your cooperation.

Sincerely,

Harvey Haddix
Environmental Manager

Enclosure

Figure 4.7 Full Block Format with Subject Line

State Health Department
1890 Clark Road
Jefferson City, MO 67220

June 6, 2006

Dale McGraw, Manager
Elmwood Mobile Home Park
Elmwood, MO 64003

Subject: Pollution Control Inspection

On April 19, 2006, Ryan Duran and I, environmental specialists from the health department, conducted an inspection of the Elmwood Mobile Home Park Wastewater Treatment Facility. The purpose was to assess compliance with the following: the state's Clean Water Law, Clean Water Commission regulations, and your facility's plan for pollution control. The inspection also would allow the state to promote proper operation of wastewater facilities and to provide technical assistance where needed to the Elmwood Mobile Homes management.

Though the Elmwood Mobile Home pollution control plan had expired in 2005, a consent judgment was issued by the state's attorney general's office. The county court stipulated a timeline for correction by connection to an available sewer system. Your mobile home park's wastewater system has continually discharged to the Little Osage River. A copy of the abatement order, which requires that monthly discharge monitoring reports (DMRs) be submitted by the 28th of the month following the reporting periods, is attached. All DMRs for the previous twelve months have been received, and reported pollution parameters are not within limits. Due to the plant's performance, the stream was placed on the 1998 303 (d) stream for impairment by the Elmwood Mobile Home.

As part of the inspection, a review of the facility's DMR was conducted. Twenty-four-hour composite samples were collected using a composite sampler. Enclosed are the results of the 24-hour composite samples collected on April 20, 2006. Every one of the problems documented is an infraction that must be addressed.

Within 30 days of receipt of this letter, please submit to the health department written documentation describing steps taken to correct each of the concerns identified in the enclosure. Also include engineering reports, and submit a timeframe to eliminate the problems. Thank you for your cooperation.

Harvey Haddix
Environmental Manager

Enclosure

Figure 4.8 Simplified Format Omitting "Dear . . ." and "Sincerely"

date, the complimentary close, or signature. The full block with subject line differs only with the inclusion of a subject line.

Another option is the simplified format. This type of letter layout is similar to the full block format in that *all text is typed margin left. The two significant omissions include no salutation* ("Dear __________:") *and no complimentary close* ("Sincerely,"). Omitting a salutation is useful in the following instances:

- You do not know your reader's name (NOTE: avoid the trite salutation "To Whom It May Concern:")
- You are writing to someone with a non-gender-specific name (Jesse, Terry, Stacy, Chris, etc.) and you do not know whether to use "Mr.," "Mrs.," or "Ms."

The Administrative Management Society (AMS) suggests that if you omit the salutation, you also should omit the complimentary close. Some people feel that omitting the salutation and the omplimentary close will make the letter cold and unfriendly. However, the AMS says that if your letter is warm and friendly, these omissions will not be missed. More importantly, if your letter's content is negative, beginning with "Dear" and ending with "Sincerely" will not improve the letter's tone or your reader's attitude toward your comments.

The simplified format includes a subject line to aid the letter's clarity.

Criteria for Different Types of Letters

Though you might write different types of letters, including letters of inquiry, cover, complaint, response, adjustment, or sales, consider using the all-purpose letter template (Figure 4.9) to format your correspondence.

Letter of Inquiry

If you want information about degree requirements, equipment costs, performance records, turnaround time, employee credentials, or any other matter of interest to you or your company, you write a letter requesting that data. Letters of inquiry require that you be specific. For example, if you write, "Please send me any information you have on your computer systems," you are in trouble. You will either receive any information the reader chooses to give you or none at all. Look at the following flawed letter of inquiry from a biochemical waste disposal company.

before

Dear Mr. Jernigan:

Please send us information about the following filter pools:

1. East Lime Pool
2. West Sulphate Pool
3. East Aggregate Pool

Thank you.

Figure 4.9 All-Purpose Letter Template

Writer's Address

Date

Reader's Address

Salutation:

A lead-in or overview stating *why* you are writing and *what* you are writing about.

Detailed development, made accessible through highlighting techniques, explaining *exactly what* you want to say. • • •

State *what* is next, *when* this will occur, and *why* the date is important.

Complimentary close,

Signed Name

Typed Name

The reader replied to the "before" sample as follows:

Dear Mr. Scholl:

I would be happy to provide you with any information you would like. However, you need to tell me what information you require about the pools.

I look forward to your response.

The first writer, recognizing the error, rewrote the letter as follows:

after

Dear Mr. Jernigan:

My company, Jackson County Hazardous Waste Disposal, Inc., needs to purchase new waste receptacles. One of our clients used your products in the past and recommended you. Please send us information about the following:

1. Lime Pool: costs, warranties, time of installation, and dimensions
2. Sulphate Pool: costs, material, and levels of acidity
3. Aggregate Pool: costs, flammability, maintenance, and discoloration

We plan to install our pools by March 12. We would appreciate your response by February 20. Thank you.

Providing specific details makes your letter of inquiry effective. You will save your reader's time by quantifying your request.

To compose your letter of inquiry, include the following:

Introduction. Clarify your intent in the introduction. Until you tell your readers why you are writing, they do not know. It is your responsibility to clarify your intent and explain your rationale for writing. Also tell your reader immediately what you are writing about (the subject matter of your inquiry). You can state your intent and subject matter in one to three sentences.

Discussion. Specify your needs in the discussion. To ensure that you get the response you want, ask precise questions or list specific topics of inquiry. You must quantify. For example, rather than vaguely asking about machinery specifications, you should ask more precisely about "specifications for the 12R403B Copier." Rather than asking, "Will the roofing material cover a large surface?" you need to quantify—"Will the roofing material cover 150′ × 180′?"

Conclusion. Conclude precisely. First, explain when you need a response. Do not write "Please respond as soon as possible." Provide dated action and tell the reader exactly when you need your answers. Second, to sell your readers on the importance of this date, explain why you need answers by the date given.

Figure 4.10 will help you understand the requirements for effective letters of inquiry.

Cover (Transmittal) Letters

In business, you are often required to send information to a client, vendor, or colleague. You might send multipage copies of reports, invoices, drawings, maps, letters, specifications, instructions, questionnaires, or proposals.

A cover letter accomplishes two goals. First, it lets you tell readers up front what they are receiving. Second, it helps you focus your reader's attention on key points within the enclosures. Thus, the cover letter is a reader-friendly gesture geared toward assisting your audience. To compose your cover letter, include the following:

Introduction. In the introductory paragraph, tell your reader why you are writing and what you are writing about. What if the reader has asked you to send the documentation? Do you still need to explain why you are writing? The answer is yes.

Figure 4.10 Letter of Inquiry Using the Simplified Format

CompuMed

8713 Hillview Reno, NV 32901 1-800-551-9000 Fax: 1-816-555-0000

September 12, 2006

Sales Manager
OfficeToGo
7622 Raintree
St. Louis, MO 66772

Writer's Insight

Jim says, "When I write letters of inquiry, I make sure that I itemize the body questions so my readers can easily access them. More than that, I get as specific as I can so I don't have to waste my time—or theirs—with follow-up e-mail questions. A real grammar challenge for me is making the bulleted items parallel!"

Subject: Request for Product Pricing and Shipping Schedules

In the introduction, explain why you are writing.

My medical technology company has worked well with OfficeToGo (OTG) for the past five years. However, in August I received a letter informing me that OTG had been purchased by a larger corporation. I need to determine if OTG remains competitive with other major office equipment suppliers in the Reno area.

Please provide the following information:

In the discussion, specify your needs. To ensure accuracy of response, ask precise questions.

1. What discounts will be offered for bulk purchases?
2. Which freight company will OTG now be using?
3. Who will pay to insure the items ordered?
4. What is the turnaround time from order placement to delivery?
5. Will OTG be able to deliver to all my satellite sites?
6. Will OTG technicians set up the equipment delivered, including desks, file cabinets, bookshelves, and chairs?
7. Will OTG be able to personalize office stationery on site, or will it have to be outsourced?

In the conclusion, state when you need a response and explain why this date is important. Providing contact information will help the reader respond.

Please respond to these questions by September 30 so I can prepare my quarterly orders in a timely manner. I continue to expand my company and want assurances that you can fill my growing office supply needs. You can contact me at the phone number provided above or by e-mail (jgood@CompuMed.com). Thank you for your help.

Jim Goodwin
Owner and CEO

Although the reader requested the information, time has passed, other correspondence has been written, and your reader might have forgotten the initial request.

Discussion. In the body of the letter, you can accomplish two things. You either will tell your reader exactly what you have enclosed or exactly what of value is within the enclosures. In both instances, you should provide an itemized list or easily accessible, short paragraphs.

Conclusion. Your conclusion should tell your readers what you want to happen next, when you want this to happen, and why the date is important.

See Figure 4.11 for an example of a cover letter from a healthcare provider.

Figure 4.11 Cover Letter in Block Format

AMERICAN HEALTHCARE
1401 Laurel Drive
Denton, TX 76201
November 11, 2006

Jan Pascal
Director of Outpatient Care
St. Michael's Hospital
Westlake Village, CA 91362

Dear Ms. Pascal:

Thank you for your recent request for information about our specialized outpatient care equipment. American Healthcare's stair lifts, bath lifts, and vertical wheelchair lifts can help your patients. To show how we can serve you, we have enclosed a brochure including the following information:

	Page
• Maintenance, warranty, and guarantee information	1–3
• Technical specifications for our products, including sizes, weight limitations, colors, and installation instructions	4–6
• Visuals and price lists for our products	7–8
• An American Healthcare order form	9
• Our 24-hour hotline for immediate service	10

Early next month, I will call to make an appointment at your convenience. Then we can discuss any questions you might have. Thank you for considering American Healthcare, a company that has provided exceptional outpatient care for over 30 years.

Sincerely,

Toby Sommers

Enclosure

A positive tone in the introduction builds rapport and informs the reader why this letter is being written: in response to a request.

An itemized body clarifies what is in the enclosure. Adding page numbers in the list helps readers find the information in the enclosed material.

Complaint Letters

You are purchasing director at an electronics firm. Although you ordinarily receive excellent products and support from a local manufacturing firm, two of your recent orders have been filled incorrectly and included defective merchandise. You don't want to have to look for a new supplier. You should express your complaint as pleasantly as possible.

To compose your complaint letter, include the following:

Introduction. In the introduction, politely state the problem. Although you might be angry over the service you have received, you want to suppress that anger. Blatantly negative comments do not lead to communication; they lead to combat. Because angry readers will not necessarily go out of their way to help you, your best approach is diplomacy.

To strengthen your assertions, in the introduction, include supporting details, such as the following: serial numbers, dates of purchase, invoice numbers, check numbers, names of salespeople involved in the purchase, and/or receipts. When possible, include copies documenting your claims.

Discussion. In the discussion paragraph(s), explain in detail the problems experienced. This could include dates, contact names, information about shipping, breakage information, an itemized listing of defect, or poor service.

Be specific. Generalized information will not sway your readers to accept your point of view. In a complaint letter, you suffer the burden of proof. Help your audience understand the extent of the problem. After documenting your claims, state what you want done and why.

Conclusion. End your letter positively. Remember, you want to ensure cooperation with the vendor or customer. You also want to be courteous, reflecting your company's professionalism. Your goal should be to achieve a continued rapport with your reader. In this concluding paragraph, include your contact information and the times you can best be reached.

Creating a Positive Tone. Audiences respond favorably to positive words. If you use negative words, you could offend your reader. In contrast, positive words will help you control your readers' reactions, build goodwill, and persuade your audience to accept your point of view.

Choose your words carefully. Even when an audience expects bad news, they still need a polite and positive response. The positive words in Table 4.2 and positive verbs in Table 4.3 will help you create a pleasant tone and build audience rapport.

TABLE 4.2 POSITIVE WORDS

advantage	efficient	meaningful
asset	enjoyable	please
benefit	favorable	positive
certain	good	profit
confident	grateful	quality
constructive	happy	successful
contribution	helpful	thank you
effective	improvement	value

TABLE 4.3 POSITIVE VERBS	
accomplish	improve
achieve	increase
assist	initiate
assure	insure (ensure)
build	maintain
coordinate	organize
create	plan
develop	produce
encourage	promote
establish	satisfy
help	train
implement	value

TABLE 4.4 NEGATIVE VS. POSITIVE SENTENCES

Before	After
1. The error is your fault. You scheduled incorrectly and cannot complain about our deliveries. If you would cooperate with us, we would work with you to solve this problem.	1. To improve deliveries, let's work together on our companies' scheduling practices.
2. I regret to inform you that we will not replace the motor in your dryer unless we have proof of purchase.	2. When you provide us proof of purchase, we will be happy to replace the motor in your dryer.
3. The accounting records your company submitted are incorrect. You have obviously miscalculated the figures.	3. After reviewing your company's accounting records, please recalculate the numbers to ensure that they correspond to the new X44 tax laws (enclosed).
4. Your letter suggesting an improvement for the system has been rejected. The reconfigurations you suggest are too large for the area specifications. We need you to resubmit if you can solve your problem.	4. Thank you for your suggestions. Though you offer excellent ideas, the configurations you suggest are too large for our area specifications. Please resubmit your proposal based on the figures provided online.
5. You have not paid your bill yet. Failure to do will result in termination of services.	5. Prompt payment of bills ensures continued service.

Table 4.4 gives you a "before and after" view of negative sentences rewritten using a positive tone.

See Figure 4.12 for a sample complaint letter to an automotive supplies company.

Figure 4.12
Complaint Letter in Block Format

1234 18th Street
Galveston, TX 77001
May 10, 2006

Mr. Holbert Lang
Customer Service Manager
Gulfstream Auto
1101 21st Street
Galveston, TX 77001

Dear Mr. Lang:

The introduction includes the date of purchase (to substantiate the claim) and the problem encountered.

On February 12, 2006, I purchased two shock absorbers in your automotive department. Enclosed are copies of the receipt and the warranty for that purchase. One of those shocks has since proved defective.

I attempted to exchange the defective shock at your store on May 2, 2006. The mechanic on duty, Vernon Blanton, informed me that the warranty was invalid because your service staff did not install the part. I believe that your company should honor the warranty agreement and replace the part for the following reasons:

In the body, explain what happened, state what you want done, and justify your demand. This letter develops its claim with warranty information.

1. The warranty states that the shock is covered for 48 months and 48,000 miles.
2. The warranty does not state that installation by someone other than the dealership will result in warranty invalidation.
3. The defective shock absorber is causing potentially expensive damage to the tire and suspension system.

Conclude your letter by providing contact information and an upbeat, pleasant tone.

I can be reached between 1 P.M. and 6 P.M. on weekdays at 763-9280 or at 763-9821 anytime on weekends. I look forward to hearing from you. Thank you for helping me with this misunderstanding.

Sincerely,

Carlos De La Torre

Enclosures (2)

Adjustment Letters

Responses to letters of complaint, also called adjustment letters, can take three different forms.

1. 100% Yes: You could agree 100 percent with the writer of the complaint letter.
2. 100% No: You could disagree 100 percent with the writer of the complaint letter.
3. Partial Adjustment: You could agree with some of the writer's complaints but disagree with other aspects of the complaint.

Table 4.5 Differences Among Adjustment Letters			
	100% Yes	100% No	Partial Adjustment
Introduction	State the good news.	Begin with a buffer, a comment agreeable to both reader and writer.	State the good news.
Discussion	Explain what happened and what the reader should do and/or what the company plans to do next.	Explain what happened, state the bad news, and provide possible alternatives.	Explain what happened, state the bad news, and provide possible alternatives—what the reader or company should do next.
Conclusion	End upbeat and positive.	Resell (provide discounts, coupons, follow-up contact names and numbers, etc.) to maintain goodwill.	Resell (provide discounts, coupons, etc.) to maintain goodwill.

Table 4.5 shows you the differences among these three types of adjustment letters.

Writing a 100% Yes response to a complaint is easy. You are telling your audience what they want to hear. The challenge, in contrast, is writing a 100% No response or a Partial Adjustment. In these letters, you must convey bad news, but you do not want to convey bad news too abruptly. Doing so might offend, anger, or cause hurt feelings. Using a buffer statement delays bad news in written communication and gives you an opportunity to explain your position.

Buffers to Cushion the Blow Use the following techniques to buffer the bad news:

- Establish rapport with the audience through positive words to create a pleasant tone. Instead of writing "We received your complaint," be positive and say, "We always appreciate hearing from customers."
- Sway your reader to accept the bad news to come with persuasive facts. "In the last quarter, our productivity has decreased by 16 percent, necessitating cost-cutting measures."
- Provide information that both you and your audience can agree upon. "With the decline of dotcom jobs, many information technology positions have been lost."
- Compliment your reader or show appreciation. "Thank you for your June 9 letter commenting on fiscal year 2005."
- Make your buffer concise with one to two sentences. "Thank you for writing. Customer comments give us an opportunity to improve service."
- Be sure your buffer leads logically to the explanation that follows. Consider mentioning the topic, as in the following example about billing practices. "Several of our clients have noted changes in our corporate billing policies. Your letter was one of several addressing this issue."
- Avoid placing blame or offending the reader. Rather than stating, "Your bookkeeping error cost us $9,890.00," write, "Mistakes happen in business. We are refining our bookkeeping policies to ensure accuracy."

See Figures 4.13 through 4.15 for sample adjustment letters.

Figure 4.13 100% Yes Adjustment Letter, Complete with Letter Essentials

Positive word usage ("Thank you") achieves audience rapport.

The introduction immediately states the good news.

The discussion explains what created the problem and provides an instruction telling the customer what to do next.

The conclusion ("We appreciate your business") resells to maintain customer satisfaction.

Gulfstream Auto

1101 21st Street
Galveston, TX 77001
(712) 451-1010
May 31, 2006

Mr. Carlos De La Torre
1234 18th Street
Galveston, TX 77001

Dear Mr. De La Torre:

Thank you for your recent letter. Gulfstream will replace your defective shock absorber according to the warranty agreement.

The Trailhandler Performance XT shock absorber that you purchased was discontinued in October 2006. Mr. Blanton, the mechanic to whom you spoke, incorrectly assumed that Gulfstream was no longer honoring the warranty on that product. Because we no longer carry that product, we either will replace it with a comparable model or refund the purchase price. Ask for Mrs. Cottrell at the automotive desk on your next visit to our store. She is expecting you and will handle the exchange.

We appreciate your business, Mr. De La Torre. I'm glad you brought this problem to my attention. If I can help you in the future, please contact me.

Sincerely,

Holbert Lang
Sales Manager

cc: Jordan Cottrell, Supervisor
Jim Gaspar, CEO

Sales Letters

You have just manufactured a new product (an electronic testing device, a fuel injection mechanism, a fiber optic cable, or a high-tech, state-of-the-art heart monitor). Perhaps you have just created a new service (computer maintenance, automotive diagnosis, home repair, or computer security). You must market your product or service. Connect with your end users, and let the public know that you exist by writing a sales letter. To compose your sales letter, include the following:

Introduction. The introductory paragraph of your sales letter tells your readers why you are writing (you want to increase their happiness or reduce their anxieties, for example). Your introduction should highlight a reader problem, need, or desire. If the readers do not need your services, then they will not be motivated to purchase your merchandise. The introductory sentences also should mention the product or service

Figure 4.14 100% No Adjustment Letter

1101 21st Street
Galveston, TX 77001
(712) 451-1010
May 31, 2006

Gulfstream Auto

Mr. Carlos De La Torre
1234 18th Street
Galveston, TX 77001

Dear Mr. De La Torre:

Thank you for your May 10 letter. Gulfstream Auto always appreciates hearing from its customers.

The Trailhandler Performance XT shock absorber that you purchased was discontinued in October 2006. Mr. Blanton, the mechanic to whom you spoke, correctly stated that Gulfstream was no longer honoring the warranty on that product. Because we no longer carry that product, we can not replace it with a comparable model or refund the purchase price. Although we can not replace the shock absorber, we want to offer you a 10 percent discount off of a replacement.

We appreciate your business, Mr. De La Torre. I'm glad you brought this problem to my attention. If I can help you in the future, please contact me.

Sincerely,

Holbert Lang,
Sales Manager

cc: Jordan Cottrell, Supervisor
Jim Gaspar, CEO

The introduction begins with a buffer. The writer establishes rapport with the audience through positive words to create a pleasant tone.

The discussion explains the company's position, states the bad news, and offers an alternative.

you are marketing, stating that this is the solution to their problems. Arouse your readers' interest with anecdotes, questions, quotations, or facts.

Discussion. In the discussion paragraph(s), specify exactly what you offer to benefit your audience or how you will solve your readers' problems. You can do this in a traditional paragraph. In contrast, you might want to itemize your suggestions in a numbered or bulleted list. Whichever option you choose, the discussion should provide data to document your assertions, give testimony from satisfied customers, or document your credentials.

Conclusion. Make your readers act. If your conclusion says, "We hope to hear from you soon," you have made a mistake. The concluding paragraph of a sales letter should motivate the reader to act. Conclude your sales letter in any of the following ways:

Figure 4.15 Partial Adjustment Letter

1101 21st Street
Galveston, TX 77001
(712) 451-1010
May 31, 2006

Mr. Carlos De La Torre
1234 18th Street
Galveston, TX 77001

Dear Mr. De La Torre:

Begin your letter with the good news. →

Thank you for your recent letter. Gulfstream will replace your defective shock absorber according to the warranty agreement.

Explain what happened, state the bad news, and provide a possible alternative. →

The Trailhandler Performance XT shock absorber that you purchased was discontinued in October 2006. Mr. Blanton, the mechanic to whom you spoke, incorrectly assumed that Gulfstream was no longer honoring the warranty on that product. However, we no longer carry that product. We will replace the shock absorber with a comparable model, but you will have to pay for installation.

We appreciate your business, Mr. De La Torre. I'm glad you brought this problem to my attention. If I can help you in the future, please contact me.

Sincerely,

Holbert Lang,
Sales Manager

cc: Jordan Cottrell, Supervisor
Jim Gaspar, CEO

- Give directions (with a map) to your business location.
- Provide a tear-out to send back for further information.
- Supply a self-addressed, stamped envelope for customer response.
- Offer a discount if the customer responds within a given period of time.
- Give your name or a customer-contact name and a phone number (toll-free if possible).

See Figure 4.16 for a successful sales letter from a computer hardware/software company.

4520 Shawnee Dr. Tulsa, OK 86221 721-555-2121

November 12, 2006

Bill Schneider
Office Manager
REM Technologies
2198 Silicon Way
Tulsa, OK 86112

The introduction arouses reader interest by asking questions. The questions highlight reader problems: *profits, costs, productivity,* and *breakdowns.*

Dear Mr. Schneider:

Are hardware and software upgrades making your profits plummet? Would you like to reduce your company's computer purchase and maintenance costs? Do computer breakdowns hurt your business productivity? Don't let technology breakdowns harm your bottom line. Many companies have taken advantage of **Office Station's** computer prices, service guarantees, and certified technicians.

The last sentence shows how Office Station will solve the problems.

Office Station, located in your neighborhood, offers you the following benefits:

The letter uses positive words to persuade: *advantage, guarantees, certified, benefits, satisfied, prompt,* and *courteous.*

- Purchase prices at least 10 percent lower than our competitors.
- IBM-trained technicians, available on a yearly contract or per-call basis.
- An average response time to service calls of under two hours.
- Repair loaners to keep your business up and running.
-  Over 5000 satisfied customers, like IBM, Ford, Chevrolet, and Boeing.
-  All-inclusive agreements that cover travel, expenses, parts, and shop work.
-  State-of-the-art technologies, featuring the latest hardware and software.

The body provides specific proof to sway the reader: *10 percent lower, IBM-trained technicians, two-hour response,* and *satisfied customers.*

Our service is prompt, our technicians are courteous, and our prices are unbeatable. For further information and a written proposal, please call us at **721-555-2121** or e-mail your sales contact, Steve Hudson (shudson@os.com). He's waiting to hear from you. Take advantage of our *Holiday Season Discounts!*

The conclusion urges action by giving contact names and numbers and seasonal discounts.

Sincerely,

Rachel Adams,
Sales Manager

Office Station
Authorized Sales and Service for
Gateway 3M Microsoft HP Apple Dell Swingline

Figure 4.16 Sales Letter in Block Format

Letters Checklist

The Letters Checklist will give you the opportunity for self-assessment and peer evaluation of your writing.

Letters Checklist
__ 1. **Letter Essentials:** Does your letter include the eight essential components (writer's address, date, reader's address, salutation, text, complimentary close, writer's signed name, and writer's typed name)?
__ 2. **Introduction:** Does the introduction state *what* you are writing about and *why* you are writing?
__ 3. **Discussion:** Does your discussion clearly state the details of your topic depending on the type of letter?
__ 4. **Highlighting/Page Layout:** Is your text accessible? To achieve reader-friendly ease of access, use headings, boldface, italics, bullets, numbers, underlining, or graphics (tables and figures). These add interest and help your readers navigate your letter.
__ 5. **Organization:** Have you helped your readers follow your train of thought by using appropriate modes of organization? These include chronology, importance, problem/solution, or comparison/contrast.
__ 6. **Conclusion:** Does your conclusion give directive action (tell what you want the reader to do next and when) and end positively?
__ 7. **Clarity:** Is your letter clear, answering reporter's questions and providing specific details that inform, instruct, or persuade?
__ 8. **Conciseness:** Have you limited the length of your words, sentences, and paragraphs?
__ 9. **Audience Recognition:** Have you written appropriately to your audience? This includes avoiding biased language, considering the multicultural/cross-cultural nature of your readers, and your audience's role (supervisors, subordinates, coworkers, customers, or vendors). Have you created a positive tone to build rapport?
__10. **Correctness:** Is your text grammatically correct? Errors will hurt your professionalism. See Appendix A for grammar rules.

The Writing Process at Work

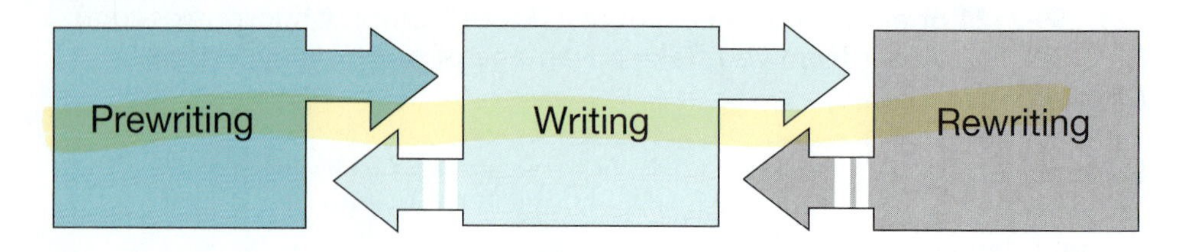

Effective writing follows a process of prewriting, writing, and rewriting. Each of these steps is sequential and yet continuous. The writing process is dynamic, with the three steps frequently overlapping. To clarify the importance of the writing process, look at how Jim Goodwin, the CEO of CompuMed, used prewriting, writing, and rewriting to write a memo to his employees.

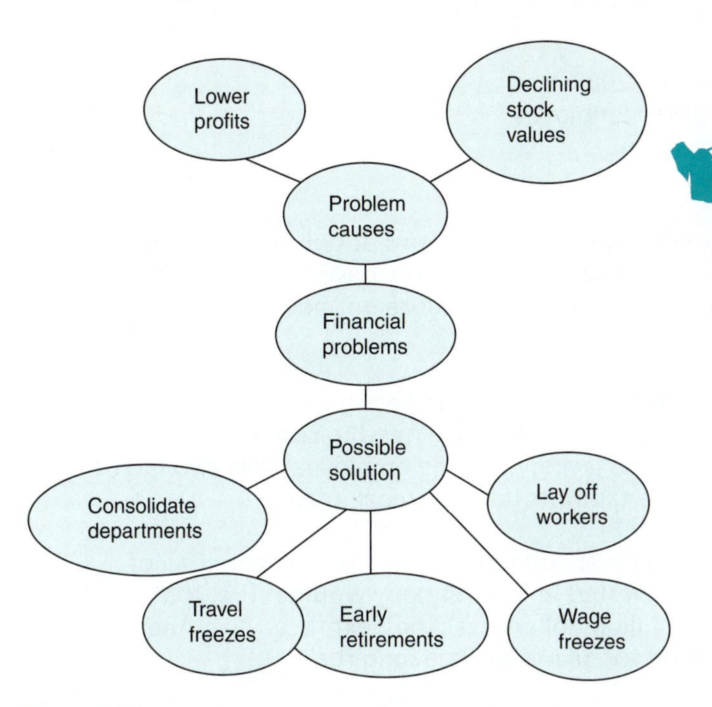

Figure 4.17 Mind Mapping/Clustering to Gather Data

Writer's Insight

Jim Goodwin says that "clustering always helps me. When I cluster, I'm able to see the different parts of any subject. Clustering helps me develop my ideas and conceptualize. Though I also use outlining, I have found that clustering is less restrictive, less structured, and this allows me to gather data more creatively."

Prewriting

No single method of prewriting is more effective than another. Throughout this textbook, you will learn many different types of prewriting techniques, geared uniquely for different types of communication. The goal of all prewriting is to help you overcome the blank page syndrome (writer's block). Prewriting will allow you to spend time before writing your memo or letter, gathering as much information as you can about your subject matter. In addition, prewriting lets you determine your objectives.

Jim used mind mapping/clustering to gather data and determine objectives (Figure 4.17).

Writing

Once you have gathered your data and determined your objectives in prewriting, your next step is to draft your memo or letter. In doing so, consider the following techniques:

- **Clarify your audience.** Before writing the draft, consider your audience. Are you writing laterally or vertically? Is your audience composed of management, subordinates, or colleagues? Is it a combination audience?
- **Organize your ideas.** If your supporting details are presented randomly, your audience will be confused. As a writer, develop your content logically. When you draft your memo or letter, choose a method of organization that will help your readers understand your objectives. This could include comparison-contrast, problem-solution, chronology, cause-effect, and more.

Jim drafted a memo, focusing on the information he discovered in prewriting (Figure 4.18).

Figure 4.18 Rough Draft Memo

Date: October 14, 2006
To: CompuMed Employees
From: Jim Goodwin
Subject: Problems

As you know, we are experiencing some problems at CompuMed. These include lower profits and stock value declines. We have a lot of unhappy stockholders. Its up to me to help everyone figure out how to solve our problems.

I have some ideas I want to share with you. I'm happy to have you share your ideas with me too. Here are my ideas: we need to consider consolidating departments and laying off some employees. We also might need to freeze wages and certainly its time to freeze travel.

The best idea I have is for some of you to take early retirement. If all of you who have over 20 years vested in the company would retire, that would save us around 2.1 million dollars over the next fiscal year. And, you know, saving money is good for all of us in the long run.

Rewriting

To edit and revise your memo or letter and make it as professional as it can be, follow these techniques:

- **Add new detail for clarity.** Reread your draft. If you have omitted any information stemming from the reporter's questions (who, what, when, why, where, how), insert answers to these questions.
- **Delete dead words and phrases for conciseness.** Chapter 2 will help you understand how to achieve conciseness. Review techniques for revising unnecessarily long words, sentences, and paragraphs.
- **Move information for emphasis.** Place your most important information first, or move information around to maintain an effective chronological order.
- **Reformat for access.** Use highlighting techniques (white space, bullets, boldface, italics, numbers, or headings) for reader-friendly appeal.
- **Correct the tone of your correspondence.** Based on audience, achieve an appropriate tone.
- **Edit for accuracy.** Check your memo or letter for punctuation, grammar, and spelling errors. You must proofread to be professional. Appendix A will help you with your grammar concerns.

After drafting this memo, Jim asked a coworker to help revise the text (Figure 4.19).

Jim factored in his coworker's suggestions and rewrote the memo. See Figure 4.20 for the finished product.

Figure 4.19
Revision Suggestions

Date: October 14, 2006
To: CompuMed Employees
From: Jim Goodwin
Subject: Problems

As you know, we are experiencing some problems at CompuMed. These include lower profits and stock value declines. We have alot of unhappy stockholders. Its up to me to help everyone figure out how to solve our problems.

I have some ideas I want to share with you. I'm happy to have you share your ideas with me too. Here are my ideas: we need to consider consolidating departments and laying off some employees. We also might need to freeze wages and certainly its time to freeze travel.

The best idea I have is for some of you to take early retirement. If all of you who have over 20 years vested in the company would retire, that would save us around 2.1 million dollars over the next fiscal year. And, you know, saving money is good for all of us in the long run.

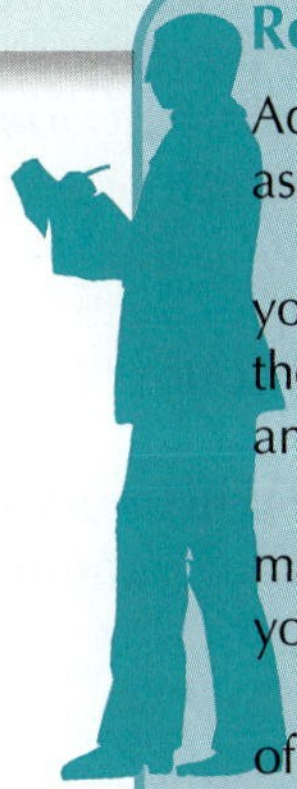

Revision Suggestions

Add a focus to the subject line, such as "Problems with . . . ."

I'd consider removing words like "as you know," "some," and "a lot." Replace them with stronger words. Also, "a lot" and "it's" are spelled wrong.

List these problems and solutions to make them more accessible. Also, could you add more details?

Jim, I think you need to alter the tone of this memo. Is there some way to avoid talking about saving money by firing people?

Date October 14, 2006
To: CompuMed Employees
From: Jim Goodwin
Subject: Suggestions for Improving Corporate Finances

CompuMed is experiencing lower profits and declining stock value. Consequently, stockholders are displeased with company performance. I have been meeting with the Board of Directors and division managers to determine the best course of action. Here are ideas to improve our company's financial situation.

1. Consolidating departments: By merging our marketing and advertising departments, for example, we can reduce redundancies. This could save CompuMed approximately $275,000 over a six-month period.
2. Reducing staff: We need to cut back employees by 15 percent. This does not necessarily mean that layoffs are inevitable. One way, for instance, to reduce staff is through voluntary retirements. We will be encouraging employees with over 20 years vested in the company to take our generous early-retirement package.
3. Freezing wages: For the next fiscal quarter, no raise increases will go into effect. Internal auditors will review the possibility of reestablishing raises after the first quarter.
4. Freezing travel: Conference attendance will be stopped for six months.

I encourage you to visit with me and your division managers with questions or suggestions. CompuMed is a strong company and will bounce back with your help. Thank you for your patience and understanding.

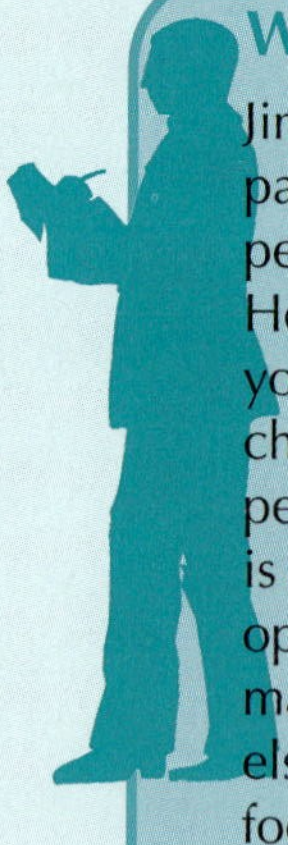

Figure 4.20
Finished Bad-News, Problem-Solution Memo Incorporating Revision Suggestions

Writer's Insight

Jim says, "I find it extremely difficult and painful to communicate bad news to people I care about—my employees. However, to run a business successfully, you sometimes have to make difficult choices that will negatively affect many people. The best way to convey bad news is to state it clearly and follow up with options. I always try to end positively to maintain good relations. Having someone else read my correspondence helps me focus on what needs to be changed and what's successful."

Developing Workplace Skills

Case Studies

After reading the following case studies, write the appropriate correspondence required for each assignment.

1. As director of human resources at CompuMed biotechnology company, Andrew McWard helps employees create and implement their Individual Development Plans (IDPs). Employees attend 360-Degree Assessment Workshops where they learn how to get feedback on their job performance from their supervisors, coworkers, and subordinates. They also provide self-evaluations.

   Once the 360-Degree Assessments are complete, employees submit them to Andrew, who, with the help of his staff, develops IDPs.

   Andrew sends the IDPs to the employees, prefaced by a cover letter. In this cover letter, he tells them why he is writing and what he is writing about. In the letter's body, he focuses their attention on the attachment's contents: supervisor's development profile, the schedule of activities that helps employees implement their plans, the courses designed to increase their productivity, the costs of each program, and guides to long-term professional development.

   In the cover letter's conclusion, Andrew ends upbeat by emphasizing how the employees' IDPs help them resolve conflicts and make better decisions.

   Based on the information provided, write this cover letter for Andrew McWard. He is sending the letter to Sharon Baker, Account Executive, 1092 Turtle Hill Road, Evening Star, GA 20091.

2. Mark Shabbot works for Apex, Inc., at 1919 W. 23rd Street, Denver, CO 80204. Apex, a retailer of computer hardware, wants to purchase 125 new flat-screen monitors from a vendor, Omnico, located at 30467 Sheraton, Phoenix, AZ 85023. The monitors will be sold to Northwest Hills Educational Cooperative. However, before Apex purchases these monitors, Mark needs information regarding bulk rates, shipping schedules, maintenance agreements, equipment specifications, and technician certifications. Northwest Hills needs this equipment before the new term (August 15). Write a letter of inquiry for Mr. Shabbot based on the preceding information.

3. Gregory Peña (121 Mockingbird Lane, San Marcos, TX 77037) has written a letter of complaint to Donya Kahlili, the manager of TechnoRad (4236 Silicon Dr., San Marcos, TX 77044). Mr. Peña purchased a computer from a TechnoRad outlet in San Marcos. The *San Marcos Tattler* advertised that the computer "came loaded with all the software you'll need to write effective letters and perform basic accounting functions." (Mr. Peña has a copy of this advertisement.) When Mr. Peña booted up his computer, he expected to access word processing software, multiple fonts, a graphics package, a grammar check, and a spreadsheet. All he got was a word processing package and a spreadsheet. Mr. Peña wants Ms. Kahlili to upgrade his software to include fonts, graphics, and a grammar check; he wants a computer technician from TechnoRad to load the software on his computer; and he wants TechnoRad to reimburse him $400 (the full price of the software) for his trouble.

Ms. Kahlili agrees that the advertisement is misleading and will provide Mr. Peña software including the fonts, graphics, and grammar check (complete with instructions for loading the software). Write Ms. Kahlili's 100% Yes Adjustment to Mr. Peña based on the information provided.

4. TechToolshop provides automotive sales and service. They install and repair automotive equipment at service sites nationwide; through an online catalog and storefront sites, they sell automotive equipment both wholesale and retail.

   TechToolshop's home office is in Big Springs, Iowa, at 11324 Elm, where over 1,200 employees work. Their phone number at this site is 212-345-6666, and their email address is *ToolHelp@TechTools.com*. TechToolshop's new local address in your city is 5110 Nueces Avenue. Their phone number is 345-782-8776.

   TechToolshop offers free product support 24 hours a day at 1-800-TechHelp. They also can guarantee arrival at your site within two hours of any automotive service emergency call. Plus—their greatest innovation—TechToolshop has installed service kiosks in every mall, library, and bank in your city where you can look up answers to frequently asked automotive questions. They warrant all products and services—money back—for 90 days, covering defects in material and workmanship.

   Write a sales letter marketing TechToolshop's.

5. Bob Ward, an account manager at HomeCare Health Equipment, has not gotten the raise that he thinks he deserves. When Bob met with his boss, Helene Koren, last Thursday for his annual evaluation, she told him that he had missed too many days of work (eight days during the year), was unwilling to work beyond his 40-hour workweek to complete rush jobs, and had not attended two mandatory training sessions on the company's new computerized inventory system.

   Bob agrees that he missed the training sessions, but he was out of town on a job-related assignment for one of those sessions. He missed eight days of work, but he was allowed five days of sick leave as part of his contract. The other three days missed were due to his having to stay home to take care of his children when they were sick. He believed that these absences were covered by the company's parental leave policy. Finally, he does not agree that employees should be required to work beyond their contractual 40 hours.

   Write a memo to Helene Koren, stating Bob's case.

Individual and Team Projects

1. Write a letter of inquiry. You might want to write to a college or university requesting information about a degree program or to a manufacturer for information about a product or service. Whatever the subject matter, be specific in your request.
2. Write a cover letter. Perhaps your cover letter will preface a report you are working on in school, a report you are writing at work, or documentation you will need to send to a client.
3. Write a letter of complaint. You might want to write to a retail store, a manufacturing company, a restaurant, or a governmental organization. Whatever the subject matter, be specific in your complaint.

4. Write an adjustment letter. Envision that a client has written a complaint letter about a problem he or she has encountered with your product or service. Write a 100% Yes letter in response to the complaint.

5. Write a sales letter. You plan to sell a new product, portable computer zip drives, that are small enough to fit on key chains.

6. Write a memo requesting office equipment. Your company plans to purchase new office equipment. Your memo will explain your office's needs. Specifically state what equipment and furniture you want and why these purchases are important.

7. Write a memo reporting on a project's progress. Draw from your experiences in one of your classes. How are you progressing on an assignment? What work have you accomplished? What problems have you encountered? What work remains on this project? In a memo to your instructor, detail this status.

8. Write a memo inviting coworkers to your company's annual picnic. In this memo, tell why the picnic is being held, when the celebration will occur, what special events are planned, where the picnic will take place, and what the guests should bring (equipment, clothing, etc.) to enjoy their outing.

9. Write a good news memo. One of your coworkers has done an outstanding job (with customer service, sales, training, or helping other colleagues in the department). You want to write a memo to your manager commending this employee.

10. Write a bad news memo. One of your subordinates has not been performing well on the job. Maybe this employee has been rude to customers, has not performed tasks up to the company's standards, has been shirking responsibilities, often has arrived late for work, or has failed to meet quotas. As manager, you must document these problems. Write the required memo to the employee. Remember to control the memo's tone—though the memo will convey bad news, you still want to be pleasant and positive.

Problem Solving Think Pieces

Northwest Regional Governmental Training Consortium (GTC) provides educational workshops for elected and appointed officials, as well as employees of city and state governmental offices.

One seminar participant, Mary Bloom, supervisor of the North Platte County Planning and Zoning Department, attended a GTC seminar entitled "Developing Leadership Skills" on February 12, 2006. Unfortunately, she was disappointed in the workshop and the facilitator. On February 16, Mary wrote a complaint letter to GTC's director, Georgia Randall, stating her dissatisfaction. Ms. Bloom said that the training facilitator's presentation skills were poor. According to Mary, Doug Aaron, the trainer, exhibited the following problems:

- Late arrival at the workshop
- Too few handouts for the participants
- Incorrect cables for his computer, so he could not use his planned PowerPoint presentation
- Old, smudged transparencies as a backup to the PowerPoint slides

Mary also noted that the seminar did not meet the majority of the seminar participants' expectations. She and the other government employees had expected a hands-on workshop with breakout sessions. Instead, Mr. Aaron lectured the entire time. In addition, his information seemed dated and ignored the cross-cultural challenges facing today's supervisors.

Neither Georgia nor her employees had ever attended this workshop. They offered the seminar based on the seemingly reliable recommendation of another state agency, the State Data Collection Department. From Doug Aaron's course objectives and resume, he appeared to be qualified and current in his field.

However, Mary Bloom deserves consideration. Not only are her complaints justified by others' comments, but she is a valued constituent. The GTC wants to ensure her continued involvement in their training program.

Assignments

1. Georgia needs to write a 100% Yes Adjustment letter. In this letter, Georgia wants to accommodate the dissatisfied client. How should she recognize Mary's concern, explain what might have gone wrong, and offer satisfaction?
2. Georgia needs to write an internal memo to her staff. In this memo, she will provide standards for hiring future trainers. What should the standards include?

Web Workshop

Many state governments provide guidelines for writing letters either to government officials or for government employees. In addition, you can go online, type something like "how to write state government letters" in a search engine, and find sites with instructions and sample letters for writing to state officials or state agencies.

Research the Internet and find letter samples and guidelines. Once you have done so,

1. review sample letters to or from governmental agencies.
2. determine whether these letters are successful, based on the criteria provided in this chapter.
3. either write a letter or memo to your instructor explaining how and why the letters succeeded.
4. if you find letters that can be improved, rewrite them, using the guidelines provided in this chapter.

Electronic Communication

OBJECTIVES

When you complete this chapter, you will be able to do the following:

1. Understand the impact and importance of electronic communication.
2. Know the characteristics of online communication.
3. Compose effective e-mail messages.
4. Know the characteristics of instant messaging.
5. Use a Web log (blog).
6. Create effective Web sites.
7. Write effective online help.
8. Use the writing process—prewriting, writing, and rewriting—to write effective electronic communication.
9. Test your knowledge of electronic communication through end-of-chapter activities:
   - Case Studies
   - Individual and Team Projects
   - Problem Solving Think Pieces
   - Web Workshops

COMMUNICATION AT WORK

The Future Promise scenario shows the importance of electronic communication.

Future Promise is a not-for-profit organization geared toward helping at-risk high school students. This agency realizes that to reach its target audience (teens age 15 to 18), it needs an Internet presence and a Web log (blog).

Future Promise's CEO, Brent Searing, has decided to form a cross-functional team to create the agency's Web site. Brent will encourage the team to work collaboratively to determine the Web site's content, its level of interactivity, and its design features. Brent wants the Web site to include the following:

- College scholarship opportunities
- After-school intramural sports programs
- Job-training skills (resume building and interviewing)
- Service learning programs to encourage civic responsibility
- An FAQ page
- Future Promise's 800-Hotline (for suicide prevention, depression, substance abuse, and peer counseling)

To accommodate these Web components, the Future Promise Web team will consist of the agency's accountant, sports and recreation director, public relations manager, counselor, training facilitator, graphic artist, and computer and information systems director. In addition to these Future Promise employees, Brent also has asked two local high school principals, two local high school students, and a representative from the mayor's office to serve on the committee. Future Promise's public relations manager, Jackie Velasquez, will chair the committee. Given this team's diverse locations, the best ways for them to communicate will be through e-mail and instant messaging.

Though the task is difficult, the end product will be invaluable for the city and the city's youth. Jackie and Brent know that by conveying information about jobs, training, scholarships, and counseling to their end users (at-risk teens), Future Promise can improve the quality of many people's lives.

Overview: The Importance of Electronic Communication

During the last decade, the written word underwent significant changes, leaping from the printed page into cyberspace. Correspondence, once limited to hard-copy letters, reports, and memos, is now often online as e-mail, Web sites, Web logs (blogs), instant messages, and online help. Corporate brochures and newsletters, once paperbound, now are online. Product and service manuals, once paperbound, now are online. Resumes are online. Research is online. Workplace communication in the twenty-first century is increasingly electronic.

The 2004 National Commission on Writing in the workplace reports about the changing role of electronic communication ("Writing: A Ticket to Work" 11). E-mail messages are the most prevalent type of communication written on the job. The 2005 National Commission on Writing follow-up report further emphasizes the importance of electronic communication. In this report, survey results state that 83.7 percent of respondents "almost always" use e-mail; 100 percent "frequently or almost always" use e-mail. These numbers more than doubled all other communication channels used in the workplace ("Writing: A Powerful Message" 17).

This trend toward electronic communication is increasing with the growing presence of "Wi/Fi" (wireless fidelity) hotspots. At Wi/Fi hotspots, people can connect to the Internet by way of laptops, Tablet PCs, cell phones, and handheld computers. This allows people to work offsite, anywhere and anytime.

Writer's Insight

Brent Searing, CEO of Future Promise, says, "Last year, our not-for-profit agency chose to give up our lease on an office space to save overhead costs. Now, our employees at Future Promise use cell phones, PDAs, and laptops to conduct business. Our organization meets weekly at a coffeehouse with wireless access. This gives the employees a chance to discuss work face to face and retrieve and reply to their e-mail messages through the Internet unplugged. It's a less stressful environment that provides my staff flexibility and enhances their creativity.

And we're not looking to rent office space. My staff's productivity has increased, and working outside of a cubicle is more convenient for their lifestyles. Electronic communication is allowing for a mobile work environment, offices without walls and employees who can do their jobs from almost anywhere."

The Characteristics of Online Communication

Electronic communication is an entirely different mode of communication than hard-copy text. What are the differences between paper text and online communication?

E-readers are topic specific. In libraries or bookstores, we wander up and down aisles, looking for any book that interests us. In contrast, e-readers tend to access the Web or blogs with specific goals in mind. You go online to search for specific information found in specific Web sites: CD prices, automobile loan rates, hotel room availability, restaurant menus, technical specifications for laser printers, the start date for your college's spring semester, and so forth (Moore).

E-readers want information quickly. E-readers often scan, skim, and skip over text, looking just for the information they want and ignoring the rest of the text. In fact, readers want to find information in "ten seconds. Your web site visitors love skim-readable pages. They're not lazy. They're just in a hurry. They don't want to waste time and money reading the wrong web page, when there are millions of other pages to choose from" (McAlpine).

Electronic communication platforms are diverse. Another difference between an e-reader and the same person who reads a hard-copy novel is the way in which he or she accesses the document. Most hard-copy text is printed on either book-sized pages or on $8^1/_2" \times 11"$ pieces of paper. In contrast, e-readers can access electronic communication "using cell phones, PDAs, and other wireless devices" (Moore 17). Thus, screen resolution and the size of type font are key elements for online readability. Furthermore, e-readers will be using Netscape, Internet Explorer, AOL, Macs and PCs, and a host of other electronic platforms to view your text. Each of these electronic platforms differs in subtle ways.

Electronic communication encourages random access. Because e-readers are unique, you must alter the way you write online communication. This means changing your mind-set as a writer. When you write a blog, a Web site, e-mail, or an instant message, you must consider hypertext links versus chronological reading. We read books from beginning to end, sequentially. Web sites, however, allow us, even encourage us, to leap randomly from screen to screen, from Web site to Web site, or from blog to blog.

Electronic communication is more casual than other forms of written communication. Another unique distinction of electronic communication is its casual tone. Letters and reports have formal formats that elevate the tone of your correspondence. In contrast, instant messaging, blogs, the Internet, and e-mail messages tend to be more informal, even conversational.

Why Is E-Mail Important?

Many companies are "geared to operate with e-mail," creating what the Harvard Business School calls "e-mail cultures" for the following reasons ("The Transition to General Management" 1998):

Time. "Everything is driven by time. You have to use what is most efficient" (Miller et al. 10). The primary driving force behind e-mail's prominence is time. E-mail is quick. Whereas a posted letter might take several days to deliver, e-mail messages can be delivered within seconds.

Convenience. With wireless communication, you can send e-mail from notebooks to handhelds. Current communication systems combine a voice phone, a personal digital assistant, and e-mail into a package that you can slip into a pocket or purse. Then, you can access your e-mail messages anywhere, anytime.

Internal/External. E-mail allows you to communicate internally to coworkers and externally to customers and vendors. Traditional communication channels, like letters and memos, have more limited uses. Generally, letters are external correspondence written from one company to another company; memos are internal correspondence transmitted within a company.

Cost. E-mail is cost effective because it is paper-free. With an ability to attach files, you can send many kinds of documentation without paying shipping fees. This is especially valuable when considering international business.

Documentation. E-mail provides an additional value when it comes to documentation. Because so many writers merely respond to earlier e-mail messages, what you end up with is a "virtual paper trail" (Miller et al. 15). When e-mail is printed out, often the printout will contain dozens of e-mail messages, representing an entire string of dialogue. This provides a company an extensive record for future reference. In addition, most companies archive e-mail messages in backup files.

Techniques for Writing Effective E-Mail Messages

To convey your messages effectively and to ensure that your e-mail messages reflect professionalism, follow these tips for writing e-mail.

Recognize Your Audience. E-mail messages can be sent to managers, coworkers, subordinates, vendors, and customers, among other audiences. Your e-mail readers will be specialists, semi-specialists, and lay people. Thus, you must factor in levels of knowledge.

If an e-mail message is sent internationally, you also might have to consider your readers' language. Remember that abbreviations and acronyms are not universal. Dates, times, measurements, and monetary figures differ from country to country. Audience concerns are discussed in greater detail in Chapter 2. In addition, your reader's e-mail system might not have the same features or capabilities that you have. Hard-copy text will look the same to all readers. E-mail platforms, in AOL, Outlook, Juno, HotMail, and Yahoo, for example, display text differently. To communicate effectively, recognize your audience's level of knowledge, unique language, and technology needs.

Identify Yourself. Identify yourself by name, affiliation, or title. You can accomplish this either in the "from" line of your e-mail or by creating a signature file or ".sig file." This .sig file acts like an online business card. Once this identification is complete, readers will be able to open your e-mail without fear of corrupting their computer systems ("Email Netiquette").

Provide an Effective Subject Line. Readers are unwilling to open unsolicited or unknown e-mail, due to fear of spam and viruses. In addition, corporate employees receive approximately 50 e-mail messages each day. They might not want to read every message sent to them. To ensure that your e-mail messages are read, avoid uninformative subject lines, such as "Hi," "What's New," or "Important Message." Instead, include an effective subject line, such as *Subject: Meeting Dates for Tech Prep Conference.*

Keep Your E-Mail Message Brief. E-readers skim and scan. To help them access information quickly, "Apply the 'top of the screen' test. Assume that your readers will look at the first screen of your message only" (Munter et al. 31). Limit your message to one screen (if possible).

Organize Your E-Mail Message. Successful writing usually contains an introductory paragraph, a discussion paragraph or paragraphs, and a conclusion. Although many e-mail messages are brief, only a few sentences, you can use the introductory sentences to tell the reader why you are writing and what you are writing about. In the discussion, clarify your points thoroughly. Use the concluding sentences to tell the reader what is next, possibly explaining when a follow-up is required and why that date is important.

Use Highlighting Techniques Sparingly. Many e-mail packages will let you use highlighting techniques, such as boldface, italics, underlining, computer-generated bullets and numbers, centering, font color highlighting, and font color changes. Many other e-mail platforms will not display such visual enhancements. To avoid having parts of the message distorted, limit your highlighting to asterisks (*), numbers, double spacing, and all-cap headings.

Proofread Your E-Mail Message. Errors will undermine your professionalism and your company's credibility. Recheck your facts, dates, addresses, and numerical information before you send the message. Try these tips to help you proofread an e-mail message.

- Type your text first in a word processing package, like Microsoft Word.
- Print it out. Sometimes it is easier to read hard-copy text than text online. Also, your word processing package, with its spell check and/or grammar check, will help you proofread your writing.
- Once you have completed these two steps (writing in Word or WordPerfect and printing out the hard-copy text), copy and paste the text from your word processing file into your e-mail.

Make Hard Copies for Future Reference. Making hard copies of all e-mail messages is not necessary because most companies archive e-mail. However, in some instances, you might want to keep a hard copy for future reference. These instances could include transmissions of good news. For example, you have received compliments about your work and want to save this record for your annual job review. You also might save a hard copy of an e-mail message regarding flight, hotel, car rental, or conference arrangements for business-related travel.

Be Careful When Sending Attachments. When you send attachments, tell your reader within the body of the e-mail message that you have attached a file; specify the file name of your attachment and the software application that you have used (HTML, PowerPoint, PDF, RTF [rich text format] Word, or Works); and use compression (Zip) files to limit your attachment size. Zip files are necessary only if an attachment is quite large.

Practice Netiquette. When you write your e-mail messages, observe the rules of "netiquette."

- Be courteous. Do not let the instantaneous quality of e-mail negate your need to be calm, cool, deliberate, and professional.
- Be professional. Occasionally, e-mail writers compose excessively casual e-mail messages. They will lowercase a pronoun like "i," use ellipses (. . .) or dashes instead of more traditional punctuation, use instant messaging shorthand language, such as "LOL" or "BRB," and depend on emoticons (☺ ☹). These e-mail techniques might not be appropriate in all instances. Don't forget that your e-mail messages represent your company's professionalism. Write according to the audience and communication goal.
- Avoid abusive, angry e-mail messages. Because of its quick turnaround abilities, e-mail can lead to negative correspondence called flaming. Flaming is sending angry e-mail, often TYPED IN ALL CAPS.

Organizing and Writing E-Mail Messages

E-mail is used to convey many types of information in business and industry. You can write an e-mail message to accomplish any of the following purposes.

- **Directive.** Tell a subordinate or a team of employees to complete a task.
- **Cover/Transmittal.** Tell a reader or readers that you have attached a document, and list the key points that are included in the attachment.
- **Documentation.** Report on expenses, incidents, accidents, problems encountered, projected costs, study findings, hiring, firings, and reallocations of staff or equipment.
- **Confirmation.** Tell a reader about a meeting agenda, date, time, and location; decisions to purchase or sell; topics for discussion at upcoming teleconferences; conclusions arrived at; or fees, costs, or expenditures.
- **Procedures.** Explain how to set up accounts, research on the company intranet, operate new machinery, use new software, apply online for job opportunities through the company intranet, create a new company Web site, or solve a problem.
- **Recommendations.** Provide reasons to purchase new equipment, fire or hire personnel, contract with new providers, merge with other companies, revise current practices, or renew contracts.
- **Feasibility.** Study the possibility of changes in the workplace (practices, procedures, locations, staffing, or equipment).
- **Status.** Provide a daily, weekly, monthly, quarterly, biannual, or yearly report about where you, the department, or the company is regarding a topic of your choice (class project, sales, staffing, travel, practices, procedures, or finances).
- **Inquiry.** Ask questions about upcoming processes, procedures, or assignments.

Many e-mail messages only require a sentence or two. If you need to convey more information than can be accomplished in only a few sentences, try this easy-to-use template (Figure 5.1):

Subject: *Topic* and *Focus* of the e-mail

Opening sentence or paragraph: Briefly tell the reader *why* you are writing and *what* you are writing about.

Discussing the topic: Develop your ideas in one or two short paragraphs (three to four sentences per paragraph) or in a list. Use this discussion section to explain *what exactly* you need to communicate. This can include meeting dates, steps to follow, a list of questions, or actions required.

Ending the e-mail: Tell the reader *what's next*. This could include a follow-up action and a due date.

Figure 5.1 E-Mail Template

Samples of E-Mail Messages

See Figure 5.2 for an example of a well written e-mail message. Figure 5.3 is a flawed e-mail message.

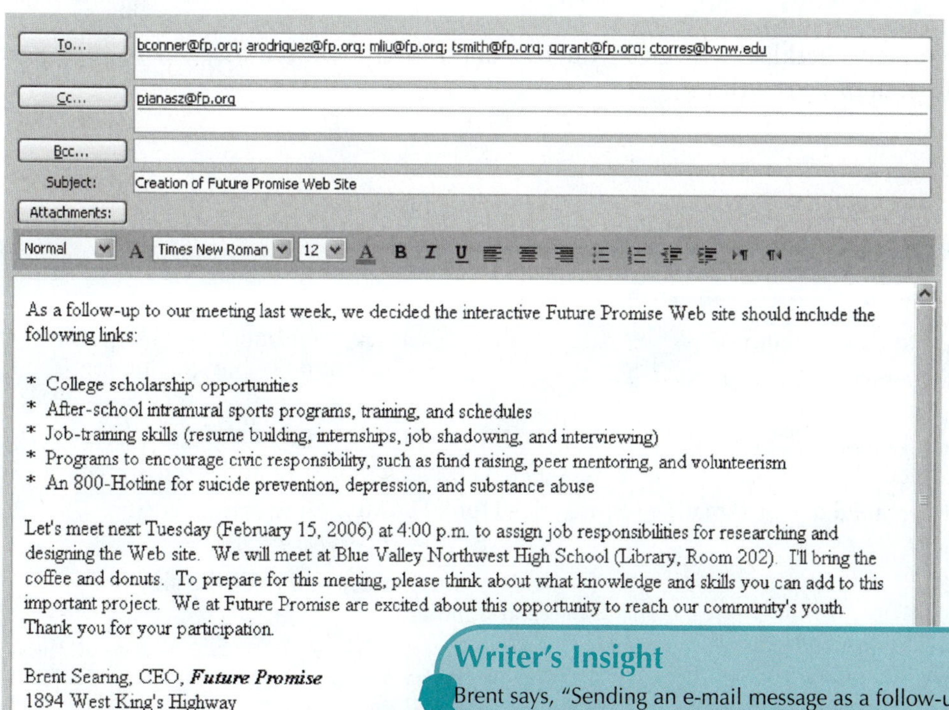

To... bconner@fp.org; arodriguez@fp.org; mliu@fp.org; tsmith@fp.org; ggrant@fp.org; ctorres@bvnw.edu

Cc... pjanasz@fp.org

Bcc...

Subject: Creation of Future Promise Web Site

Attachments:

As a follow-up to our meeting last week, we decided the interactive Future Promise Web site should include the following links:

* College scholarship opportunities
* After-school intramural sports programs, training, and schedules
* Job-training skills (resume building, internships, job shadowing, and interviewing)
* Programs to encourage civic responsibility, such as fund raising, peer mentoring, and volunteerism
* An 800-Hotline for suicide prevention, depression, and substance abuse

Let's meet next Tuesday (February 15, 2006) at 4:00 p.m. to assign job responsibilities for researching and designing the Web site. We will meet at Blue Valley Northwest High School (Library, Room 202). I'll bring the coffee and donuts. To prepare for this meeting, please think about what knowledge and skills you can add to this important project. We at Future Promise are excited about this opportunity to reach our community's youth. Thank you for your participation.

Brent Searing, CEO, ***Future Promise***
1894 West King's Highway
San Antonio, TX 78532
Phone: 814-236-5482

Figure 5.2 Successful E-Mail Message About the Status of a Company's Web Site

Writer's Insight

Brent says, "Sending an e-mail message as a follow-up to meetings is one of the best ways I know to keep people informed. I take notes during the meetings. Then, when I get back to my office, I review the notes and decide what to highlight in the e-mail. By the time employees get back to their offices or homes, I will have already sent the e-mail message.

I use a specific subject line so that my readers will not delete the message assuming it's spam. And I always try to itemize points for easy reading.

E-mail allows me to communicate quickly and ensure that all of the meeting participants get the same message. E-mail has been a real time-saver for me and for everyone I work with. This communication channel helps me do my job efficiently and almost effortlessly."

This e-mail is flawed due to its use of Instant Messaging abbreviations, lowercase text, ellipses (. . .), slang, and emoticons. Though these techniques might be appropriate for personal communication, avoid overly casual e-mail messages for job-related communication.

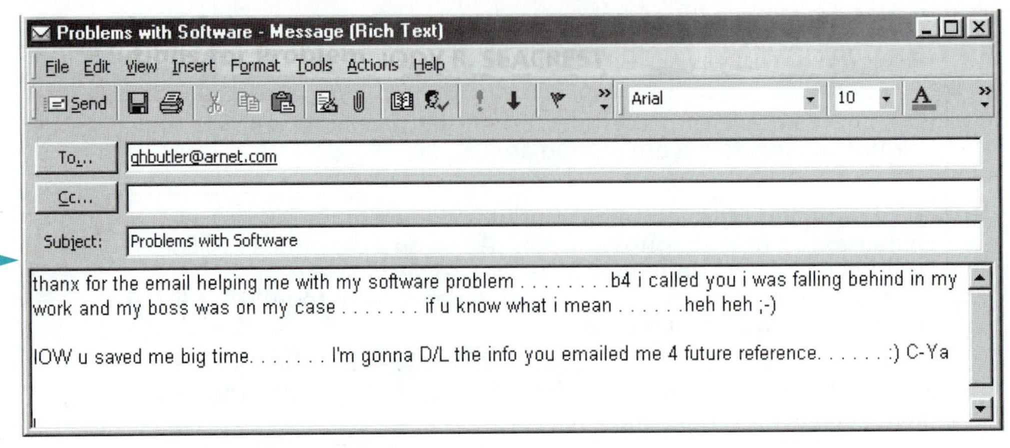

Figure 5.3 Flawed E-Mail Message

E-Mail Checklist

The e-mail checklist will give you the opportunity for self-assessment and peer evaluation of your writing.

E-Mail Checklist

__ 1. **Does the e-mail use the correct address?**

__ 2. **Have you identified yourself?** Provide a "sig" (signature) line.

__ 3. **Did you provide an effective subject line?** Include a *topic* and a *focus.*

__ 4. **Have you effectively organized your e-mail?** Consider including the following:
- an opening sentence(s) telling *why* you are writing and *what* you are writing about.
- a discussion unit with itemized points telling *what exactly* the e-mail is discussing.
- a concluding sentence(s), *summing* up your e-mail message or telling your audience what to do next.

__ 5. **Have you used highlighting techniques sparingly?**
- Avoid boldface, italics, color, or underlining.
- Use asterisks (*) for bullets, numbers, all-cap headings, and double spacing for access.

__ 6. **Did you practice netiquette?**
- Be polite, courteous, and professional.
- Don't **FLAME.**

__ 7. **Is the e-mail concise?**

__ 8. **Did you identify and limit the size of attachments?**
- Tell your reader(s) if you have attached files and what types of files are attached (PPT, PDF, RTF, Word, etc.).
- Limit the files to 750 K.

__ 9. **Does the memo recognize the audience?**
- Define acronyms or abbreviations where necessary.
- Consider a diverse audience (factoring in issues such as multiculturalism).

__ 10. **Did you avoid grammatical errors?**

Instant Messaging

E-mail could be too slow for today's fast-paced workplace. Instant messaging (IM) could replace e-mail in the workplace within the next five years, studies suggest. IM pop-ups are already providing businesses many benefits.

Benefits of Instant Messaging

Following are benefits of instant messaging:

- Increased speed of communication.
- Improved efficiency for geographically dispersed workgroups.
- Collaboration by multiple users in different locations.
- Communication with colleagues and customers at a distance in real time, like the telephone.
- The avoidance of costly long distance telephone rates. (Note: Voice-over IP [VoIP] services, which allow companies to use the Internet for telephone calls, could be more cost-efficient than IM).
- A more "personal" link than e-mail.
- A communication channel that is less intrusive than telephone calls.
- A communication channel that allows for multitasking. With IM, you can speak to a customer on the telephone or via an e-mail message *and simultaneously* receive product updates from a colleague via IM. (Hoffman; Shinder)

Challenges of Instant Messaging

However, IM used for business purposes is so new that corporate standards have not been formalized. Software companies have not yet redesigned IM home versions for the workplace. This leads to numerous potential problems, including security, archiving, monitoring, and employee misuse (Hoffman; Shinder):

- **Lost productivity.** Use of IM on the job can lead to job downtime. First, we tend to type more slowly than we talk. Next, the conversational nature of IM leads to "chattiness." If employees are not careful, or monitored, a brief IM conversation can lead to hours of lost productivity.
- **Employee abuse.** IM can lead to personal messages rather than job-related communication with coworkers or customers.
- **Distraction.** With IM, a bored colleague easily can distract you with personal messages, online chats, and unimportant updates.
- **Netiquette.** As with e-mail, due to the casual nature of IM, people tend to relax their professionalism and forget about the rules of polite communication. IM can lead to rudeness or just pointless conversations.
- **Spim.** IM lends itself to "spim," instant messaging spam—unwanted advertisements, pornography, pop-ups, and viruses.
- **Security issues.** This is the biggest concern. IM users are vulnerable to hackers, electronic identity theft, and uncontrolled transfer of documents. With unsecured IM, a company could lose confidential documents, internal users could download copyrighted software, or external users could send virus-infected files.

Techniques for Successful Instant Messaging

To solve potential problems, consider these ten suggestions:

1. **Choose the correct communication channel.** Use IM for speed and convenience. If you need length and detail, other options—e-mail messages, memos, reports, letters—are better choices. In addition, sensitive topics or bad news should never be handled through IM. These deserve the personal attention provided by telephone calls or face-to-face meetings.
2. **Document important information.** For future reference, you must archive key text. Therefore, copy and paste IM text into a word processing tool for long-term documentation.
3. **Summarize decisions.** IM is great for collaboration. However, all team members might not be online when decisions are made. Once conclusions have been reached that affect the entire team, the designated team leader should e-mail everyone involved. In this e-mail, the team leader can summarize the key points, editorial decisions, timetables, and responsibilities.
4. **Tune in, or turn off.** The moment you log on, IM software tells everyone who is active online. Immediately, your IM buddies can start sending messages. IM pop-ups can be distracting. Sometimes, in order to get your work done, you might need to turn off your IM system. Your IM product might give you status options, such as "on the phone," "away from my desk," or "busy." Turning on IM could infringe upon your privacy and time. Turning off might be the answer.
5. **Limit personal use.** Your company owns the instant messaging in the workplace. IM should be used for business purposes only.
6. **Create "buddy" lists.** Create limited lists of IM users, including legitimate business contacts (colleagues, customers, and vendors).
7. **Avoid public directories.** This will help ensure that your IM contacts are secure and business-related.
8. **Disallow corporate IM users from installing their own IM software.** A company should require standardized IM software for safety and control.
9. **Never use IM for confidential communication.** Use another communication channel if your content requires security.
10. **Use IM software that allows you to archive and record IM communications.** As with e-mail, IM programs can let systems administrators log and review IM conversations. Some programs create reports that summarize archived information and let users search for text by key words or phrases. (Hoffman; Shinder)

Blogging for Business

Jonathan Schwartz, president and chief operating officer of Sun Microsystems, says that blogging is a "must-have tool for every executive. It'll be no more mandatory that they have blogs than that they have a phone and an e-mail account," says Schwartz (Kharif). Bill Gates, Microsoft's CEO, says that blogs could be a better way for firms to communicate with customers, staff, and partners than e-mail and Web sites. "More than 700 Microsoft employees are already using blogs to keep people up to date with their projects" ("Gates backs blogs for business"). Marc Cuban, owner of the Dallas Mavericks basketball team, has a blog (Ray). IBM is planning "the largest corporate blogging initiative" by encouraging its 320,000 employees to become active bloggers with the goal of achieving "thought leadership" in the global information technology market (Foremski). Nike has launched an "adverblog" to market its products.

What is "blogging," and why are so many influential companies and business leaders becoming involved in this new communication channel?

Blogging: A Definition

Web logs, also knows as "blogs," are web-based journals posted online. Blogs allow individuals to create diary entries, consisting of observations, thoughts, insights, "news, opinions, ideas and brainstorms" (Wuoria "5 Ways").

Blogs are a very unique type of communication channel. Blogs are more informal than reports and letters. Thus, in tone, blogs are similar to e-mail messages or instant messaging. Like Web sites, blogs appear online. However, blogs are different from Web sites, e-mail messages, and instant messaging because blogs are even more casual in tone and style.

Though blogging is not yet commonplace, it is a growing phenomenon. Blog readership grew by 58 percent in 2004 (Blumberg). As of late 2004, "an estimated 4.8 million blogs" existed, with up to 6 million Americans reading blogs, "up from just 100,000" in 2002 (Gard). Blogging is growing at a faster pace than the Internet did in its early years (Cross). To put that number in perspective, four million bloggers equals more people than readers of the popular newspaper *USA Today* per day (Bruner).

Ten Reasons for Blogging

The rapid growth of blogging might be reason enough for businesses to consider using this online communication channel. However, following are ten more reasons why companies are beginning to blog:

1. **Communicate with colleagues.** Chris Winfield, president of an online marketing company, says that he uses blogging to improve "communication flow to his employees" (Ray). Many companies encourage their employees to use blogging for project updates, issue resolutions, and company announcements. The engineering department at Disney ABC Cable Networks Group uses blogs "to log help desk inquiries" (Li).
2. **Communicate with customers.** In contrast to private, intranet-based blogs used for internal corporate communication, a company also can have public blogs. Through public blogs, a company can initiate question/answer forums, respond to customer concerns, allow customers to communicate with each other, create interactive newsletters, and build rapport with customers, vendors, and stakeholders.
3. **Introduce new product information.** Because blogs are quick and current, they allow companies to provide up-to-date information about new products and services.
4. **Reach an influential audience segment.** One reason that companies are turning to blogging is that blogs appeal to an influential and emerging audience—youth. Young adults "between the ages of 18 and 24" read blogs "three times more frequently than older adults" (Li). This age segment of online consumers provides an attractive target for companies.

5. **Improve search engine rankings.** Marketing is a key attraction of corporate blogs. A blog post using keywords, allowing for comments and responses, and providing references and links to other sites, tends to rank in the top 10 to 20 listings in Internet search engines (Ray).
6. **Network through "syndication."** To access a Web site, a reader must know the URL or use a search engine, such as Google. Blogs can be distributed directly to the end users through a blog "feed." By using feed programs, such as RSS ("Really Simple Syndication" or "Rich Site Summary"), Atom, or Current Awareness, bloggers can syndicate their blogs or be notified when topics of interest are published. Thus, blogs can be very personalized, essentially delivered to your door on a moment's notice.
7. **Facilitate online publishing.** A key value to blogging is its ease and affordability. Blogging services and software, such as Blogger, Movable Type, TypePad, Xanga, Live Journal, and Weblogger, give bloggers access to easy-to-use Web-based forms.
8. **Encourage "bubble-up" communication.** Blogging promotes brainstorming and public forums. A corporate blog builds networks, allows people to dialogue, and encourages conversation.
9. **Track public opinion.** "Trackback" features, available from many blog services, let companies track blog usage. Tracking lets companies monitor their brand impact and learn what customers are saying (good news or bad news) about their products or services.
10. **Personalize your company.** Personal responses to customer comments help personalize a company. Blogging offers a refreshing option that reaches out to customers, offering a human dimension.

Ten Guidelines for Effective Corporate Blogging

If you and your company decide to enter the world of blogging, follow these guidelines:

1. **Determine if you need to blog.** Not every company needs a blog, any more than every company needs a Web site. If blogging does not fit into your corporation's culture, you should avoid the blogosphere.
2. **Identify your audience.** As with all workplace communication, audience recognition and involvement are crucial. Before blogging, decide what topics you want to focus on, what your unique spin will be, what your goals are in using a blog, and who your blog might appeal to.
3. **Achieve customer contact.** Blogs are innately personal. Take advantage of this feature. Make your blogs fun and informal. You can give your blog personality and encourage customer outreach by including "interesting news of the day, jokes," personnel biographies, question/answer forms, updates, an opportunity to add comments, as well as information about products and services (Ray). In addition, make sure the blog is interactive, allowing for readers to comment, check out new links, or add links.
4. **Determine where to locate the blog.** You can locate and market blogs in many different places: Add a blog link to your existing Web site; create a totally independent blog site; distribute your blog through RSS aggregators; add a blog link to your e-mail signature; or mention your blog in sales literature (Wuorio "Blogging"; Li).
5. **Start "blogrolling."** Once you have determined audience and blog location, it's time to start blogrolling. Start talking. Not only do you need to start the

dialogue by adding content to your blog, but also you want to link your blog to other sites (Wuorio "Blogging").

6. **Emphasize keywords.** By mentioning keywords in your blog titles and text, you can increase the number of hits your blog receives from Internet search engines.
7. **Keep it fresh.** Avoid a stale blog. By posting frequently (daily or weekly), you encourage bloggers to access your site and return to it often.
8. **Respond quickly to criticism.** In a communication channel like the blogosphere where dialogue is the desired end, companies inevitably will receive criticism about products or services. Blogs allow companies to respond quickly to and manage bad news.
9. **Build trust.** Don't use the blog only to promote new corporate products. Transparent marketing tends to backfire on bloggers who want sincere content, an opportunity to learn more about a company's culture, and a chance to engage in dialogue.
10. **Develop guidelines for corporate blog usage.** Because blogs encourage openness from customers as well as employees, a company must install guidelines for corporate blog usage. One potential problem associated with blogging, for example, is divulging company information, such as financial information or trademark secrets. Figure 5.4 offers a sample code of ethics created by Forrester Research (Li).

Web Sites

Web sites are created by companies, organizations, schools, and government agencies. That's why a Web site's URL (Uniform Resource Locator—the Web site's address) reads *.com* (commercial), *.org* (organization), *.edu* (education), *.gov* (government), and so on.

Blogging Code of Ethics

1. I will tell the truth.
2. I will write deliberately and with accuracy.
3. I will acknowledge and correct mistakes promptly.
4. I will preserve the original post, using notations to show where I have made changes so as to maintain the integrity of my publishing.
5. I will never delete a post.
6. I will not delete comments unless they are spam or off topic.
7. I will reply to e-mails and comments when appropriate and will do so promptly.
8. I will strive for high quality with every post—including basic spell-checking.
9. I will stay on topic.
10. I will disagree with other opinions respectfully.
11. I will link to online references and original source materials directly.
12. I will disclose conflicts of interest.
13. I will keep private issues and topics private because discussing private issues would jeopardize my personal and work relationships.

Figure 5.4 Blogging Code of Ethics

In December 1996, the Internet search engine Yahoo listed 161,068 company Web sites in its Business and Economy link. By March 1999, this number had almost tripled to 431,034 companies with Web sites listed in Yahoo's Business and Economy page. As of 2006, Yahoo listed close to 800,000 Business and Economy Web sites. And that number increases every day. A Web site might be your company's most powerful communication channel in reaching customers, potential employees, franchisors, investors, stakeholders, and vendors.

Criteria for a Successful Web Site

Follow these criteria when creating your site:

1. **Home Page.** The home page sets the tone for your site. A successful home page should consist of the following components:

   Identification Information. A good home page clearly names the company, service, or product. Provide the reader access to a corporate phone number, an e-mail address, a fax number, an address, and customer service contacts.
   Graphic. An informative, attractive, and appealing graphic depicting your product or service could convey more about your company than words.
   Lead-in Introduction. In addition to a graphic, provide a phrase or sentence that tells your reader who and what you do. Applied Communications Group uses the phrase "A Software Consulting Firm" on its home page to define itself. Creative Courseware's home page includes the following lead-in: "Creative Courseware specializes in performance analysis and customized training materials."
2. **Linked Pages.** Once your reader clicks on the hypertext links from the home page, he or she will jump to the designated linked pages. These linked pages should contain the following:

   Headings and Subheadings. To ensure that readers know where they are in the context of the Web site, you need to use headings. These give the readers visual reminders of their location. Successful headings on linked pages are consistently in the same location, font size, and font type.
   Development. As in all correspondence, you need to develop your ideas thoroughly. Each linked page will develop a new idea. Prove your points precisely.
3. **Navigation.** Help the readers navigate through cyberspace by using the following:

   Home Buttons. The reader needs to be able to return to the home page easily from any page of a Web site. Remember, the home page acts as a table of contents or index for all the pages within the site. By returning to the home page, the reader can access any of the other pages. To ensure this easy navigation, you need to provide a hypertext-linked home button on each page.
   Links between Web pages. If each page has a navigation bar with a hypertext link to all pages within the site, then the reader can access any page, in any order of discovery.
4. **Document Design.** Document design should enhance your text and promote your product or service, not distract from your message. On a Web site, less is best. Choose your background, font type, and color carefully.

   Background. When choosing a background, consider your corporate image and intended audience. While a black background with red font, for example, might not be a wise advertising tool for a day care center, this color scheme could be effective if your product is online gaming. In addition, to achieve readability, you want the best contrast between text and background. Despite the vast selection of backgrounds at your disposal, the best contrast is black text on a white background.

Font. Use sans serif fonts, such as Arial, or one specifically designed for the Web—Verdana. Serif and sans serif fonts are discussed in Chapter 3. Use 12- to 14-point type for the text and a larger font for headings. Avoid using underlining, which readers will confuse with a hypertext link. Avoid using all caps. As with e-mail, text typed in all caps makes you look as if you are shouting.

Color. A primary concern is contrast. Red, blue, and black font colors on a white background are very legible because their contrast is optimum. Other combinations of color don't offer this contrast. Thus, the reader will have trouble deciphering your words. Test your use of color on several different monitors to ensure readability.

5. **Style.** Conciseness is important in all workplace communication. Conciseness is even more important in a Web site. A successful Web page should be limited to one viewable screen, and a line of text should rarely exceed two-thirds of the screen. Personalized tone is another key feature of Web style. Web sites encourage pronouns, contractions, and positive words. Create a personalized tone that engages your readers.

See Figure 5.5 for a well-designed Web page. Notice how the Missouri Department of Transportation's (MoDOT) Web page is divided approximately into thirds. The left

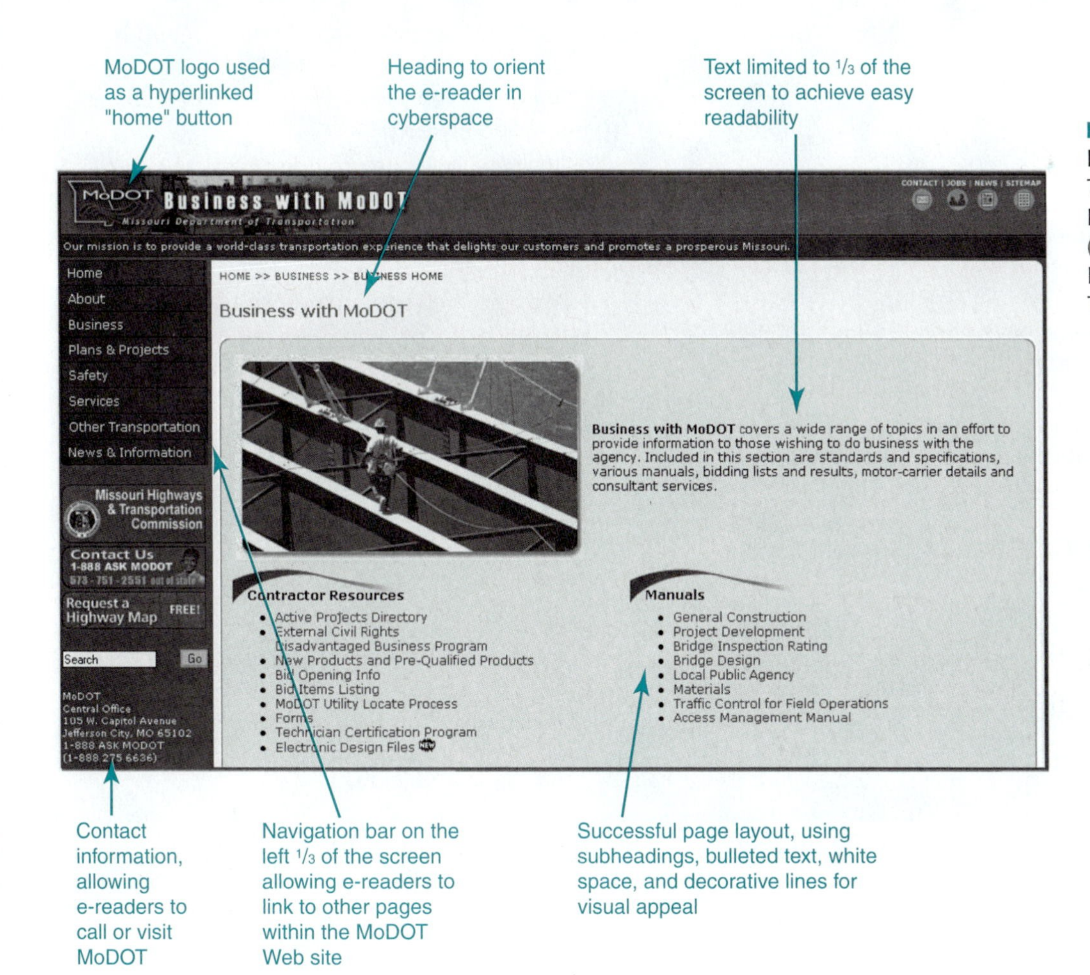

Figure 5.5 Missouri Department of Transportation Web Page. (Courtesy of Missouri Department of Transportation)

third contains the navigation bar. The middle third contains a visual to catch the reader's attention. The final third provides text for content.

Web Site Usability Checklist

The Web site usability checklist will give you the opportunity for self-assessment and peer evaluation of your writing.

Web Site Usability Checklist

Audience Recognition and Involvement

__ 1. Does the Web site meet your reader's needs?

__ 2. Does the Web site solicit user feedback for interactivity, display customer comments, or provide a frequently asked question page?

__ 3. Does the Web site make it easy for your audience to purchase online through online forms?

__ 4. Does the Web site use pronouns to engage the reader?

Home Page

__ 1. Does the home page provide identification information (name of service or product, company name, e-mail, fax, city, state, street address, etc.)?

__ 2. Does the home page provide an informative and appealing graphic that represents the product, service, or company?

__ 3. Does the home page provide a welcoming and informative introductory phrase, sentence, or paragraph?

__ 4. Does the home page provide hypertext links connecting the reader to subsequent screens?

Linked Pages

__ 1. Do the linked pages provide headings clearly indicating to the reader which screen he or she is viewing?

__ 2. Do the linked pages develop ideas thoroughly (appropriate amount of detail, specificity, valuable information)?

__ 3. Are the linked pages limited to one or two primary topics?

Navigation

__ 1. Does the Web site allow for easy return from linked pages to the home page?

__ 2. Does the Web site allow for easy movement between linked pages?

__ 3. Does the Web site provide access to other Web sites for additional information?

Document Design

__ 1. Does the Web site provide an effective background, suitable to the content and creating effective contrast for readability?

__ 2. Does the Web site use color effectively, in a way suitable to the content and creating effective contrast for readability?

__ 3. Does the Web site use a consistent document design (colors, background, graphics, font) carried throughout the entire site?

__ 4. Does the Web site use headings and subheadings for easy navigation?

__ 5. Does the Web site vary font size and type to create a hierarchy of headings?

__ 6. Does the Web site use graphics effectively—suitable to the content, not distracting, and loading quickly?

Web Site Usabillity Checklist (Continued)

__ 7. Does the Web site use highlighting techniques effectively (white space, lines, bullets, icons, audio, video, frames, font size and type, etc.) in a way that is suitable to the content and not distracting?
- Line length limited to approximately two-thirds of the screen (40–60 characters per line)
- Text per Web page limited to one viewable screen, minimizing the need to scroll

Style

__ 1. Is the Web site concise? (Remember that reading online is a challenge to end users.)
- Short words (1–2 syllables)
- Short sentences (10–12 words per sentence)
- Short paragraphs (4 typed lines maximum)

Accuracy

__ 1. Does the Web site avoid grammatical errors?

__ 2. Does the Web site ensure that information (phone numbers, e-mail, fax numbers, addresses, content) is current and correct?

__ 3. Do the hypertext links work?

Online Help

Online help screens are commonplace in word processing software, Web help and support sites, and corporate intranet sites for several reasons, including the following:

- the increased use of computers in business, industry, education, and the home
- the reduced dependence on hard-copy manuals by consumers
- the need for readily available online assistance
- proof that people learn more effectively from online tutorials than from printed manuals (Pratt)

Online Help: A Definition

Online help systems, which employ computer software to help users complete a task, include procedures, reference information, wizards, and indexes. Typical online help navigation provides readers hypertext links, tables of contents, and full-text search mechanisms.

Help menus on your computer are excellent examples of online help systems. As the computer user, you click on Help to search for a topic of your choice. When you select your topic, you could get a *pop-up* (a small window superimposed on your text), or your computer might *link* to another full-sized screen layered over your text. In either instance, the pop-up or hyperlink gives more information about your topic:

- Overviews. Explanations of why a procedure is required and what outcomes are expected.
- Processes. Discussions of how something works.
- Definitions. An online glossary of terms.
- Procedures. Step-by-step instructions for completing a task.

- **Examples.** Feedback verifying the completion of a task or graphic depictions of a completed task. These could include a screen capture or a description with call-outs.
- **Cross-References.** Hypertext links to additional information.
- **Tutorials.** Opportunities to practice online.

Online help systems allow workplace communicators to create interactive training tools and informational booths within a document. These online systems can be created using a wide variety of authoring tools. Some popular ones include *RoboHELP, HelpBreeze, ForeHelp, Help Magician, Visual Help*, and *WYSI-Help* (Zubak).

Techniques for Writing Effective Online Help

To create effective online help screens, consider these suggestions.

- **Organize your information for easy navigation.** To help your readers access information, provide an online contents Menu so your audience can cross-reference help screens within the system. Provide a back button or a home button to allow the readers to return to a previous screen.
- **Recognize your audience.** Find out what information your readers need. You can accomplish this goal through usability testing, focus groups, brainstorming sessions, surveys, and your company's hotline help desk logs.
- **Achieve a positive, personalized tone.** Users want to be encouraged, especially if they are trying to accomplish a difficult task. Your help screens should be constructive, not critical. Text should be written positively. The messages should be personalized, including pronouns to involve the reader.
- **Design your document.** To achieve effective document design, consider these points:
  - Use color sparingly. Color causes several problems in online documents. Bright colors and too many colors strain your reader's eyes. Your primary goal is contrast. To help your audience read your text, you want to maximize the contrast between the text color and the background color. Black text on a light background offers optimum contrast.
  - Be consistent. Pick a color scheme and stick with it. Your headings should be consistent, along with your word usage, tone, placement of help screen links and pop-ups, graphics, wizards, and icons.
  - Use a 10-point sans serif font. A 12-point type size is standard for most printed documents, but 10-point type will save you valuable space online. Serif fonts (like Times New Roman) are the standard for most workplace communication, but sans serif fonts (like Arial or Verdana) provide contrast with a text's primary content.
  - Use white space. Don't clutter your help screens. Minimize your reader's overload by adding ample horizontal and vertical white space.
- **Be concise.** Limit word and sentence length. A help screen should avoid horizontal and vertical scrolling. Each screen should include one self-contained message.
- **Be clear.** Clarity online could include tutorials to guide the reader through a task, graphics that depict the end result, cross-references, and pop-up definitions.

See Figure 5.6 for an example of effective online help.

The Microsoft Online Help screen lets you search for a term (such as "margins") and then provides a list of options regarding that topic. The Help screen uses a sans serif font and black text on a light yellow background to enhance readability. In addition, the Help screen minimizes text by limiting words to no more than three or four words per line.

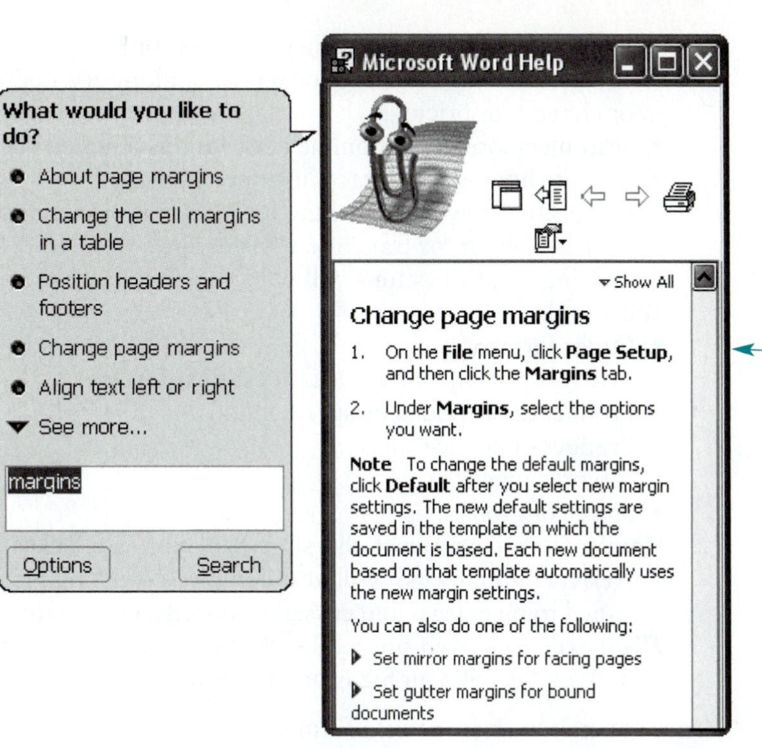

Figure 5.6 Effective Online Help (Courtesy Microsoft Word)

Once a topic has been chosen (such as "Change page margins"), the "pop-up" box uses black text on a white background. Text is personalized through pronouns (you usage).

The Writing Process at Work

Effective writing follows a process of prewriting, writing, and rewriting. Each of these steps is sequential and yet continuous. The writing process is dynamic, with the three steps frequently overlapping. Follow the writing process to create an effective Web site.

Prewriting

Before you begin to construct your Web site, gather data to determine your Web site's goal, audience, focus, and content. Answer these **reporter's questions.**

1. *Why* are you creating a Web site? Are you writing to inform, instruct, persuade, or build rapport? Are you creating the Web site to
   - inform the public about products and services?
   - sell products via online order forms?
   - encourage new franchisees to begin their own businesses?
   - advertise new job openings within the company?
   - provide a company profile, including annual reports and employee resumes?
   - list satisfied clients?
   - provide contact information?
2. *Who* is your audience? Are you writing to specialists, semi-specialists, prospective employees, the general lay public, or stockholders? Will your audience be

- customers looking for product information?
- current users of a product or service looking for upgrades, enhancements, or changes in pricing?
- customers looking for online help with instructions or procedures?
- vendors hoping to contact customer service?
- prospective new hires looking for information about your company?
- people looking for franchise information?

3. Answering these questions will help you decide *what* to include in your site. You could include
   - Order forms
   - Answers to frequently asked questions (FAQs)
   - Product descriptions or specifications
   - Product or service support
   - Job openings
   - Storefront locations
4. *How* will you construct the site? Will you use HTML coding or HTML converters and editors? Will you download existing graphics that have been modified to meet legal and ethical standards, or will you create new graphics?
5. *Where* will your Web site reside? Will it be an Internet site open to the World Wide Web, or will you focus on an intranet or extranet site?

Once you have gathered this information, consider using **storyboarding** for a preliminary sketch of your Web site's layout.

- Sketch the home page. How do you want the home page to look? What color scheme do you have in mind? How many graphics do you envision? Where will you place your graphics and text—margin left, centered, margin right?
- Sketch the linked page(s). How many linked pages do you envision? Will they have graphics, online forms, e-mail links, audio or video, headings and subheadings, or links to other Internet, intranet, or extranet sites? Will you use frames or tables?

See Figure 5.7 for a sample Web site storyboard.

Writing

To create a rough draft of your Web site, follow this procedure:

1. *Study the Web site usability checklist*. This will remind you what to include in your rough draft and how best to structure the Web site.
2. *Review your prewriting*. Have you answered all the reporter's questions? How does your Web site storyboarding sketch look? Do you see any glaring omissions or unnecessary items? If so, add what's missing and omit what's unneeded.
3. *Draft the Web site*. You can create your rough draft in several ways.
   - Write a rough draft in any word processing program. Then, after revision, you can save your text as a Web page.
   - Draft your text directly in an HTML editor, such as FrontPage, Dreamweaver, Ant HTML, HotDog Web Editor, Netscape Navigator 4.0 and above, Microsoft Internet Assistant for Word 6.0 for Windows and above, Corel WordPerfect 8.0, and others.
   - Write your draft using Hypertext Markup Language (HTML) coding.

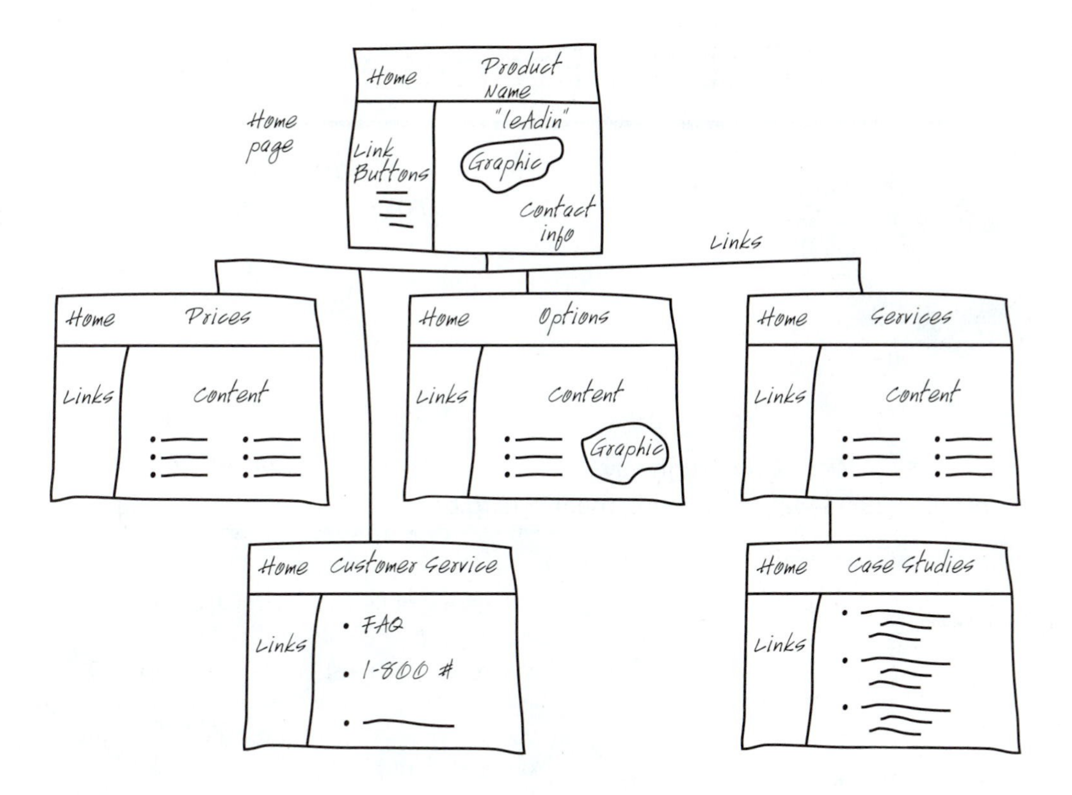

Figure 5.7 Web Site Storyboard

Rewriting

Once you have completed drafting your Web site, is it what you had in mind? If not, edit and revise the site to ensure its usability. Usability is a way to determine whether your reader can use the Web site effectively, whether your site meets your user's needs and expectations. Not only does the reader want to find information that helps him or her better understand the topic, but also the audience wants the Web site to be readable, accurate, up-to-date, and to easily access. Thus, web usability focuses on three key factors [Dorazio 2000]:

1. *Retrievability*. The user wants to find specific information quickly and be able to navigate easily between the screens.
2. *Readability*. The user wants to be able to read and comprehend information quickly and easily.
3. *Accuracy*. The user wants complete, correct, and up-to-date information.

The goal of usability testing is to solve problems that might make a Web site hard for people to use. By successfully testing a site's usability, a company can reduce customer and colleague complaints; increase employee productivity (reducing the time it takes to get work done); increase sales volume because customers can purchase products or services online; and decrease help desk calls and their costs. To achieve user satisfaction, revise your Web site to meet the goals listed in our Web Site Usability Checklist. In revising a Web site, consider these options:

- **Add new detail for clarity.** Have you said all you need to say? What additional information should you include (contact information, forms to help your customers order goods online, job openings, an online catalog, etc.)?

Future Promise

Future Promise

Scholarships

Intramurals

Volunteerism

Job Links

FAQ

"Preparing Today's Youth for Tomorrow's Challenges"

1894 West King's Highway San Antonio, TX 78532

Phone: 814-236-5482 E-mail: bsearing@fp.org Fax: 814-236-5480

Figure 5.8 Rough Draft Home Page

- **Delete anything unnecessary for conciseness.** Do your text and graphics all fit on one viewable screen? Does your Web site take too long to load due to excessively large graphics, java scripting, animation, or audio and video plug-ins? Delete the unnecessary add-ons, dead words and phrases, and excessively large graphics.
- **Simplify words and phrases.** Web sites are very friendly types of correspondence. Revise your text, striving for easy-to-understand words, short sentences, and short paragraphs.
- **Move information for emphasis.** As in real estate, location, location, location sells. Decide what is most important on your Web site and then place this information where it can be seen clearly.
- **Reformat for access.** No communication medium depends more on layout than a Web site. Your Web site's design will make or break it. How does your

Writer's Insight

Brent says, "I like the overall layout for your home page. The color link bar goes great with our logo. I just have a few questions: Are you going to use the logo for a home button? Why have we listed the links in the order shown? Did someone on the Web team do a survey or create a questionnaire to find out what our target audience really wants the most? For example, maybe jobs should be listed first.

Where did the photo come from? Since I don't recognize any of the kids, I'm guessing we downloaded the graphic from the Internet. Do we need permission for its use? Talk to FP lawyers about that. Should we have a separate link for contact information? If we did, we could use the bottom of the home page for scrolling announcements. What do you think?

Finally, there seems to be a large gap of wasted space below our slogan. What about if we centered the slogan beneath our name and added additional photos across the page? In fact, once we get our programs going, we should use original photos of our student participants (with their permission, of course)."

Future Promise

Intramurals

Scholarships

Intramurals

Volunteerism

Job Links

FAQ

- Flag Football—BVNW District Field
- Volleyball—SMN Gym

- Soccer—BVNW District Field
- Tennis—San Antonio Clear Creek Courts
- Badmitton—San Antonio Civic Center

Figure 5.9 Rough Draft Intramurals Page

Writer's Insight

Brent says, "Again, I like this page (colors especially). Notice that 'Badminton' is spelled incorrectly. Also, I don't think we have enough information on this page.

Where's our contact info? Are the sports co-ed? Does our target audience know how to find the various sites? And do the readers understand the abbreviations 'BVNW' and 'SMN'? Maybe we need to create Mapquest links, for example, for each venue. Most importantly, we don't list the dates and times for the intramural events and whether equipment is provided (tennis racquets, for example).

As for layout, I'd move the graphic to the right and line up the bullets on the left.

Still, good first draft! Let me see your next draft, please, before the site goes online."

Web site look? Is there ample white space? Do the graphics integrate well with the text? Have you used headings and subheadings for easy navigation? Have you used color effectively, making sure to achieve optimum contrast between text and background?

- **Enhance the tone and style of your Web site.** Web sites tend to be informal. Don't write a Web site using the same tone that you would use when writing a report. Enhance the tone of your site by using contractions, positive words, and pronouns.
- **Correct errors.** On the World Wide Web, millions of people might read your Web site. Grammatical errors, therefore, are magnified. You must proofread so your company or organization looks professional.
- **Make sure it works.** Test your Web site. How do your graphics and colors look on multiple browsers, platforms, and monitors? Do your hypertext links work? Test links to external sites frequently. No matter how beautiful your Web site looks on your monitor, the site will not be successful unless all the components work.

The Future Promise Web design team created a rough draft home page and second page mock-up for Brent Searing to review. Once he read the Web pages shown in Figures 5.8 and 5.9, he made revision suggestions.

CASE STUDIES

1. Future Promise is a not-for-profit organization geared toward helping at-risk high school students. This agency realizes that to reach its target audience (teens aged 15 to 18), it needs an Internet presence.

   To do so, it has formed a 12-person team consisting of the agency's accountant, sports and recreation director, public relations manager, counselor, technical writer, graphic artist, computer and information systems director, two local high school principals, two local high school students, and a representative from the mayor's office. Jackie Velasquez, the public relations manager, is acting as team leader.

   The team needs to determine the Web site's content, design, and levels of interactivity. Jackie's boss, Brent Searing, has given the team a deadline and a few components that must be included in the site:

   - College scholarship opportunities
   - After-school intramural sports programs
   - Job-training skills (resume building and interviewing)
   - Service learning programs to encourage civic responsibility
   - Future Promise's 800-hotline (for suicide prevention, STD information, depression, substance abuse, and peer counseling)
   - Additional links (for donors, sponsors, educational options, job opportunities, etc.)

Assignment

Form a team (or become a member of a team as assigned by your instructor) and design Future Promise's Web site. To do so, follow the criteria for Web design provided in this chapter. Then prewrite, write, and rewrite as follows:

Prewrite:

- Research the topics listed above (either in a library, online, or through interviews) to gather details for your Web site.
- Consider your various audiences and their respective needs and interests.
- Focus on your Web site's purpose. Are you writing to inform, persuade, instruct, or build rapport?
- Draw a storyboard of how you would like the Web site to look or use an organizational chart to lay out the Web design.
- Divide your labors among the team members (who will research, write, create graphics, etc.?).

Write:

- Draft your text either in a word processing program (you can save as an HTML file) or directly in FrontPage (or any other Web design program of your choice).

Rewrite:

- Review the Web criteria in this chapter.
- Add, delete, simplify, move, enhance, and correct your Web site.
- Test the site to make sure all links work and that it meets your audience's needs.

2. Barney Allis Stores (BAS), home-based in Seattle, WA, has department stores in all 50 states and in the following cities: Toronto, Vancouver, London, Paris, Berlin, Barcelona, Rome, Beijing, Hong Kong, Sidney, Mexico City, and Rio de Janeiro.

   It's time for BAS's marketing department to inform all BAS stores which products they should "push" for Christmas sales. This year, BAS has created a new product line called "BASK"—Barney Allis Stores-Kids. The focus will be on winter and spring children's wear, for ages Pre-K through 5th grade. Marketing needs to highlight three item lines:

   - Corduroy pants
   - Reversible jackets
   - Long-sleeved T-shirts. These will be customized to each country with slogans (in the country's language) or cartoon characters decorating the shirt-fronts. The cartoon characters will also be customized, stemming from each country's cartoon television shows.

Assignment

As BAS' director of marketing, your job is to write an e-mail message to all stateside and international marketing managers. The e-mail not only will provide the above information but also act as a cover correspondence, transmitting an attached product catalog. The e-mail must direct the marketing managers to create print, radio, and television advertisements for their local markets. Direct them to the BAS intranet site: *http://www.bas.com/BASK/salestalk.html.*

In writing this e-mail, follow the criteria provided in this chapter.

3. Suburban Day Surgery (SDS) publicized an RFP (request for proposal) requesting proposals from architectural/engineering companies. SDS wanted to expand its facility, adding new surgical rooms, recovery rooms, offices, waiting rooms, and a cafeteria.

   After reviewing several proposals, SDS has narrowed its choices. One company that SDS is especially interested in is Lamb & Sons Construction Company. This company has a wonderful reputation, as shown in their proposal's client references. In addition, SDS has seen Lamb's work, visiting other sites the company has built.

   However, SDS management has a few remaining questions. The proposal did not specify when Lamb & Sons could begin work, what their timeframe was for completing the project, and the credentials of their subcontractors.

Assignment

Write an e-mail to Bill Lamb (blamb@l&sco.com). In the e-mail, explain why you are writing, specify your requests, and give a deadline for the response. Follow the criteria for e-mail provided in this chapter.

4. The following short report recommends that upper management approve the construction of a corporate Web site. Java Lava's management has agreed with the proposal. You and your team have received an e-mail directing you to build the Web site. Using the report shown below, build your company's Web site.

Date: November 11, 2006
To: Distribution
From: Shannon Conner
Subject: RECOMMENDATION REPORT FOR NEW CORPORATE WEB SITE

Impact of the Internet
In response to your request, I have researched the impact of the Internet on corporate earnings. Companies are striving to position themselves on the World Wide Web and maximize their earnings potential. The time is right for our company to go online.

Reasons to Go Online
We can maximize our profits by making our local product a global product.

- *International bean sales.* Currently, coffee bean sales account for only 27 percent of our company's overall profit. These coffee bean sales depend solely on walk-in trade. The remaining 73 percent stems from over-the-counter beverage sales. If we go global via the Internet, we can expand our coffee bean sales dramatically. Potential clients from every continent will be able to order our coffee beans online.
- *International promotional product sales.* Mugs, T-shirts, boxer shorts, jean jackets, leather jackets, key chains, paperweights, and calendars could be marketed. Currently we give away items imprinted with our company's logo. By selling these items online, we could make money while marketing our company.
- *International franchises.* We now have three coffeehouses, located at 1200 San Jacinto, 3897 Pecan Street, and 1801 West Paloma Avenue. Let's franchise. On the Internet, we could offer franchise options internationally.
- *Online employment opportunities.* Once we begin to franchise, we'll want to control hiring practices to ensure that Java Lava standards are met. Through a Web site, we could post job openings internationally and list the job requirements. Then potential employees could submit their résumés online.

What to Include in Our Internet Site

- A map showing our three current sites.
- Our company's history—founded in 1898 by Hiram and Miriam Coenenburg, with money earned from their import/export bean business. "Hi" and "Mam," as they were affectionately called, handed their first coffeehouse over to their sons (Robert, John, and William), who expanded to our three stores. Now a third generation, consisting of Rob, John, and Bill's six children (Henry, Susan, Andrew, Marty, Blake, and Stefani), runs our business.
- Sources of our coffee beans: Guatemala, Costa Rica, Colombia, Brazil, Sumatra, France, and the Ivory Coast.
- Freshness guarantees: posted ship dates and ground-on dates; 100 percent money-back guaranteed.
- Corporate contacts (addresses, phone numbers, e-mail, fax numbers, etc.).

Recommended Action
Coffee is a "hot" commodity now. The international market is ours for the taking. We can maximize our profits and open new venues for expansion. The Web is our tarmac for takeoff. Let's meet in your office on Monday, at 2:00 P.M. to discuss this exciting venture.

Figure 5.10 Recommendation Report for Web Site

Individual and Team Projects

E-Mail Messages

1. E-mail is used to convey many types of information in business and industry. Write an e-mail message to accomplish any of the following purposes (pick any topic you would like to complete the assignment).

- **Directive.** Tell a subordinate or a team of employees to complete a task.
- **Cover/Transmittal.** Tell a reader or readers that you have attached a document and list the key points that are included in the attachment.
- **Documentation.** Report on expenses, incidents, accidents, problems encountered, projected costs, study findings, hiring, firings, and reallocations of staff or equipment.
- **Confirmation.** Tell a reader about a meeting agenda, date, time, and location; decisions to purchase or sell; topics for discussion at upcoming teleconferences; conclusions arrived at; fees, costs, or expenditures.
- **Procedures.** Explain how to set up accounts, research on the company intranet, operate new machinery, use new software, apply online for job opportunities through the company intranet, create a new company Web site, or solve a problem.
- **Recommendations.** Provide reasons to purchase new equipment, fire or hire personnel, contract with new providers, merge with other companies, revise current practices, or renew contracts.
- **Feasibility.** Study the possibility of changes in the workplace (practices, procedures, locations, staffing, or equipment).
- **Status.** Provide a daily, weekly, monthly, quarterly, biannual, or yearly report about where you, the department, or the company is regarding a topic of your choice (class project, sales, staffing, travel, practices, procedures, or finances).
- **Inquiry.** Ask questions about upcoming processes, procedures, or assignments.

2. Visit five to ten local companies. When meeting with these companies' employees, ask them how much time they spend writing e-mail. Ask them what they generally write about, who their audiences are, and whether their e-mail messages are important parts of their jobs. Then, report your findings in an e-mail message to your instructor.

Online Help

1. In a small group, have each individual ask a software-related question regarding a word processing application. These could include questions such as "How do I print?" and "How do I set margins?" and so on. Then, using a help menu, find the answers to these questions. Rewrite any of these examples to improve their conciseness or visual layout.
2. Take an existing document (one you have already written in your class or writing from your work environment) and rewrite it as online help. To do so, find a word or phrase in the text and expand on these as either pop-up windows or hyperlinks. The hyperlinks or pop-ups could provide definitions or procedural steps. Then highlight new terms within the expanded pop-ups or hyperlinks for additional hypertext documents. Create four or five such layers.

Web Sites

1. Create a corporate Web site. To do so, make up your own company and its product or service. Your company's service could focus on dog training, computer repair, basement refinishing, vent cleaning, Web site construction, child care, auto repair, personalized aerobic training, or online haute cuisine. Your company's product could be paint removers, diet pills, interactive computer

games, graphics software packages, custom-built engines, flooring tiles, or duck decoys. The choice is yours.

To create this Web site, follow our writing process. Prewrite by listing answers to reporter's questions. Then sketch your site through storyboarding. Next, draft your Web site. Finally, rewrite by adding, deleting, simplifying, moving, reformatting, enhancing, correcting, and making sure your links work.

2. Research several Web sites, either corporate or personal. Use our Web Site Usability Checklist to determine which sites excel and which sites need improvement. Then write an e-mail message justifying your assessment. In this e-mail, clarify exactly what makes the sites successful or why the unsuccessful sites fail. For an unsuccessful Web site, suggest ways the site could be improved.

Problem Solving Think Pieces

1. You are the manager of a human resources department. You are planning a quarterly meeting with your staff (training facilitators, benefits employees, personnel directors, and company counselors). The staff works in three different cities and twelve different offices. You need to accomplish four goals: get their input regarding agenda items, find out what progress they have made on various projects, invite them to the meeting, and provide the final agenda.

   How should you communicate to them? Should you write an e-mail, an instant message, or a Web log? Write an e-mail message to your instructor explaining your answer. Be sure to give reasons for and against each option.

2. Access any company's Web site and study the site's content, layout (color, graphics, headings, use of varying font sizes/types, etc.), links, ease of navigation, tone, and any other considerations you think are important. Then, determine how the Web site could be improved. Once you have made this determination, write an e-mail message to your instructor recommending the changes that you believe will improve the site. In this e-mail message,

   - analyze the Web site's current content and design, focusing on what is successful and what could be improved.
   - provide feasible alternatives to improve the site.
   - recommend changes.

Web Workshop

1. Most city, county, and state governments have written guidelines for their employees' e-mail. For example, see "Electronic and Voice Mail: Connecticut's Management and Retention Guide for State and Municipal Government Agencies" (*http://www.cslib.org/email.htm*). In addition, see the e-mail policy provided by Hennepin County, Minnesota (*http://www.co.hennepin.mn.us/webmail.html*). Similarly, most companies and college/universities have written guidelines for their employees' and students' e-mail use. For example, see what Florida Gulf Coast University says about e-mail use (*http://admin.fgcu.edu/compservices/policies3.htm*).

Research your city, county, and state e-mail policies. Research your college/university's e-mail policies. Determine what they say, what they have in common, and how they differ.

Write an e-mail to your instructor reporting your findings.

2. Bloggers provide up-to-date information on newsbreaking events and ideas.

   - For e-commerce news, visit *http://blog.clickz.com/* (ClickZ Network—Solution for Marketers).
   - For business blogging, visit *http://www.businessweek.com/the_thread/blogspotting/* (Business Week Online) to learn where the "worlds of business, media and blogs collide."
   - For technology news (information technology, computer information systems, biomedical informatics, and more), visit *http://blogs.zdnet.com/*. Once in this site, use their search engine to find a technology topic that interests you.
   - For information about criminal justice, visit *http://lawprofessors.typepad.com/crimprof_blog/criminal_justice_policy/* (Law Professor Blog Network).
   - For information about science, visit *http://www.scienceblog.com/cms/index.php* (ScienceBlog), which has links to blogs about aerospace, medicine, anthropology, geoscience, and computers.
   - The Car Blog (*http://thecarblog.com/*) has news items about automotive topics.
   - Every news agency, such as ABC, NBC, CBS, CNBC, has a news blog.

   To see what's new in your field of interest, check out a blog. Then, report key findings to your instructor in an e-mail message.

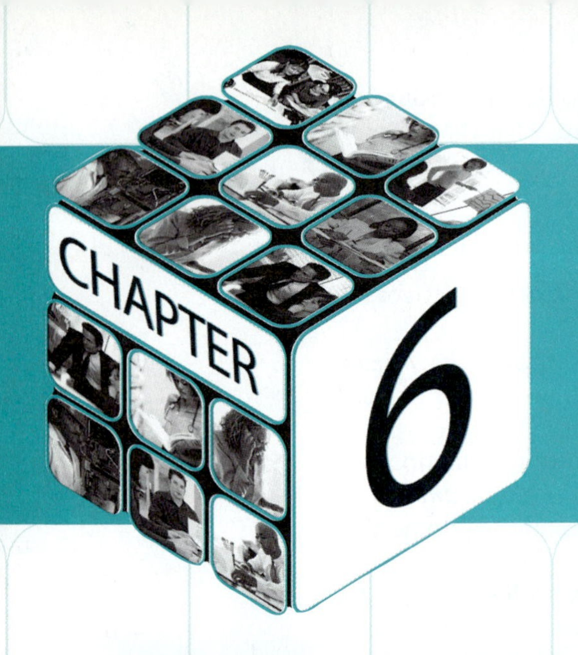

Short Reports

OBJECTIVES

When you complete this chapter, you will be able to do the following:

1. Write a short report, using headings and visual aids.
2. Answer the reporter's questions (who, what, when, where, why, and how) to develop ideas in your report.
3. Write different types of short reports, including the following:
   - Trip Reports
   - Progress Reports
   - Feasibility/Recommendation Reports
   - Incident Reports
4. Use the writing process—prewriting, writing, and rewriting—to prepare effective short reports.
5. Test your knowledge of report writing through end-of-chapter activities:
   - Case Studies
   - Individual and Team Projects
   - Problem Solving Think Pieces
   - Web Workshop

COMMUNICATION AT WORK

In the following scenario, administrator Cindy Kaye spends a significant amount of time writing short reports to document her activities during a year-long project.

Cindy Kaye is director of administrative technology at EFA Incorporated (Education For All), a proprietary university system with branches in 35 cities and a home office in Philadelphia, PA.

EFA has 49,000 students nationwide and offers degrees in accounting, networking, computer security, biomedical technology, electronic engineering technology, computer information systems, computer engineering technology, and business.

Cindy has been traveling to Philadelphia from her university site in Miami, FL, one week a month for a year. She is part of a team being trained on EFA's new administrative software. Her team consists of faculty, staff, and administrators. The software they are learning will be used to manage systems nationwide. It will allow for

- electronic registration
- online grading
- e-mail for faculty, staff, and students
- employee benefits
- online coursework

Cindy must record her travel expenses and the team's achievements. She does this by writing monthly **trip reports,** submitted to her dean at her home school site.

Cindy also studies technology options and the extent to which changes will influence EFA's academic procedures. She will write a **feasibility/recommendation report** following her study.

Finally, when her project is completed, Cindy will collaborate with her team members to write a **progress report** for EFA's board of directors.

Though Cindy's area of expertise is computer information systems, her job requires much more than programming or overseeing the networking of her corporation's computer systems. Cindy's primary job has become communication with colleagues and administrators. Writing reports to document her activities is a major component of this job requirement.

Reasons for Writing Short Reports

At one time or another as a workplace communicator, you will be asked to write a short report. Short reports can vary in length. Your reports will be read by one or several individuals in the business and will satisfy one or all of the following needs:

- Supply a record of work accomplished
- Record and clarify complex information for future reference
- Present information to a large number of people
- Record problems encountered
- Document schedules, timetables, and milestones
- Recommend future action
- Document current status
- Record procedures

Criteria for Writing Short Reports

Although report formats will vary from company to company, even department to department, all short reports share certain generic similarities in organization, development, and style.

Organizing Short Reports

Almost every report should contain four basic units: heading, introduction, discussion, and conclusion/recommendations.

Heading. The heading includes the following (you can place these heading components in any order you choose).

- **Date.** When you wrote the report
- **To.** The name(s) of the people to whom the report is written
- **From.** The name(s) of the people from whom the report is sent
- **Subject.** The *topic* of the report and a *focus*

In this subject line, the *topic* is "Trip Report" (the generic subject matter) and the *focus* is "Southwest Regional Conference."

DATE: August 13, 2006
TO: Helene Mittleman
FROM: Joan O'Brien
SUBJECT: Trip Report on Southwest Regional Conference for Workplace Communication (Fort Worth, TX)

Introduction. The introduction supplies an overview of the report. In this introductory section, use headings or talking headings to summarize the content (see Chapter 3 for a discussion of "talking headings"). These can include headings for organization, such as "Overview," "Purpose," or more informative talking headings, such as "Purpose of Report," "Committee Members," "Conference Dates," "Needs Assessment," and "Problems with Machinery." The headings will depend on the type of report and the topic of discussion. Note how headings are used in the following example.

Report Objectives

I attended the Southwest Regional Conference on Workplace Communication in Fort Worth, TX, to learn more about how our company can communicate effectively. This report addresses the workshops I attended, consultants I met with, and pricing for training seminars.

Conference Dates

August 5–8, 2006

Committee Members

Susan Lisk and Larry Rochelle

Discussion. The discussion section of the report can summarize many topics, including your activities, the problems you encountered, costs of equipment, warranty information, and more. This is the largest section of the report requiring detailed development. Use subheadings or talking headings to organize your content.

Ways to Develop Content. To develop your ideas, answer the reporter's questions.

1. *Who* did you meet or contact, who was your liaison, who was involved in the incident, or who was on your support team?
2. *When* did the documented activities occur (dates of travel, milestones, incidents, investigation, or meeting)?
3. *Why* are you writing the report, or why were you involved in the activity (rationale, justification, or objectives)?
4. *Where* did the activity take place?
5. *What* were the steps in the procedure, what conclusions have you reached, what are your findings, what occurred during the meeting, or what are your recommendations?

Following is an example of vague, imprecise writing and an improved version. In the "before" example, the heading "Discussion" tells the reader what section of the report you are in, but the heading is not very informative. What topics of "discussion" will this section cover? In addition, the text is vague: how can "PowerPoint usage" be revised? What "new techniques for headings" should be used? How can the company pay more "attention to audience"? The "after" version provides more informative headings and specifies the content. We now know what is meant by "revised PowerPoint usage," "techniques for headings," and "attention to audience."

before

Discussion

Workplace communication can be improved in several ways: revised PowerPoint usage, new techniques for headings, and attention to audience.

after

Revised PowerPoint Usage

A presentation I attended suggested that all PowerPoint screens be limited to six words per line and six lines per screen.

Techniques for Headings

In addition to using organizational headings, such as "Introduction," "Discussion," and "Conclusion," I learned that more informative "talking headings" clarify content.

Attention to Audience

We need to consider diversity issues, such as gender, race, and culture when we write our letters, e-mail messages, and reports.

Conclusion/Recommendations. The conclusion section of the report allows you to sum up, to relate what you have learned, or to state what decisions you have made regarding the activities reported. The recommendation section allows you to suggest future action, such as what the company should do next. Not all reports require recommendations. You can use headings or talking headings to organize the content.

Benefits of the Conference

The conference was beneficial. Not only did it teach me how to improve my workplace communication but also it provided me contacts for workplace communication training consultants.

Proposed Next Course of Action

To ensure that all employees benefit from the knowledge I acquired, I recommend hiring a consultant to provide workplace communication training.

Talking headings, such as "Benefits of the Conference" and "Proposed Next Course of Action" provide more focus than simple headings, such as "Conclusion" and "Recommendation."

The conclusion shows how the writer benefited.

The recommendation explains what the company should do next and why.

Types of Short Reports

Four common types of reports include the following:

- Trip
- Progress

- Feasibility/recommendation
- Incident

Trip Report

Purpose. A *trip report* documents job-related travel. When you leave your work site and travel for job-related purposes, your supervisors will require that you document your expenses and time while off-site. In addition, they will want a report on your work activities for future reference and to share with other project stakeholders. For example, you might be engaged in work-related travel as follows.

- Information Technology. You go to a conference to learn about the latest hardware and software technologies for the workplace. There, you meet with vendors, participate in hands-on technology workshops, and learn what other companies are doing to manage their technology needs. When you return, you write a trip report documenting your activities.
- Heating/Ventilating/Air Conditioning. One of your clients is building an office site. Your company has been hired to install their HVAC system. You travel to your client's home office to meet with other contractors (engineering and architecture) so all team members can agree on construction plans. At the conclusion of your job-related travel, you will write a trip report about your meeting.
- Biomedical Equipment. Four days a week, every week, you are on the road repairing biomedical equipment. Each month, you must document your job-related travel and receive recompense for travel expenditures.

Your trip report should include the following information:

Introduction. A trip report's introduction may include but is not limited to the following components.

Purpose. Document the date(s) and destination of your travel. Then comment on your objectives or rationale. If this is the first job-related travel, you might want to explain what motivated the trip, what you planned to achieve, or why you were involved in the job-related travel. If you are involved in an ongoing project where travel is regularly a part of the job, you could omit an explanation of the trip's objectives.

Personnel. With whom did you travel? Who were your contacts?

Authorization. Who recommended or suggested that you leave your work site for job-related travel?

Discussion (Body, Findings, or Agenda). Using subheadings, document your activities and costs incurred. You might include a review of your observations, contacts, seminars attended, or difficulties encountered. A day-by-day agenda is not necessary. Provide enough information to justify your expenses. In addition, develop your content so management can understand your conclusions and make decisions based on your recommendations.

Conclusion/Recommendations

Conclusion. What did you accomplish—what did you learn, whom did you meet, what sales did you make, or what occurred that might benefit you, your colleagues, or your company?

Recommendations. What do you suggest next? Should the company continue on the present course (status quo)? Should changes be made in personnel or in the approach to a particular situation? Would you suggest that other colleagues attend this conference in the future, or was the job-related travel not effective? In your opinion, what action should the company take?

Figure 6.1 provides a sample trip report for an electronics company.

Progress Report

Purpose. A *progress report* lets you document the status of an activity, explaining what work has been accomplished and what work is remaining. Supervisors or customers want to know what progress you are making on a project, whether you are on schedule, what difficulties you might have encountered, or what your plans are for the next reporting period. Progress reports also let you document your work accomplished for billing purposes. Because of this, your reader might ask you to write progress (or activity or status) reports—daily, weekly, monthly, quarterly, biannually, or annually.

- Biomedical Technology. You and your team are developing a new heart monitor. This entails researching, patenting, building, testing, and marketing. A progress report tells your investors and supervisors where you stand, if you are on schedule, and when the project will conclude.
- Hospitality Management. The city's convention center is considering new catering options. Your job has been to compare and contrast catering companies to see which one or ones would be best suited for the convention center's needs. The deadline is arriving for a decision. Submit a progress report so management can determine the next steps.
- Project Management. Your company is renovating its home office. Many changes have occurred. These include new carpeting, walls moved to create larger cubicles, the construction of larger conference rooms, a new cafeteria and fitness center, and improved lighting. Other changes are still in progress, such as increased parking spaces, exterior landscaping, a child-care center, and handicapped accessibility. Write a progress report to quantify what has occurred, what work is remaining, and when work will be finished.
- Automotive Technology. Your automobile dealership recently suffered negative publicity due to problems in the service bay area. As service manager, you have initiated new procedures to improve your product quality and customer service. Write a progress report to update your dealership's CEO.

The progress report should include the following:

Introduction

Objectives. These can include the following: why you are working on this project (the rationale), what problems motivated the project, what you hope to achieve, and who initiated the activity.

Personnel. List others working on this project (work team, liaison, contractors, contacts).

Previous activity. If this is one report in a series of reports, remind your readers what work has already been accomplished. Bring them up to date with background data or a reference to previous reports.

DATE: February 26, 2006
TO: Debbie Rulo
FROM: Oscar Holloway
SUBJECT: Trip Report—Unicon West Conference on Electronic Training

The "focus"

The "topic"

Purpose of the Meeting: On Tuesday, February 23, 2006, I attended the Unicon West Conference on Electronic Training held in Ruidoso, NM. My goal was to acquire hands-on instruction and learn new techniques for electronic training, including the following:

- Online discussion groups
- E-based tutorials
- Intranet instruction
- Videoconference lecture formats

The introduction section answers the "reporter's questions"—who, what, when, where, and why.

Conference Participants: My coworkers Bill Cole and Gena Sebree also attended the conference.

Presentations at the Conference:

Different "heading levels" and highlighting techniques are used to make the information more accessible and to help the readers navigate your text.

Gena, Bill, and I attended the following sessions:

- ***Online Discussion Groups***
  This two-hour workshop was presented by Dr. Peter Tsui, a noted instructional expert from Texas State University, San Marcos, TX. During Dr. Tsui's presentation, we reviewed how to develop online questions for discussion, post responses, interact with colleagues from distant locales, and add to streaming chats. Dr. Tsui worked individually with each seminar participant.

- ***E-based Tutorials***
  This hour-long presentation was facilitated by Debbie Gorse, an employee of Xenadon E-Learning Inc. (Colorado Springs, CO). Ms. Gorse used video and overhead screen captures to give examples of successful E-based, computer-assisted instructional options.

- ***Intranet Instruction***
  Dr. Randy Towner and Dr. Karen Pecis led this hour-long presentation.
  Both are professors at the University of Nevada, Las Vegas. Their workshop focused on course development, online instructional methodologies, customizable company-based examples, and firewall-protected assessment. The professors provided workbooks and hands-on learning opportunities.

- ***Videoconferencing***
  Denise Pakula, Canyon E-Learning, Tempe, AZ, spoke about her company's media tools for teleconferenced instruction. These included lapel microphones, multidirectional pan/tilt video cameras, plasma display touch screens, wideband conference phones, recessed lighting ports with dimming preferences, and multimedia terminals.

Presentation Benefits

Every presentation we attended was beneficial. However, the following information will clarify which workshop(s) would benefit our company the most:

Figure 6.1 Trip Report

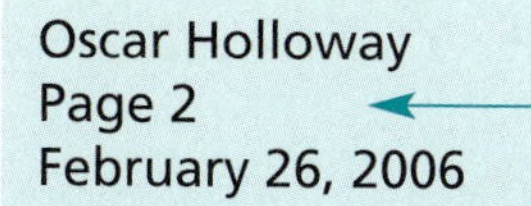
Oscar Holloway
Page 2
February 26, 2006

A "new-page notation" helps your readers avoid losing or misplacing pages.

The conclusion sums up the report by focusing on primary findings.

1. Dr. Tsui's program was the most useful and informative. His interactive presentation skills were outstanding and included hands-on activities, small-group discussions, and individual instruction. In addition, his online discussion techniques offer the greatest employee involvement at the most cost-effective pricing. We met with Dr. Tsui after the session, asking about his fees for on-site instruction. He would charge only $90 per person (other people we researched charged at least $150 per person). Dr. Tsui's fees should fit our training budget.

2. E-based tutorials will not be a valid option for us for two reasons. First, Xenadon's products are prepackaged and allow for no company-specific examples. Second, Ms. Gorse's training techniques are outdated. Videos and overhead projections will not create the interactivity our employees have requested in their annual training evaluations.

3. The Towner/Pecis workshop was excellent. Intranet instruction would be ideal for our needs. We will be able to customize the instruction, provide participants individualized feedback, and ensure confidentiality through firewall protection. Furthermore, Drs. Towner and Pecis used informative workbooks and hands-on learning opportunities in their presentation.

4. Ms. Pakula's presentation on videoconferencing focused more on state-of-the-art equipment rather than on instruction. We believe that the price of the equipment both exceeds our budget and our needs. Our current videoconference equipment is satisfactory. If the purchasing department is looking for new vendors, they might want to contact Canyon.

Recommendations

Gena, Bill, and I suggest that you invite Dr. Tsui to our site for further consultation. We also think you might want to contact Drs. Towner and Pecis for more information on their training.

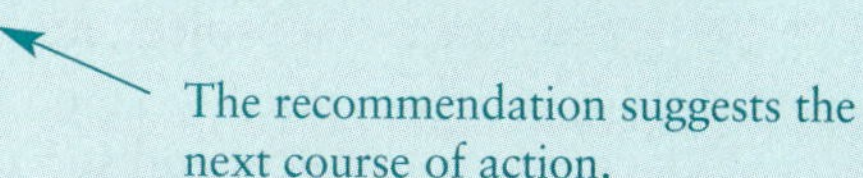
The recommendation suggests the next course of action.

Figure 6.1 *continued*

Discussion

Work accomplished. Using subheadings, itemize your work accomplished either through a chronological list or a discussion organized by importance. You might need to include information about expenditures, budgets, deadlines, personnel, marketing, travel needs, hours worked on a project, or governmental regulations.

Problems encountered. Inform your reader(s) of any difficulties encountered (late shipments, delays, poor weather, or labor shortages). This not only explains why you are behind schedule but also shows the readers where you will need help to complete the project.

Work remaining. Tell your reader what work you plan to accomplish next. List these activities, if possible, for easy access. A visual aid, such as a Gantt chart or a pie chart, fits well after these two sections. The charts will graphically depict both work accomplished and work remaining. (See Chapter 3 for discussions of graphics, including Gantt and pie charts.)

Conclusion/Recommendations

Conclusion. Sum up what you have achieved during this reporting period and provide your target completion date.

Recommendations. If you presented problems in the discussion, you can recommend changes in scheduling, personnel, budget, or materials that will help you meet your deadlines.

Figure 6.2 is an example of a progress report from an engineering company.

Feasibility/Recommendation Report

Purpose. A *feasibility/recommendation report* accomplishes two goals: it studies the practicality of a proposed plan and recommends action. Occasionally, your company plans a project but is uncertain whether the project is feasible. Will the plan work, does the company have the correct technology, will the idea solve the problem, or is there enough money? One way a company decides if a project is practical is by performing a feasibility study. The company then writes a report to document the findings and to recommend the course of action.

- Manufacturing. Your company is considering the purchase of new equipment but is concerned that the machinery will be too expensive, the wrong size for your facilities, or incapable of performing the desired tasks. Research and analyze the options, determine which equipment best suits your company's needs, and recommend purchase.
- Accounting. Your company wants to expand and is considering new locations. The decision makers, however, are uncertain whether the market is right for expansion. The accounting department needs to study the feasibility of expansion, assess P&L (profit and loss) ramifications of new monies invested in a project, and report recommendations.
- Web Design. Your company wants to create a Web site to market your products and services globally. The CEO wants this Web site to be different from the competitors' sites. The CEO wants to be sure that online checkout is easy, that pricing is cost-effective, that products are depicted in a visually appealing way, and that the site loads quickly. Write a feasibility report to present the options as well as to offer your recommendations.
- Health Information Management. It is time to update your health information system. With increasingly complex insurance and regulatory challenges, your current system is outdated. You need to install software to help code and classify patient records. A feasibility report is needed to study the options before you recommend changes.

A feasibility report should include the following:

Introduction

Objectives (Overview, Purpose, Current Status, etc.). Answer any of the following questions:

- What is the purpose of this feasibility report? To answer this question, you might need to provide background data, explaining the current status of a

TO: Buddy Ramos
FROM: Pat Smith
DATE: April 2, 2006
SUBJECT: First Quarterly Report—Project 80 Construction

The introduction explains *why* the report has been written and *what* topic will be discussed.

Purpose of Report

In response to your December 20, 2005, request, following is our first quarterly report on Project 80 Construction (Downtown Airport). Department 93 is in the start-up phase of our company's 2006 building plans for the downtown airport and surrounding site enhancements. These construction plans include the following:

1. *Airport construction*: terminals, runways, feeder roads, observation tower, parking lots, maintenance facilities.

2. *Site enhancements*: northwest and southeast collecting ponds, landscaping, berms, and signage.

Work Accomplished

In this first quarter, we have completed the following:

1. *Subcontractors*: Toby Summers (project management) and Karen Kuykendahl (finance) worked with our primary subcontractors (Apex Engineering and Knoblauch and Sons Architects). Toby and Karen arranged site visitations and confirmed construction schedules. This work was completed January 12, 2006.

2. *Permits*: Once site visitations were held and work schedules agreed upon, Toby Summers and Wilkes Berry (public relations) acquired building permits from the city. They accomplished this task on January 20, 2006.

3. *Core Samples*: Core sample screening has been completed by Department 86 with a pass/fail ratio of 76.4 percent pass to 23.6 percent fail. This meets our goal of 75 percent. Sample screening was completed January 30, 2006.

4. *Shipments*: Timely concrete, asphalt, and steel beam shipments this quarter have provided us a 30-day lead on scheduled parts provisions. Materials arrived February 8, 2006.

5. *EPA Approval*: Environmental Protection Agency (EPA) agents have approved our construction plans. We are within guidelines for emission controls, pollution, and habitat endangerment concerns. Sand cranes and pelicans nest near our building site. We have agreed to leave the north plat (40 acres) untouched as a wildlife sanctuary. This will cut into our parking plans. However, since the community will profit, we are pleased to make this concession. Our legal department also informs us that we will receive a tax break for creating this sanctuary. EPA approval occurred on February 15, 2006.

The discussion provides quantified data and dates for clarity, such as "76.4 percent pass" and "January 30, 2006." The discussion also clarifies who worked on the project and lists other primary contacts.

(*continued*)

Figure 6.2 Progress Report

Pat Smith
Page 2
April 2, 2006

A "Problems Encountered" section helps justify delays and explain why more time, personnel, or funding might be needed to complete a project.

Problems Encountered

Core samples are acceptable throughout most of our construction site. However, the area set aside for the northwest pond had a heavy rock concentration. We believed this would cause no problem. Unfortunately, when Anderson Brothers began dredging, they hit rock, which had to be removed with explosives.

Since this northwest pond is near the sand crane and pelican nesting sites, EPA told us to wait until the birds were resettled. The extensive rock removal and wait for wildlife resettlement have slowed our progress. We are behind schedule on this phase.

This schedule delay and increased rock removal will affect our budget.

Numbered points with italicized sub-headings help readers access the information more readily. This is especially valuable when an audience needs to refer to documents at a later date.

Work Remaining

To complete our project, we need to accomplish the following:

1. *Advertising*: Our advertising department is working on brochures, radio and television spots, and highway signs. Advertising's goal is to make the construction of a downtown airport a community point of pride and civic celebration.

2. *Signage*: With new roads being constructed for entrance and exit, our transportation department is working on street signage to help the public navigate our new roads. In addition, transportation is working with advertising on signage designs for the downtown airport's two entrances. These signs will juxtapose the city's symbol (a flying pelican) with an airplane taking off. The goal is to create a logo that simultaneously promotes the preservation of wildlife and suggests progress and community growth.

3. *Landscaping*: We are working with Anderson Brothers Turf and Surf to landscape the airport, roads, and two ponds. Our architectural design team, led by Fredelle Schneider, is selecting and ordering plants, as well as directing a planting schedule. Anderson Brothers also is in charge of the berms and pond dredging. Fredelle will be our contact person for this project.

4. *Construction*: The entire airport must be built. Thus, construction comprises the largest remaining task.

The conclusion sums up the overall status of the project: "15 percent" complete.

Project Completion

Though we have just begun this project, we have completed approximately 15 percent of the work. We anticipate a successful completion, especially since deliveries have been timely.

Only the delays at the northwest pond site present a problem. We are two weeks behind schedule and $3,575.00 over cost. With approximately 10 additional personnel to speed the rock removal and with an additional $2,500, we can meet our target dates. Darlene Laughlin, our city council liaison, is the person to see about corporate investors, city funds, and big-ticket endowments. With your help and Darlene's cooperation, we should meet our schedules.

The recommendation explains what is needed to complete the job: "additional personnel" and "increased funds." It also states who to contact for help, "Darlene Laughlin."

Figure 6.2 *continued*

Pat Smith
Page 3
April 2, 2006

The Gantt chart in Figure 1 clarifies our status at this time.

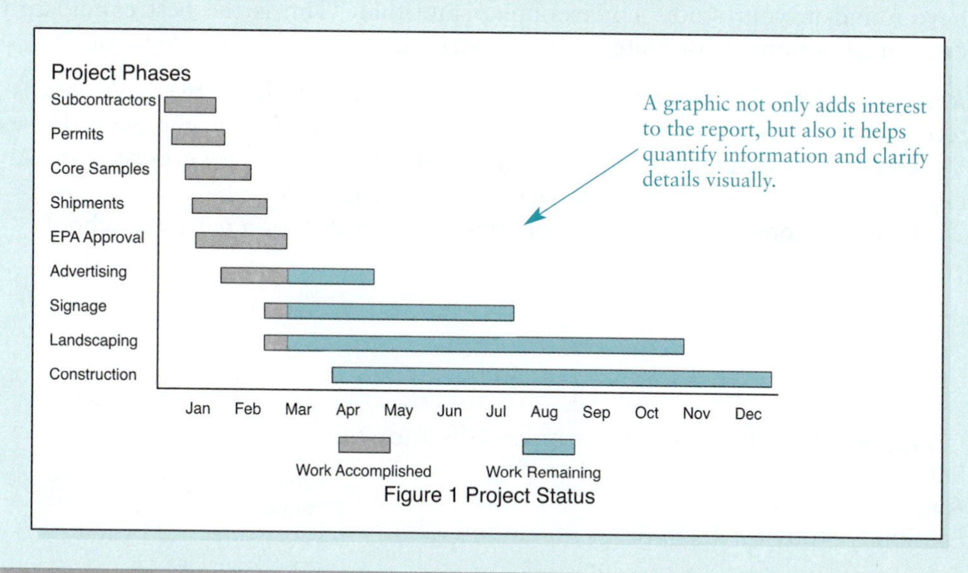

Figure 1 Project Status

Figure 6.2 *continued*

project or problem. For example, what problems cause doubt about the feasibility of the project? Is there a market for your new project? Is there a piece of equipment available that would meet the company's needs? Is land available for expansion? What problems led to the proposed project (i.e., current equipment is too costly or time-consuming, current facilities are too limited for expansion, current net income is limited by an insufficient market)?

- Who initiated the feasibility study? List the name(s) of the manager(s) or supervisor(s) who requested this report.

Personnel. Document the names of your project team members, your liaison between your company and other companies involved, and your contacts at these other companies.

Discussion. Provide accessible and objective documentation, including the following.

Criteria. State the criteria upon which your recommendation will be based. Criteria are established so you have a logical foundation for comparison of personnel, products, vendors, costs, options, schedules, and so on.

Analysis. Compare your findings against the criteria. This might require a critical analysis of the company's needs, the cost of new equipment or facilities, the credentials of required personnel, the time frame required to complete a proposed project, vendor qualifications, or consultant fees. In objectively written paragraphs, develop the points being considered. You might want to use a visual, such as a table, to organize the criteria and to provide easy access.

Conclusion. In this section, you go beyond the mere facts as evident in the discussion section. State the significance of your findings. Draw a conclusion from what you have found in your study. For example, state that "Tim is the best candidate for director of personnel" or "Site 3 is the superior choice for our new location."

Recommendations. Once you have drawn your conclusions, the next step is to recommend a course of action. What do you suggest that your company do next? Which piece of equipment should be purchased, where should the company locate its expansion, or is there a sufficient market for the product?

Figure 6.3 presents an example of a feasibility report for a technical school.

FROM: Cindy Kaye, Director of Administrative Technology
TO: Pat Hobby, Registrar
DATE: August 13, 2006
SUBJECT: Feasibility Study for Technology Purchases

The "Purpose" reminds the reader why this report has been written and what the report's goal is.

Purpose of the Report

The purpose of this report is to study which technology will best meet your communication needs and budget. After analyzing the feasibility of various technologies, we will recommend the most cost-effective technology options.

The "Problem" details what issues have led to this report.

Technology Problems

According to your memo dated August 1, 2006, your department needs new communication technologies for the following reasons:

- Your department has hired three new employees, increasing your head count to ten.
- Currently, your department has only five laptops. This slows down the registration process because we can not have all ten employees registering students simultaneously.
- We also must update training because EFA has purchased new registration software.
- The registrar's current printer is also insufficient due to your increased head count.

Vendor Contacts

Our vendor contacts for the laptops, printers, and software are as follows:

Electek	**Tech On the Go**	**Mobile Communications**
Steve Ross	Jay Rochlin	Karen Allen
stever1@electek.com	jrochlin@tog.com	karen.allen@mobcom.net

The "Vendor" section provides contact information (names and e-mail addresses).

Figure 6.3 Feasibility Report

Cindy Kaye
Page 2
August 13, 2006

Criteria for Vendors
The following criteria were considered to determine which communication technology would best meet the registrar's needs.

1. *Trainers*: Because of the unique aspects of the registration software, we need trainers who are familiar with educational requirements (privacy, 24/7 access, schedules, faculty use, etc.).
2. *Maintenance*: We need to purchase equipment and software complete with either quarterly or biannual service agreements (at no extra charge).
3. *Service Personnel*: The service technicians should be certified to repair and maintain whatever hardware we purchase. In addition, the vendors also must be able to train our personnel in hardware usage.
4. *Warranties*: The warranties should be for at least one year with options for renewal.
5. *Cost*: The total budgeted for your department is $15,000.

Needs Assessment
Purchasing agrees that the registrar's hardware and software needs exceed their current technology. Not only are the department's laptops and printer insufficient in number, but also they do not allow the personnel to access corporate e-mail, the Internet, or word processing packages. Updated equipment is necessary.

Vendor Evaluation

- **Electek**. Having been in business for 10 years, this company is staffed by highly trained technicians and sales staff. All Electek employees are certified for software training. The company promises a biannual maintenance package and subcontracted personnel if employees can not repair hardware problems. They offer manufacturer's guarantees with extended service warranties costing only $100 a year for up to 5 years. Electek offers 20 percent customer incentives for purchases of over $2,000.
- **Tech On the Go (TOG)**. This company has been in business for two years. TOG provides only subcontracted service technicians for hardware repair. TOG's employees are certified in software training. The owners do not offer extended warranty options beyond manufacturers' guarantees. No special customer pricing incentives are offered though TOG sells retail at a wholesale price.
- **Mobile Communications**. Having been in business for five years, Mobile has certified technicians and sales representatives. All repairs are provided in-house. The company offers quarterly maintenance at a fee of $50 ($200 per year). Mobile offers a customer incentive of 10 percent discounts on purchases over $5000.

(*continued*)

Writer's Insight

Cindy says, "In this feasibility report, I had to establish criteria before I could perform my research. I needed to base my decisions on certain specific needs, including cost, warranties, personnel certifications, and more. The criteria provided me direction not only for performing the research but also for determining feasible solutions to the problem.

While I was in college, I was constantly asked to write research papers. I never fully understood why I was doing this. I certainly never realized how frequently I would be doing research on the job, including surveys, interviews, and online searches. Now I know that research allows me to understand a topic and to justify the decisions I make. By clarifying the criteria for this feasibility report and researching the topic, I could make my case more convincingly."

Figure 6.3 *continued*

Cindy Kaye
Page 3
August 13, 2006

The discussion provides specific details to prove the feasibility of the plan or project.

Cost Analysis

- Laptops: The registrar requests one laptop per departmental employee. Our analysis has determined that the most affordable laptops we can purchase (with the requested software and wireless Internet connections) would cost $1,500 per unit. Thus, 10 laptops would cost $15,000. Even with discounts, this exceeds your department's budget.
- Printers: The registrar requests three additional printers with the capability to print double-side pages, staple, collate, print in color, and print three different sizes of envelopes and three different sizes of paper. Our analysis has found that the most affordable printers meeting your specifications would cost $2,500 each. Again, when combining this cost with that of laptops, you exceed your budget.
- Training: The manufacturer can provide training on our new registration software. Training can be offered Monday through Friday, starting in September. The manufacturer says that effective training must entail at least 20 hours of hands-on practice. The cost for this would be $5,000.

The following table compares the three vendors we researched on a scale of 1–3, 3 representing the highest score.

Graphics depict the findings more clearly and more concisely than a paragraph of text.

Table 1: Criteria Comparison

Criteria	**Electek**	**TOG**	**Mobile**
Maintenance	3	2	3
Personnel	3	3	3
Warranties	3	2	2
Cost	3	2	2
Total	**12**	**9**	**10**

Summary of Findings

We can not purchase the number of laptops and printers you have requested. Doing so exceeds the budget. Training is essential. Your department must adjust its budget accordingly to accommodate this need.

All three vendors have the technology you require. However, TOG and Mobile do not meet the criteria. In particular, these companies do not provide either the maintenance packages, warranties, or pricing required.

The conclusion sums up the findings, explaining the feasibility of a course of action—why a plan should or should not be pursued.

Writer's Insight

Cindy says, "When I write reports, especially feasibility reports in response to a department's request, I know that I have to prove my recommendations. This is particularly difficult when I have to refuse a portion of a request. After all, my refusal will make some of my colleagues unhappy. To clarify my findings and to maintain rapport with my co-workers, I always use tables to make the text more understandable and to emphasize the facts and figures. I don't make decisions based on just my opinion; I recommend according to quantifiable findings."

Figure 6.3 *continued*

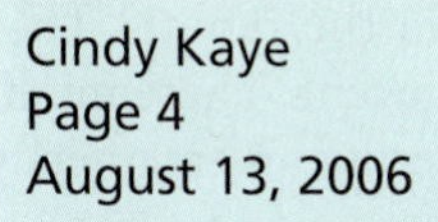

Cindy Kaye
Page 4
August 13, 2006

The recommendation explains what should happen next and provides the rationale for this decision.

Recommended Action
Given the combination of cost, maintenance packages, warranties, and service personnel, Electek is our best choice.

We suggest the following options for printers and laptops: purchasing five laptops instead of ten; purchasing one additional printer instead of three; and/or sharing printers with nearby departments.

Figure 6.3 *continued*

Incident Report

Purpose. An *incident report* documents an unexpected problem that has occurred. This could be an automobile accident, equipment malfunction, fire, robbery, injury, or problems with employee behavior. In this report, you will document what happened during and after the incident. If a problem occurs within your work environment that requires analysis (fact finding, review, or study) and suggested solutions, you might prepare an incident report. Incident reports also are called trouble reports or accident reports.

- **Engineering.** An historic, 100-year-old bridge crossing your city's river is buckling. The left lane is now two inches higher than the right lane, and expansion joints are separating beyond acceptable specifications. You must visit the bridge site, inspect the damage, and report on the causes for this incident.
- **Biomedical Technology.** A CAT scan in the radiology department is not functioning correctly. This has led to the department's inability to read X-rays and delays in patient care. To avoid similar problems, report this incident.
- **Computer Information Systems.** A company might need to report on security incidents involving the Internet, intranet, or extranet. For example, someone has attempted to gain unauthorized access to a system, a computer system has experienced unwanted disruption of service, or someone has illegally used a company's system for the processing or storage of data. In such incidents, a systems analyst must document the event to identify weaknesses that could lead to major corporate liabilities.
- **Hospitality Management.** An oven in your restaurant caught fire. This injured one of your cooks; the equipment must be replaced with an oven less prone to damage. A report needs to document this incident.

To write your incident report, include the following:

Introduction

Overview. In this section, provide an overview of the incident. Answer reporter's questions: *what* was the incident, *when* did the event occur (time and date), and *where* did the incident take place.

Personnel. *Who* was involved, and *what* role do you play in the report? That could entail listing all of the people involved in the accident or event. These might be people injured, as well as police or medical personnel answering an emergency call.

In addition, *why* are you involved in the activity? Are you a supervisor in charge of the department or employee? Are you a police officer or medical person writing the report? Are you a maintenance employee responsible for repairing the malfunctioning equipment?

Discussion. Using subheadings or itemization, quantify what you saw (the problems motivating the report). You should list your findings in chronological order. Be specific. Include the following:

- make or model of the equipment involved
- police departments or hospitals contacted
- names of witnesses
- witness testimonies (if applicable)
- extent of damage—financial and physical
- graphics (sketches, schematics, diagrams, layouts, etc.) depicting the incident visually
- follow-up action taken to solve the problem

Conclusion/Recommendation. Explain what caused or might have caused the problem. Then, detail future preventive action to avoid similar problems.

Figure 6.4 presents an example of an incident report from a wastewater treatment facility.

To: Bowstring City Council; Arrowhead School District 234
From: Mike Moore, Frog Creek Wastewater Treatment Plant Director of Public Relations
Date: September 15, 2006
Subject: Investigative Report on Frog Creek Wastewater Pollution

The subject line provides a topic (Frog Creek Wastewater Pollution) and a focus (Investigative Report).

Background of Incident

On September 7, 2006, teachers at Arrowhead Elementary School reported that over a five-day period (September 2–6), approximately 20 students complained of nausea, light-headedness, and skin rashes. On the fifth day, the Arrowhead administration called 911 and the Arrowhead School District (ASD 234) in response to this incident.

Bowstring City paramedics treated the children's illnesses, suggesting that the problems might be due to airborne pollutants. The Bowstring City Council contacted the Frog Creek Wastewater Treatment Plant (FCWTP) to investigate the causes of this problem. This report is submitted in response to Bowstring City's request.

In the introduction, reporter's questions clarify when and where the event occurred, who was involved, and why the report is being written.

Team Members

Mike Moore (Director of Public Relations), Sue Cottrell (Wastewater Engineer), and Fred Mittleman (Wastewater Engineer)

Committee Findings

Impact on Students: Arrowhead Elementary School administrators reported the following.

- Monday, September 2: two children reported experiencing nausea.
- Tuesday, September 3: two children reported experiencing nausea, and one child experienced light-headedness.
- Wednesday, September 4: three children experienced skin rashes.
- Thursday, September 5: two children complained of nausea, one child was light-headed, and two children showed evidence of skin rashes.
- Friday, September 6: two children reported nausea, three reported skin rashes, and two incidents of light-headedness.

The findings use chronological organization to document the incidents.

After Friday's occurrences, Arrowhead Elementary School administrators called 911. Bowstring paramedics reported that (with parental approval) the children were treated with antacids for nausea, antihistamines for skin rashes, and oxygen for their light-headedness. No other incidents were reported in the neighborhood surrounding the school.

Pollutants in Creek: Frog Creek is usually characterized by low alkalinity (generally <30 mg/l) (milligrams per liter). Inorganic fertilizer nutrients (phosphorous and nitrogen) are also generally low (<20 mg/l), with limited algae growth.

Despite normally low readings, in late summer, with heat and rain, these readings can escalate. Higher algae-related odors above the 3–6 picometer thresholds, along with increased alkalinity (<50 mg/l) can create health problems for youth, elderly, or anyone with respiratory illnesses.

Wastewater Engineers Sue Cottrell and Fred Mittleman took samples of Frog Creek on September 7–10. These studies showed that algae, alkalinity, and fertilizer nutrients were higher than usual.

- Algae readings: 4 picometers
- Alkalinity readings: <45 mg/l
- Phosphorous and nitrogen readings: <25 mg/l

On September 11–14, our engineers rechecked Frog Creek, finding that the chemical levels had returned to a normal, acceptable range.

(*continued*)

Figure 6.4 Incident Report

Mike Moore
Page 2
September 15, 2006

The findings not only investigate the causes of the incident but also document with specific details. To achieve a readable format, the text is made accessible through highlighting techniques—subheadings and bulleted details.

Atmospheric Factors: The above elevated readings were caused by three factors: heat, rain, and northeasterly winds.

- Heat: On the days of the Arrowhead Elementary School incident, the temperature ranges were 92 degrees F to 95 degrees F, unusually high for early September. Algae and chemical growth increases in temperatures above 84 degrees F.
- Rain: In addition, on September 4–6, Bowstring City received 2" of rain, swelling Frog Creek to 3' above its normal levels. Studies show that rain-swollen creeks and rivers lead to increased pollutants, as creek bottom silt rises.
- Wind: On September 4–5, a prevailing northeasterly wind blew from Frog Creek toward Arrowhead Elementary School's playground.

Conclusions

Frog Creek normally has acceptable levels of algae, alkalinity, and fertilizer nutrient levels. The heat and higher water levels temporarily led to elevated pollutant readings. These levels subsequently returned to normal. Wind direction during the school incidents also had an impact on the children's illnesses. On follow-up questionnaires, FCWTP employees found that the children's ailments had subsided.

The Arrowhead Elementary School situation appears to have been an isolated incident due to atmospheric changes.

Recommendations for Future Course of Action

Frog Creek is constantly monitored for safety. However, rain, wind, and heat will continue to affect its chemical levels. In unusual situations (children playing outside, the wind blowing from Frog Creek toward the elementary school's playground, and a combination of heat and rain), similar results could occur. Children especially susceptible to airborne pollutants could experience nausea, rashes, and/or light-headedness.

FCWTP HAZMAT employees would be happy to work with parents and teachers. In a one-hour workshop, presented during the school day or at a Parent-Teacher Organization meeting, FCWTP will provide the following information:

- Scientific data about stream and creek pollutants
- The effects of rain, wind, and heat on creek chemicals
- Useful preventive medical emergency techniques

This information would explain real-world applications for science classes, as well as provide valuable health tips for parents and teachers.

Please let us know if you would like to benefit from this free-to-the-public workshop. We would be happy to schedule one at your convenience.

The conclusion provides options for the readers and a positive tone appropriate for the intended audience.

Figure 6.4 *continued*

Short Report Checklist

The short report checklist will give you the opportunity for self-assessment and peer evaluation of your writing.

Short Report Checklist

__ 1. **Subject Line:** Does your subject line contain a *topic* and *focus*? If you write only "Subject: Trip Report" or "Subject: Feasibility/Recommendation Report," you have not communicated precisely to your audience. Such a subject line merely presents the focus of the correspondence. But what's the topic? To provide both *topic* and *focus*, write "Subject: Trip Report on Solvent Training Course, Arco Corporation, 3/15/06."

__ 2. **Introduction:** Does the introduction explain the *purpose* of the report, list the *personnel* involved, and/or state *when* and *where* the activities occurred?

__ 3. **Discussion:** When you write the discussion section of the report, do you quantify what occurred? In this section, you must clarify precisely. Specify dates, times, prices, locations, people involved, calculations, or problems encountered.

__ 4. **Accessibility:** Is your report accessible? To create reader-friendly ease of access, use headings, boldface, italics, bullets, numbers, underlining, or graphics (tables and figures). These add interest and help your readers navigate your report.

__ 5. **Organization:** Have you selected an appropriate method of organization in your discussion? These include chronology, importance, problem/solution, or comparison/contrast.

__ 6. **Conclusion:** Does your conclusion present a value judgment regarding the findings presented in the discussion? The discussion presents the facts, and the conclusion explains what the facts mean and their implications.

__ 7. **Recommendation:** In the recommendation, have you suggested what should happen next? What is the appropriate follow-up, when should this occur, and why is that date important?

__ 8. **Conciseness:** Have you limited the length of your sentences, words, and paragraphs?

__ 9. **Audience:** Have you effectively recognized your audience's level of understanding (specialists, semi-specialists, lay, multicultural or cross-cultural, supervisors, subordinates, coworkers, customers, or vendors) and written accordingly?

__10. **Accuracy:** Are the facts (numbers, data, research, etc.) correct? Is your report's grammar accurate? Correct errors in spelling, punctuation, grammar, research, or mathematics.

The Writing Process At Work

To clarify the importance of the writing process, look at how one student used prewriting, writing, and rewriting to construct her progress report.

Prewriting

The student used a simple topic outline to gather data and determine her objectives (Figure 6.5).

Writing

The student drafted her progress report, focusing on the information she discovered in prewriting (Figure 6.6).

Rewriting

After drafting this report, the student submitted her copy to a peer review group, which helped her edit and revise her text (Figure 6.7).

After discussing the suggested changes with her peer review group, the student revised her draft and submitted her finished copy (Figure 6.8).

Figure 6.5 Simple Topic Outline for Prewriting

Sales Proposal to Round Rock Hospital

I. Introduction
- A. When: Oct. 10
- B. What: Sales proposal to hospital
- C. Why: In response to your memo

II. Discussion
- A. What points to cover?
  1. Containers
  2. Sterilization
  3. Documentation
  4. Cost
  5. Disposal
- B. What's next?
  1. Future work requirements

III. Conclusion/Recommendations
- A. Conclusion: When? On schedule, 50 percent completed
- B. Recommendations: Comply with regulations

November 6, 2006
To: Carolyn Jensen
From: Shuan Wang.
Subject: Progress Report.

Purpose: This is a progress report on the status of a sales proposal you requested in your memo of October 10, 2006. The objective of this proposal is the sale of a total program for infectious waste control and disposal to a Round Rock Hospital. The proposal will cover the following five things.

1. Containers
2. Steam sterilization
3. Cost savings
4. Landfill operating
5. Computerized documentation

Work Completed: The following is a list of items that are finished on the project.

Containers
I have finished a description of the specially designed containers with disposable biohazard bag.

Steam sterilization
I have set a instructions for using the Biological Detector, model BD 12130 as a reliable indicator of sterilization of infectious wastes.

Cost savings
Central Hospital sent me some information on installing a pathological incinerator as well as the constant personal and maintenance costs to operate them. I ran this information through our computer program "Save-Save". The results show that a substantial savings will be realized by the use of our services. I have data to support this, since more and more of our customer base is online with the system proving that expenses to other companies has been reduced.

Future Work: I have an appointment to visit Round Rock Hospital on November 7, 2006. At that time I will make on-site evaluations of present waste handling practices at Round Rock Hospital. After carefully studing this evaluation, I will report to you on needed changes to comply with governmental guidelines.

Conclusion: The project is proceeding on schedule. Approximately 50 percent of the work is done (see graph). I don't see any problems at this time and should be able to meet the target date. I will update you on our continuing progress in the near future.

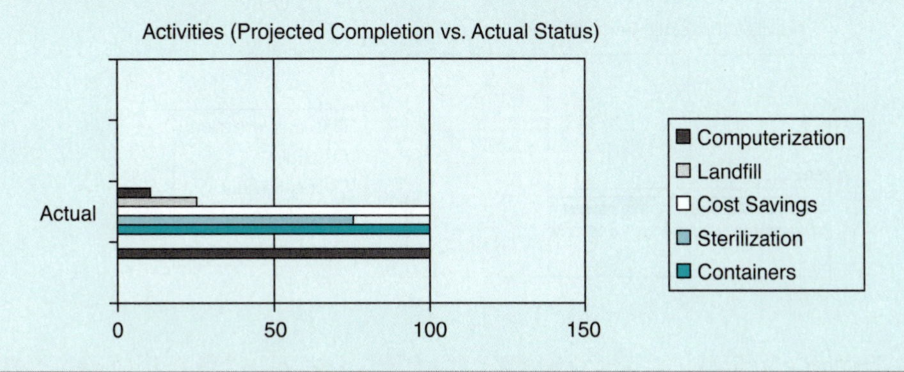

Figure 6.6 Student Rough Draft

November 6, 2006
To: Carolyn Jensen
From: Shuan Wang.
Subject: Progress Report.

Omit the period and clarify the topic of this report.

Purpose: This is a progress report on the status of a sales proposal you requested in your memo of October 10, 2006. The objective of this proposal is the sale of a total program for infectious waste control and disposal to a Round Rock Hospital. The proposal will cover the following five things.

"things" = weak word!

1. Containers
2. Steam sterilization
3. Cost savings
4. Landfill operating
5. Computerized documentation

Boldface the headings for emphasis.

The second sentence above is long. Remember, our teacher told us to limit sentences to 10-15 words.

Work Completed: The following is a list of items that are finished on the project.

Containers
I have finished a description of the specially designed containers with disposable biohazard bag.

Steam sterilization
I have set a instructions for using the Biological Detector, model BD 12130 as a reliable indicator of sterilization of infectious wastes.

Correct the typo = "a set of instructions"

Cost savings
Central Hospital sent me some information on installing a pathological incinerator as well as the constant personal and maintenance costs to operate them. I ran this information through our computer program "Save-Save". The results show that a substantial savings will be realized by the use of our services. I have data to support this, since more and more of our customer base is online with the system proving that expenses to other companies has been reduced.

Correct the spelling error = personnel

Future Work: I have an appointment to visit Round Rock Hospital on November 7, 2006. At that time I will make on-site evaluations of present waste handling practices at Round Rock Hospital. After carefully studing this evaluation, I will report to you on needed changes to comply with governmental guidelines.

Correct the spelling error = studying

Conclusion: The project is proceeding on schedule. Approximately 50 percent of work is done (see graph). I don't see any problems at this time and should be able to meet the target date. I will update you on our continuing progress in the near future.

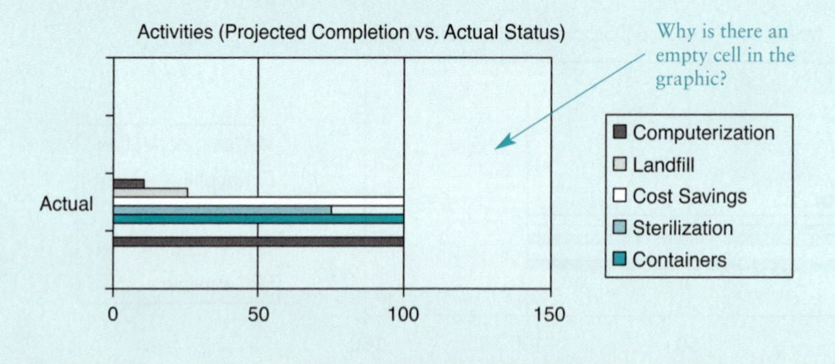

Figure 6.7 Peer Group Revision Suggestions

Date: November 6, 2006
To: Carolyn Jensen
From: Shuan Wang
Subject: Progress Report on Round Rock Hospital

Purpose of Report:

In response to your October 10, 2006, request, following is a progress report on our Round Rock Hospital sales proposal. This proposal will present Round Rock Hospital our total program for infectious waste control and disposal. The proposal will cover the following five topics:

- Containers
- Steam sterilization
- Cost savings
- On-site evaluations
- Landfill disposal

Work Completed:

1. *Containers*: On October 5, I finished a description of the specially designed containers with biohazard bags. These were approved by Margaret Chase, Round Rock's quality control administrator.

2. *Steam sterilization*: On November 4, I wrote instructions for using our Biological Detector, model BD 12130. This mechanism measures infectious waste sterilization levels.

3. *Cost savings*: Round Rock Hospital's accounting manager, Lenny Goodman, sent me their cost charts for pathological incinerator expenses and maintenance costs. I used our computer program "Save-Save" to evaluate these figures. The results show that we can save the hospital $15,000 per year on annual incinerator costs after a two-year break-even period. I have data to support this from other customers (see the attachment).

Work Remaining:

1. *On-site evaluation*: When I visit Round Rock Hospital on November 7, I will evaluate their present waste handling practices. After studying these evaluations, I will report necessary changes for governmental compliance.

2. *Landfill disposal*: I will contact the Environmental Protection Agency on November 13, 2006, to receive authorization for disposing sterilized waste at our sanitary landfill.

(*continued*)

Figure 6.8 Finished Product

Wang
Page 2
November 6, 2006

Completion of Project:

The project is proceeding on schedule. Approximately 50 percent of the work is done (see Figure 1).

Figure 1: Completion Status of Planned Activities

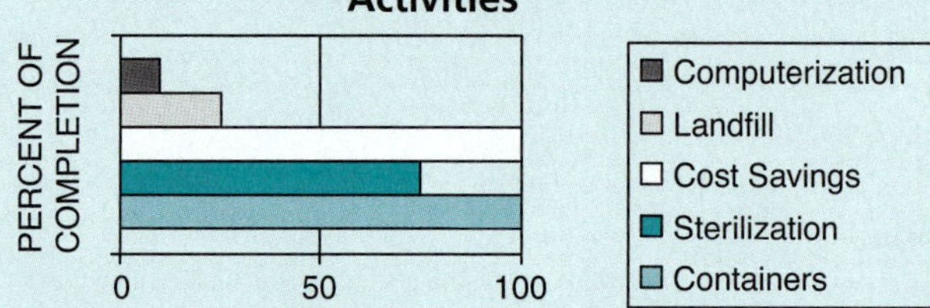

I see no problems at this time. We should meet our December 31, 2006, target date. I will update you on our continuing progress in two weeks.

Figure 6.8 *continued*

CASE STUDIES

1. Cindy Kaye, director of administrative technology at EFA Incorporated, needs to write a trip report. She just returned from a meeting in Philadelphia and must report on her activities. While in Philadelphia, Cindy accomplished many of her regular duties. These included attending the update meeting (where all involved parties report on their earlier activities), the ongoing technology training sessions (how to use the college's intranet system for grade reporting, online enrollment and counseling, and benefits information), breakout sessions on problems encountered, and meetings with vendors to assess options for new hardware and software purchases.

   Cindy also was involved in two new initiatives: discussions on how to train faculty on the new technology improvements and ways to increase revenue to pay for technology changes.

   Cindy encountered one new problem in her last trip to Philadelphia. The airline carrier she typically traveled on announced its bankruptcy. She had been traveling for free on this airline due to her banked air miles. She will lose this advantage if she can no longer travel on her preferred carrier. Cindy must find affordable options.

Assignment

Based on the information provided above, write a trip report for Cindy, addressed to her supervisor, Dr. Susan Hart, Vice President of Administrative Services. In this report, document Cindy's activities. In addition, through research, provide technology training options for her faculty and suggest ways to increase revenue. You can go online to a search engine and type in "technology training" and "how to increase revenue" (or similar synonyms) to find options for Cindy. In addition, brainstorm with your classmates reasonable ways that Cindy can travel affordably from Miami to Philadelphia.

2. You manage an engineering department at Acme Aerospace. Your current department supervisor is retiring. You must recommend the promotion of a new supervisor to the company's executive officer, Kelly Adams. You know that Acme seeks to promote individuals who have the following traits:

   - Familiarity with modern management techniques and concerns, such as teamwork, global economics, and workplace diversity.
   - An ability to work well with colleagues (subordinates, lateral peers, and management).
   - Thorough knowledge of one's areas of expertise.

   You have the following candidates for promotion.

   - Pat Jefferson. Pat has worked for Acme for 12 years. Pat, in fact, has worked up to a position as a lead engineer by having started as an assembler, then working in test equipment, quality control, and environmental safety and health (ESH). As an engineer in ESH, Pat was primarily in charge of hazardous waste disposal. Pat's experience is lengthy although Pat has only taken two years of college coursework and one class in management techniques. Pat is well liked by all colleagues and is considered to be a team player.

- Kim Kennedy. Kim is a relatively new employee at Acme, having worked for the company for two years. Kim was hired directly out of college after earning an M.B.A. degree from the Mountaintop College School of Management. As such, Kim is extremely familiar with today's management climate and modern management techniques. Kim's undergraduate degree was a B.S. in business with a minor in engineering. Currently, Kim works in the engineering department as a departmental liaison, communicating the engineering department's concerns to Acme's other departments. Kim has developed a reputation as an excellent coworker who is well liked by all levels of employees.
- Chris Clinton. Chris has a B.S. degree in engineering from Poloma College and an M.B.A. degree from Weatherford University. Prior to working for Acme, Chris served on the IEEE (Institute of Electrical and Electronic Engineering) Commission for Management Innovation, specializing in global concerns and total quality management. In 2000, Chris was hired by Acme and since then has worked in various capacities. Chris is now lead engineer in the engineering department. Chris has earned high scores on every yearly evaluation, especially regarding knowledge of engineering. When assistance is needed with new management techniques, Chris has been a valued resource. Chris's only negative points on evaluations have resulted from difficulties with colleagues, some of whom regard Chris as haughty.

Assignment

Which candidate will you recommend for supervisor? Write a feasibility/recommendation report stating your decision.

3. You are the accountant at AAA Computing, a chain of retail stores specializing in computer hardware. Your boss states that all new computers sold should be accompanied by an optional service contract. This service will be provided by an outside vendor. Your job is to research several vendors and write a feasibility report recommending the best choice for your company.

   To do so, you know your boss will emphasize the following criteria:

   - Years in business/expertise. To be sure that customers receive quality service, the vendor should have a good track record and be familiar with computer hardware innovations.
   - Quick response/turnaround. Since many of your customers depend on their computers for daily business operations, the vendor should provide on-site service for minor repairs and 24-hour turnaround service on major repairs.
   - Cost-effective pricing. The less the vendor charges, the more you can mark up the service cost. Thus, AAA can make more money.

   The following vendors have proposed their services. Using the information provided about each company and the criteria for feasibility reports provided in this chapter, write your report recommending a vendor for AAA's service contracts.

   - QuickBit. This company has 16 months' experience in computer hardware service. QuickBit, although without a lengthy track record, is co-owned by three graduates of the Silicon Valley Institute of Technology's renowned B.A. program in computer information services. The three individuals are very knowledgeable about computers, having received superior instruction and hands-on training using today's most up-to-date technology. Because QuickBit is small, it can provide immediate, 24-hour on-call service. Furthermore, QuickBit owns a 12-wheeler truck fully stocked with parts. Therefore, all service can be handled on-site. QuickBit charges $50 per hour plus parts.

- ROM on the Run. This company has been in business for 10 years. It employs 50 servicepeople, has a lengthy track record, and has provided service for ARC Telecommunications, Capital Bank, Helping Hand Hospital, and the State Penitentiary at Round Rock. Because of their years in business and the fact that all their employees have at least two-year certificates from vocational colleges, they charge $85.50 per hour plus parts. ROM on the Run promises on-site service within 24 hours on all service requests.
- You Bet Your Bytes (YBYB). YBYB has been in business for four years, employs 10 servicepeople (all of whom have at least a B.S. degree in computer systems), and specializes in retail outlets. YBYB charges $70 per hour plus parts. The company promises to respond within 30 minutes for on-site service calls. Furthermore, YBYB advertises that most parts are carried on their service trucks. Any repairs requiring unavailable parts can be made within 24 hours if the service call is received by 3:30 p.m. Calls after that time require two working days.

Assignment

Which company do you recommend? Write the feasibility report stating your choice.

4. Your company, Telecommunications R Us Enterprises (TRUE), has experienced a 45 percent increase in business, a 37 percent increase in warehoused stock, and a 23 percent increase in employees. You need more room. Your executive officer, Polina Gertsberg, has asked you to research existing options. To do so, you know you must consider these criteria:

   - Ample space for further expansion. Ms. Gertsberg suggests that TRUE could experience further growth upwards of 150 percent. You need to consider room for parking, warehouse space, additional offices, and a cafeteria—approximately 20,000 sq. ft. total.
   - Cost. Twenty million dollars should be the top figure, with a preferred payback of five years at 10 percent.
   - Location. Most of your employees and customers live within 15 miles of your current location. This has worked well for deliveries and employee satisfaction. A new location within this 15-mile radius is preferred.
   - Aesthetics. Ergonomics suggest that a beautiful site improves employee morale and increases productivity.

After research, you have found three possible sites. Based on the following information and on the criteria for feasibility reports discussed in this chapter, write your report recommending a new office site.

Site 1 (11717 Grandview). This four-story site, located 12 miles from your current site, offers three floors of finished space equaling 18,000 sq. ft. The fourth floor is an unfinished shell equaling an additional 3,000 sq. ft. As is, the building will sell for $19 million. If the current owner finishes the fourth floor, the addition would cost $4 million more. For the building as is, the owner asks for payment in five years at 12 percent interest. If the fourth floor is finished by the owner, payment is requested in seven years at 10 percent. The building has ample parking space but no cafeteria, although a building next door has available food services. Site 1 is nestled in a beautifully wooded area with hiking trails and picnic facilities.

Site 2 (808 W. Blue Valley). This one-story building offers 21,000 sq. ft. that includes 100 existing offices, a warehouse capable of holding storage bins that measure 20 ft. tall × 60 yds. long × 8 ft. wide, and a full-service cafeteria. Because the complex is one story, it takes up 90 percent of the lot, leaving only 10 percent for parking. Additional parking is located across an eight-lane highway which can be crossed via a footbridge. The building, located 18 miles from your current site, has an asking price of $22 million at 8.75 percent interest for five years. Site 2 has a cornfield to its east, the highway to its west, a small lake to its north where flocks of geese nest, and a strip mall to its south.

Site 3 (1202 Red Bridge Avenue). This site is 27 miles from your current location. It has three stories offering 23,000 sq. ft., a large warehouse with four-bay loading dock, and a cafeteria with ample seating and vending machines for food and drink. Because this site is located near a heavily industrialized area, the asking price is $15 million at 7.5 percent interest for five years.

Assignment

Which site do you recommend? Write your feasibility report stating your choice.

Individual and Team Projects

1. *Write a progress report.* The subject of this report can involve a project or activity at work. If you have not been involved in job-related projects, write about the progress you are making in this class or another course you are taking. Write about your progress toward completing your degree program. Write about the progress you are making on a home improvement project (refinishing a basement, constructing a deck, painting and papering a room). Write about the progress you are making on a hobby (rebuilding an antique car, constructing a computer, making model trains). Whatever your topic, first plan (using outlining), then write a draft, and, finally, rewrite, perfecting the text. Abide by all the criteria presented in this chapter regarding progress reports.
2. *Write a feasibility/recommendation report.* You can draw your topic from your work environment, school, or home. For example, if you are considering a new purchase (car, home, computer system, entertainment system), the implementation of a new procedure at work, expansion to a new location, or the marketing of a new product, you could study options and then write a feasibility report on your findings.

   If nothing at work lends itself to this topic, then consider plans at school. Are you and your roommates thinking about redecorating your apartment? Is your sorority or fraternity planning an event? As a member of an on-campus club, are you considering ways to get other students involved? Study your options and then write a feasibility report on your findings.

   What about your home? For example, are you and your family planning a vacation, the renovation of your basement, or a new business venture? If so, study this situation. Study vacation options, research the market for a new business, and get bids for the renovation. Then write a feasibility report to your family documenting your findings. Whether your topic comes from business, school, or home, gather your data in planning (using outlining), draft your text, and then revise. Follow the criteria for feasibility reports provided in this chapter to help you write the report.

3. *Write an incident report.* You can select a topic either from work or home. If you have encountered a problem at work, write an incident report documenting the problem and providing your solutions to the incident. This could include thefts, injuries, or damage to equipment or products. Follow the criteria in this chapter to help you write this report.

4. *Write a trip report.* Drawing from either job-related travel or a trip you have taken with friends and family, document your activities in a trip report. Follow the criteria in this chapter, and use the writing process (prewriting, writing, and rewriting).

5. *Revise a report.* An example of a flawed report that needs revising is shown in Figure 6.9. To improve this report, form small groups and first decide what is missing according to this chapter's criteria for good reports. Then use the rewriting techniques presented in this textbook to revise the report.

Problem Solving Think Pieces

1. Angel Guerrero, computer information systems technologist at HeartHome Insurance, has traveled from his home office to a branch location out of town. While on his job-related travel, he encountered a problem with his company's remanufactured laptop computer. He realized that the problem had been ongoing not only for this laptop but also for six other remanufactured laptops that the company had recently purchased.

   Angel thinks he knows why the laptops are malfunctioning and plans to research the issue. When he returns to the home office from his travel, he needs to write a report. What type of report should he write? Explain your answer, based on the information provided in this chapter.

2. Minh Tran is a special events planner in the marketing department at Thrill-a-Minute Entertainment Theme Park. Minh and her project team are in the middle of a long-term project. For the last eight months, they have been planning the grand opening of the theme park's newest sensation ride—*The Horror*—a wooden roller coaster that boasts a 10 g gravitational drop.

   During one of the team's weekly project meetings, the team has hit a roadblock. The rap group "Bite R/B Bit" originally slated to play at the midnight unveiling has cancelled at the last minute. The team needs to get a replacement band. One of Minh's teammates has researched the problem and has presented six alternative bands (at varying prices and levels of talent) for consideration. Minh needs to write a report to her supervisor.

   What type of report should she write? Explain your decision based on the criteria for reports provided in this chapter.

3. Toby Hebert is human resource manager at Crab Bayou Industries (Crab Bayou, LA), the world's largest wholesaler of frozen Cajun food. She and her staff traveled to New Orleans to attend meetings held by five insurance companies that wanted to explain and promote their employee benefits packages.

Site Visit—Alamo Manufacturing
November 1

Sam, I visited our Alamo site and checked on the following:

1. Our plant facilities suffered some severe problems due to the recent wind and hail storms. The west roof lost dozens of shingles, leading to water damage in the manufacturing room below. The HVAC unit was flooded by several feet of water, shorting out systems elsewhere in the plant. In addition, our north entryway awning was torn off its foundation due to heavy winds. This not only caused broken glass in our front entrance door and a few windows bordering the entrance, but also the entryway driveway now is blocked for customer access.

2. Maintenance and security failed to handle the problems effectively. Security did not contact local police to secure the facilities. Maintenance responded to the problems far too late. This led to additional water and wind damage.

3. Luckily, the storm hit early in the day before many of our employees had arrived at work. Still, a few cars were in the parking lot, and they suffered hail damage. Are we responsible?

4. The storm, though hurting manufacturing, will not affect sales. Our 800-lines and e-mail system were unaffected. But we might have problems with delivery if we can't fix the entryway impediment. I think I have some solutions. We could reroute delivery to one of our new plants, or maybe we should consider direct delivery to our sales staff (short term at least). Any thoughts?

5. Meanwhile, I have gotten on Maintenance's case, and they are working on repairs. Brownfield HVAC service has given us a quote on a new sump pump system that has failsafe programs (not too bad an extra cost, given what we can save in the long run). Plainview Windows & Doors is already on the front entrance problem. They promise replacement soon.

6. As for future plans, I think we need to reevaluate both Maintenance and Security. Training is an option. We also could add new personnel or reconsider our current management in those departments, or maybe just a few, good, hard, strongly-stated comments from you would do the trick.

Anyway, we're up and running again. Upfront repair costs are covered by insurance, and long-term costs are minimal since sales weren't hurt. Our only remaining challenges are preventive, and that depends on what we do with our Maintenance and Security staff. Let me know what you think.

Figure 6.9 Flawed Report Needing Revision

During the trip, Toby's company van was sideswiped by an uninsured driver. When Toby returned to Crab Bayou, she had to write a report. What type of report did she write? Base your decision on the criteria for reports provided in this chapter.

4. Yasser El-Akiba is a member of his college's FBLA club (Future Business Leaders of America). Riverbend Community College (RCC), Riverbend, Idaho, has

sponsored the FBLA team's travel to a regional convention. At this convention, Yasser and his teammates will compete in FBLA events, meet world-renowned business leaders, and attend meetings.

Yasser is the RCC secretary responsible for reporting on the team's activities. These include expenditures, contacts, team successes during the competitions, and activities during the meeting the students must attend. During one meeting, Yasser learns about challenges unique to international business. These include cultural differences, language barriers, monetary exchanges, and international legal difficulties regarding franchising.

When he returns to RCC, what kind of report should he submit to his FBLA faculty sponsor? Defend your decision based on criteria for reports provided in this chapter.

WEB WORKSHOP

1. More and more, companies and organizations are putting report forms online. The reason for doing this is simple—ease of use.

   Go online and access online report forms. All you need to do is use any Internet search mechanism, and type in "online __________ report form." (In the space provided, type in "trip," "progress," "incident," "investigative," or "feasibility/recommendation.") Once you find examples, evaluate how they are similar to and different from the written reports shown in this chapter. Share your findings with others in your class, either through oral presentations or written reports.

2. Every company writes reports, and you can find examples of them online. Use the Internet to access "Report Gallery," found at *http://www.reportgallery.com/*. Report Gallery bills itself as "the largest Internet publisher of annual reports." From this Web site, you can find the annual reports from 2,200 companies, including most Fortune 500 companies.

   Study the annual reports from 5 to 10 different companies. What do these reports have in common? How do they differ? Discuss the page layout, readability, audience involvement, content, tone, and development of the reports. How are the reports similar to and different from those discussed in this chapter?

   Share your findings with others in your class, either through oral presentations or written reports.

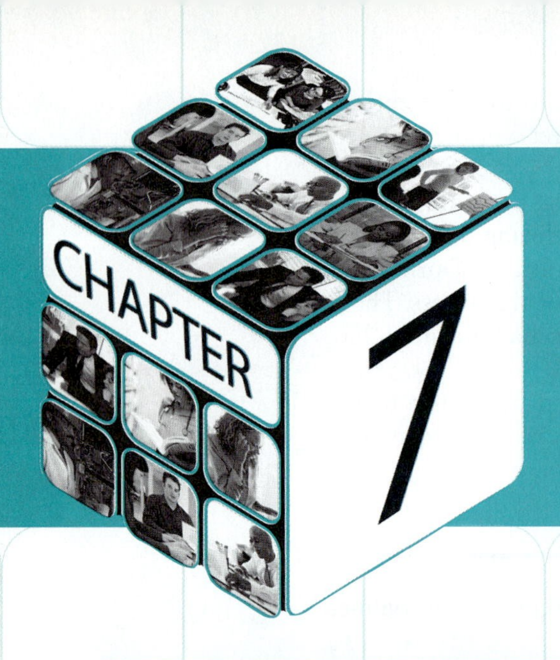

Proposals and Long Reports

OBJECTIVES

When you complete this chapter, you will be able to do the following:

1. Format proposals and long reports effectively using components such as a title page, cover letter, table of contents, list of illustrations, abstract, introduction, discussion, conclusion/recommendation, and glossary.
2. Write effective internal proposals to persuade corporate decision makers to address issues and provide resources.
3. Write effective external proposals to sell a new service or product to a potential customer.
4. Distinguish among common proposal terms, including RFP, T&C, SOW, boilerplate, solicited proposals, and unsolicited proposals.
5. Apply research techniques to gather information for proposals and long reports.
6. Test your knowledge of proposals and long reports through end-of-chapter activities:
   - Case Studies
   - Individual and Team Projects
   - Problem Solving Think Pieces
   - Web Workshop

COMMUNICATION AT WORK

In the following BioStaffing scenario, Jonathan Bacon communicates both internally and externally by writing proposals.

BioStaffing provides temporary staffing for hospitals and medical offices needing part-time nurses, medical laboratory technicians, medical records personnel, physical therapists, and occupational therapists. In addition, BioStaffing manufactures and markets biomedical equipment.

Jonathan Bacon, manager of information technology, needs to write two proposals:

1. **An internal proposal to update BioStaffing's technology.** The IT department has researched this topic thoroughly and will recommend a revised technology policy, improved hardware and software, and a new vendor. He will work with his staff of writers, as well as the IT department's technicians, to write the internal proposal to management seeking approval and funding.
2. **An external proposal to market the company's new product line—MediFactFinder.** This handheld computer, which can be hot synced to desktops and provides wireless access to BioStaffing's intranet, offers medical providers instantaneous information about medications, dosages, and treatments for over 3,500 illnesses.

To write both proposals, Jonathan will work with a diverse team from legal, marketing, manufacturing, engineering, and accounting. To accompany his written proposals, Jonathan also will work with the corporate communication staff to create PowerPoint presentations.

Though Jonathan's tasks are challenging, his proposals promise to generate income and improve the company's internal communication abilities.

LONG REPORTS

In some instances, your subject matter might be so complex that a short report will not thoroughly cover the topic. For example, your company asks you to write a report about the possibility of an impending merger. This merger will require significant commitments regarding employees, schedules, equipment, training, facilities, and finances. Only a long report, such as a proposal complete with research, will convey your content sufficiently and successfully.

Since short reports run only a few pages, you can assume that your readers will be able to follow your train of thought easily. Thus, short reports merely require that you use headings such as "Introduction," "Discussion," and "Conclusion/Recommendation" to guide your readers through the document or talking headings (see Chapter 3 for a discussion of headings and talking headings) to summarize the content more thoroughly. Long reports, however, place a greater demand on readers. Your readers could be overwhelmed with numerous pages of information and research. A few headings won't be enough to help your readers wade through the data. Figure 7.1 shows you the key components of a long report.

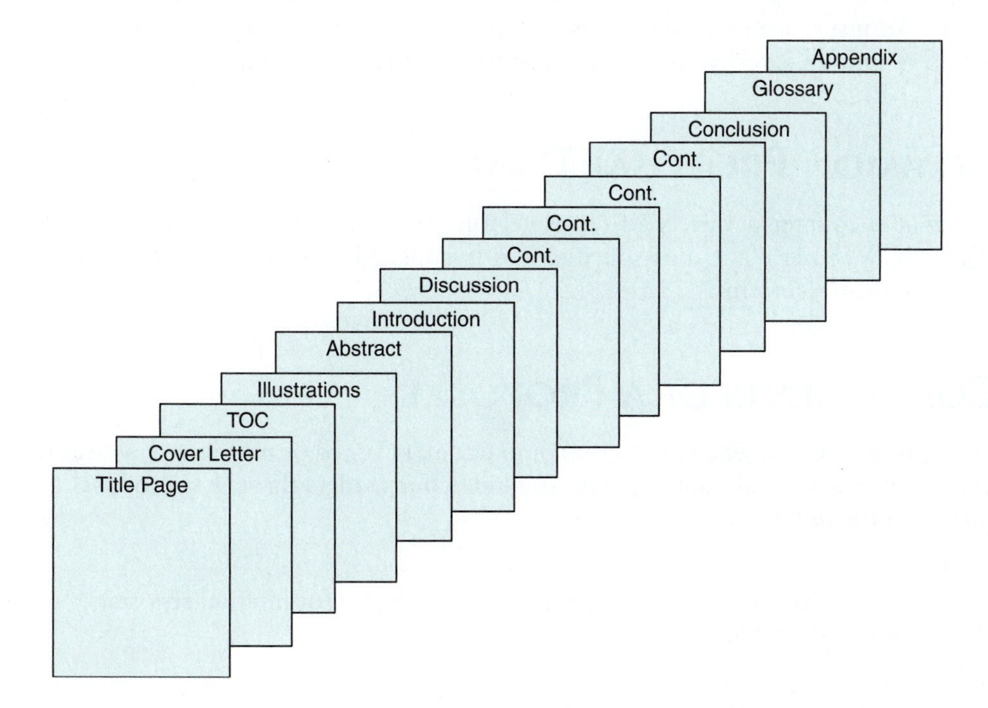

Figure 7.1 Components of a Long Report

In addition to the basic components of a short report (discussed in Chapter 6), a long report includes the following:

- Front matter (title page, cover letter, a table of contents or TOC, a list of illustrations, and an abstract or executive summary)
- Text (introduction, discussion, and conclusion/recommendation)
- Back matter (glossary, works cited or reference page, and an appendix)

Types of Proposals

Whether writing an internal or external proposal, you ask readers to make significant commitments regarding employees, schedules, equipment, training, facilities, and finances.

Internal Proposals

Your company needs a new and expanded facility to accommodate increased inventory, clientele, and workforce. Your department needs updated equipment (computers, copiers, scanners, or telephone systems) to stay competitive. Your team needs improved training to meet governmental regulations.

To persuade decision makers within your company to address these issues and provide resources, you must write an *internal proposal*, a type of long report.

External Proposals

Your company is offering a new service or manufacturing a new product. Your responsibility is to write an *external proposal*. This external proposal, addressed to potential vendors or customers, will sell the benefits of your new corporate offering. You also could write an external proposal in response to a request for proposal (RFP). A governmental agency, for example, needs Internet security systems for its offices. A hospital asks engineering companies to submit proposals about facility improvements. An insurance company needs to buy a fleet of cars for its adjusters. Vendors would submit an external proposal in response to these requests.

Common Proposal Terms

You will encounter a variety of terms relating to proposal writing, including RFP, T&C, SOW, boilerplate, and solicited and unsolicited proposals. See Table 7.1 for common proposal terms.

Components of a Proposal

Typically, proposals, whether internal or external, solicited or unsolicited, share common components to all long reports. To guide your readers through a proposal, provide them the following:

- Title page
- Cover letter for external reports (or cover memo for internal reports)
- Table of contents

TABLE 7.1 COMMON PROPOSAL TERMS

Proposal Terms	Definitions
RFP	**Request for Proposals**: means by which external companies and agencies ask for proposals
T&C	**Terms and Conditions**: the exact parameters of the request and expected responses
SOW	**Scope of Work** ***or*** **Statement of Work**: costs, dates, deliverables, personnel certifications, and/or company history
Boilerplate	Any content (text or graphics) that can be used in many proposals
Solicited Proposal	A proposal written in response to a request
Unsolicited Proposal	A proposal written on your own initiative

- List of illustrations
- Abstract
- Executive summary
- Introduction
- Discussion (the body of the proposal)
- Conclusion/recommendation
- Glossary
- Works cited information (optional; only needed if you are documenting research)
- Appendix

Title Page

The title page provides your reader the following:

- Title of the proposal
- Name of the company, organization, or reader of the proposal
- Name of the company or writer(s) submitting the proposal
- Date on which the proposal was completed

If the internal proposal is being submitted within your company to peers, subordinates, or supervisors, you might want to include a routing list of individuals who must sign off or approve the proposal. Following are two sample title pages. Figure 7.2 is for an internal proposal; Figure 7.3 is for an external proposal.

Cover Letter (or Memo)

Your cover letter for an external proposal or cover memo for an internal proposal prefaces the proposal and provides the reader with an overview of what is to follow. See Chapter 4 for information about memos and cover letters.

Table of Contents

The table of contents should be a complete and accurate listing of the main headings and subheadings covered in the proposal. Avoid providing only an outline of main headings. This could lead to page gaps. Your readers would be unable to find key ideas of interest.

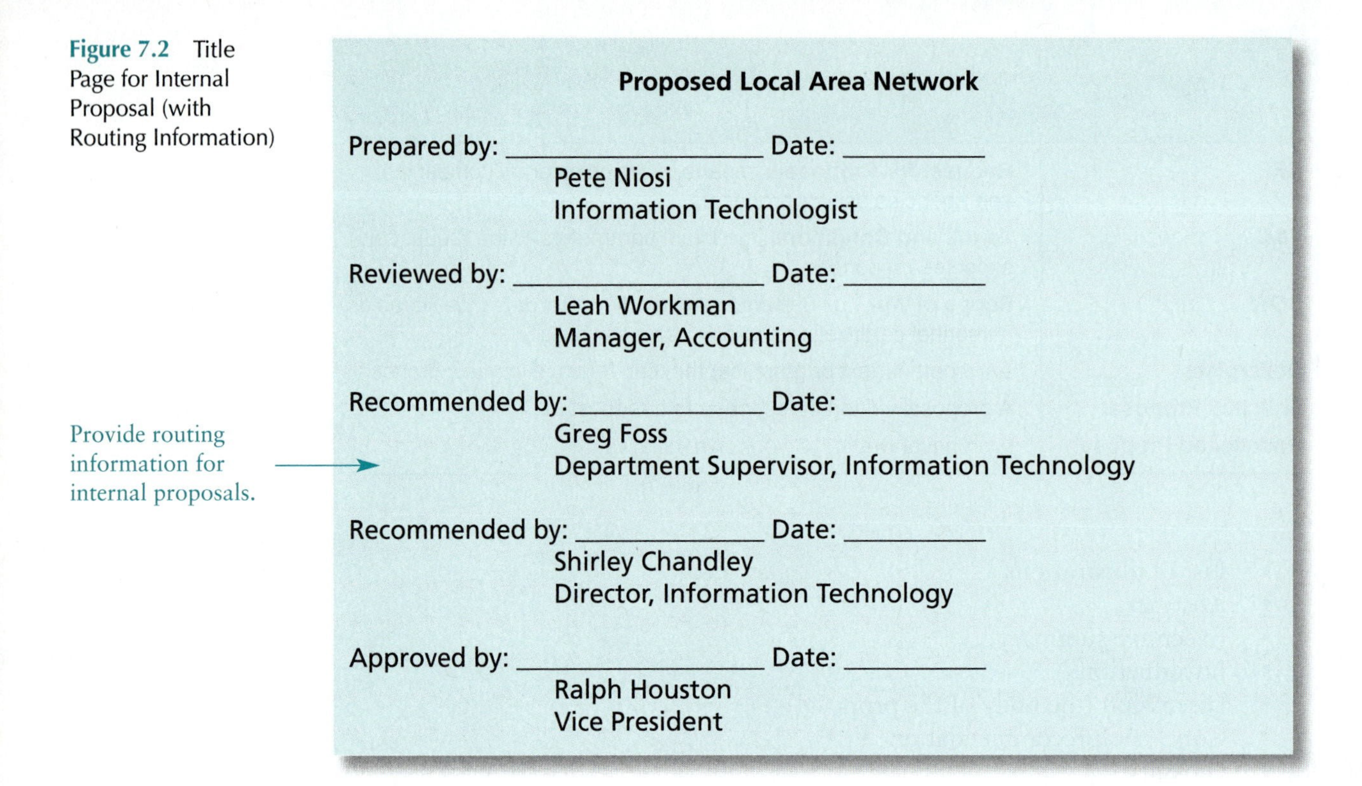

Proposed Local Area Network
Prepared by: ____________________ Date: ___________
Pete Niosi
Information Technologist
Reviewed by: ____________________ Date: ___________
Leah Workman
Manager, Accounting
Recommended by: _______________ Date: ___________
Greg Foss
Department Supervisor, Information Technology
Recommended by: _______________ Date: ___________
Shirley Chandley
Director, Information Technology
Approved by: ____________________ Date: ___________
Ralph Houston
Vice President
Provide routing information for internal proposals.

Figure 7.2 Title Page for Internal Proposal (with Routing Information)

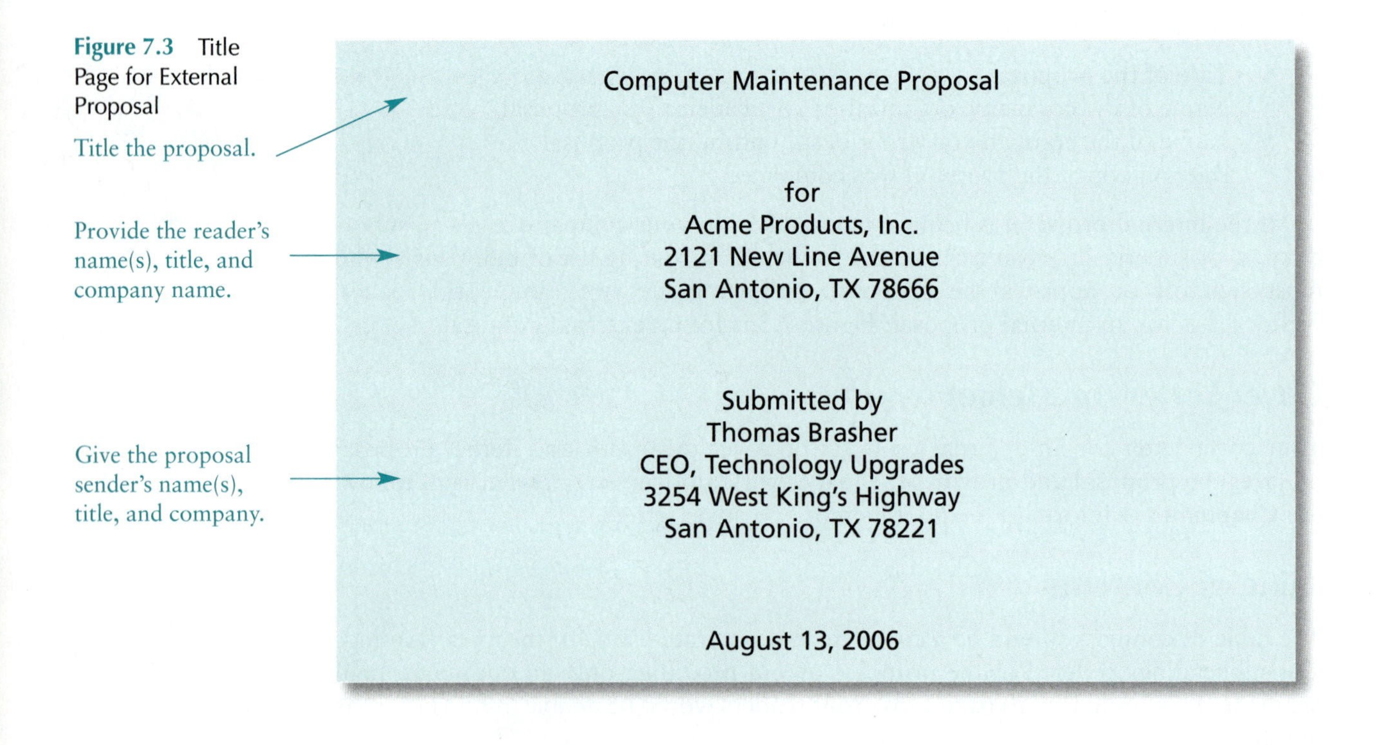

Computer Maintenance Proposal
for
Acme Products, Inc.
2121 New Line Avenue
San Antonio, TX 78666
Submitted by
Thomas Brasher
CEO, Technology Upgrades
3254 West King's Highway
San Antonio, TX 78221
August 13, 2006
Title the proposal.
Provide the reader's name(s), title, and company name.
Give the proposal sender's name(s), title, and company.

Figure 7.3 Title Page for External Proposal

	Table of Contents	
	List of Illustrations	iv
	Abstract	v
1.0	Introduction	1
2.0	Proposal	3
3.0	Conclusion	19
4.0	Recommendations	21
5.0	Glossary	22

Figure 7.4 Flawed Table of Contents

This poorly written table of contents omits second-level headings and has too many page gaps.

Figure 7.4 is a flawed table of contents. In contrast, Figure 7.5 is an effective table of contents.

List of Illustrations

If your proposal contains several tables or figures, provide a list of illustrations. This list can be included below your table of contents, if there is room on the page, or on a separate page. As with the table of contents, your list of illustrations must be clear and informative. Figure 7.6 is a flawed list of illustrations. Figure 7.7 is an effective list of illustrations.

Abstract

The abstract is a brief overview of the proposal's key points geared toward decision makers. Abstracts often focus on but are not limited to the following: the problem necessitating the proposal, the suggested solutions, and the benefits derived when the proposed suggestions are implemented. These three points work for external as well as internal proposals. In the abstract, provide readers with an easy-to-understand summary of the entire proposal's focus.

Abstract

Due to deregulation and the recent economic recession, we must reduce our workforce by 12 percent.

Our plan for doing so involves

- Freezing new hires
- Promoting early retirement
- Reassigning second-shift supervisors to our Desoto plant
- Temporarily laying off third-shift line technicians

Achieving the above will allow us to maintain production during the current economic difficulties.

An effective abstract highlights the problem, possible solutions, and benefits in the proposal.

Figure 7.5 Effective Table of Contents

Second-level and third-level headings help the audience access all parts of the proposal in any order.

Table of Contents

Figure 7.6 Flawed List of Illustrations

This list of illustrations omits figure and table titles. The reader will not know what each graphic illustrates.

List of Illustrations

List of Illustrations

Figure 7.7 Effective List of Illustrations

Executive Summary

The terms "abstract" and "executive summary" are often used interchangeably. However, the distinction between the two terms is usually dependent upon length. Whereas an effective abstract should be very brief, an executive summary might be longer. Not every proposal requires an executive summary. In some instances, the more concise abstract is sufficient. If you choose to include a longer and more detailed executive summary in addition to the abstract, this summary will accomplish the following goals:

- Highlight the proposal's objectives.
- Detail the proposal's significance and scope.
- Outline the proposal's main points.
- Help the readers understand the recommendations.
- Let the reader know what to do next.

Introduction

The introduction to your internal or external proposal should include two key components:

Purpose Statement. In one to three sentences or in a short paragraph, tell your readers the purpose of your proposal. This purpose statement informs your readers why you are writing and what you hope to achieve. It is synonymous with a paragraph's topic sentence, an essay's thesis, the first sentence in a letter, or the introductory paragraph in a shorter report.

> **Purpose Statement** The purpose of this report is to propose the immediate installation of the 102473 Numerical Control Optical Scanner. This installation will ensure continued quality checks and allow us to meet agency specifications.

A purpose statement expresses the goal of the proposal.

Problem (Needs Analysis). Though the purpose statement is limited to only a few sentences, your discussion of the problem or reader's need must be much more detailed. Your introduction's analysis of reader needs, which could average several pages, is important for two reasons. First, it highlights the importance of your proposal.

It emphasizes for your readers the proposal's priority. In this problem section, you persuade your readers that a problem truly exists and needs immediate attention.

Second, by clearly stating the problem, you also reveal your knowledge of the situation. The problem section reveals your expertise. After reading this section of the introduction, the audience should recognize the severity of the problem and trust you to solve it.

Figure 7.8 is a sample introduction stating purpose and needs analysis.

Discussion

The discussion section of a proposal constitutes its body. In this section, you sell your product, service, or suggested solution. As such, the discussion section represents the major portion of the proposal, perhaps 85 percent of the text. Though the content of one proposal will differ from another depending on subject matter, consider including any or all of the following:

- Analyses
  - Existing situation
  - Solutions
  - Benefits
- Product specifications of mechanisms, facilities, or products
- User instructions
- Optional approaches or methodologies for solving the problems
- Managerial chains of command (organizational charts)
- Biographical sketches of personnel
- Corporate and employee credentials
  - Years in business
  - Satisfied clients
  - Certifications
  - Previous accomplishments
- Schedules
  - Implementation schedules
  - Reporting intervals
  - Maintenance schedules
  - Delivery schedules
  - Completion dates
  - Payment schedules
  - Projected milestones (forecasts)
- Cost analyses
- Profit and loss potential
- Warranties
- Maintenance agreements
- Online help
- Training options

Conclusion/Recommendations

Sum up your proposal, providing your readers closure. The conclusion can restate the problem, your solutions, and the benefits to be derived. Your recommendation will suggest the next course of action. Specify when this action will or should occur and why that date is important. Figure 7.9 is a conclusion/recommendation from an internal proposal.

Figure 7.8
Introduction with Purpose Statement and Needs Analysis

1.0 INTRODUCTION

1.1 Purpose Statement

This is a proposal for a storm sewer survey for Yakima, WA. First, the survey will identify storm sewers needing repair and renovation. Then it will recommend public works projects that would control residential basement flooding in Yakima.

1.2 Needs Analysis

1.2.1. Increased Flooding

Residential basement flooding in Yakima has been increasing. Fourteen basements were reported flooded in 2005, whereas 83 residents reported flooded basements in 2006.

Provide specific details to explain the problem. Doing so shows that you understand the reader's need and highlights the proposal's importance.

1.2.2. Property Damage

Basement flooding in Yakima results in thousands of dollars in property damage. The following are commonly reported as damaged property:

- Washers
- Dryers
- Freezers
- Furniture
- Furnaces

Major appliances cannot be repaired after water damage. Flooding also can result in expensive foundation repairs.

1.2.3. Indirect Costs

Flooding in Yakima is receiving increased publicity. Flood areas, including Yakima, have been identified in newspapers and on local newscasts. Until flooding problems have been corrected, potential residents and businesses may be reluctant to locate in Yakima.

1.2.4. Special-Interest Groups

Citizens over 55 years of age represent 40 percent of the Yakima, WA, population. In city council meetings, senior citizens with limited incomes expressed their distress over property damage. Residents are unable to obtain federal flood insurance and must bear the financial burden of replacing flood-damaged personal and real property. Senior citizens (and other Yakima residents) look to city officials to resolve this financial dilemma.

Figure 7.9
Conclusion/Recommendation for Internal Proposal

Summarize the key elements of the proposal.

Recommend follow-up action and show the benefits derived.

3.0 Solutions for Problem

Our line capability between San Marcos and LaGrange is insufficient. Presently, we are 23 percent under our desired goal. Using the vacated fiber cables will not solve this problem because the current configuration does not meet our standards. Upgrading the current configuration will improve our capacity by only 9 percent and still present us the risk of service outages.

4.0 Recommended Actions

We suggest laying new fiber cables for the following reasons. They will

- Provide 63 percent more capacity than the current system
- Reduce the risk of service outages
- Allow for forecasted demands when current capacity is exceeded
- Meet standard configurations

If these new cables are laid by September 1, 2006, we will predate state tariff plans to be implemented by the new fiscal year.

Glossary

Because you will have numerous readers with multiple levels of expertise, you must be concerned about your use of abbreviations, acronyms, and specialized terms. Although some of your readers will understand your terminology, others might not. You must define your terms.

You can define terms parenthetically after each use. If you define your terms each time you use them, however, you delay your audience as they read the text. To avoid this problem, use a glossary. A glossary is an alphabetized list of specialized terminology placed before the introduction, as a part of the introduction, or after your conclusion/recommendation. Figure 7.10 is a sample glossary.

Works Cited (or References)

If you conducted research to write your proposal, you need to include a works cited or reference page. This page documents the sources (books, periodicals, interviews, computer software, or online sources) you have researched and quoted or paraphrased.

Appendix

An optional component to a proposal is an appendix. Appendices allow you to include any additional information (survey results, tables, figures, previous report findings, examples, or relevant correspondence) that you have not used in the discussion of your proposal.

GLOSSARY

BSCE:	Bachelor of Science, Civil Engineering
Drainage Studies:	The study of moving surface or surplus water
Gb:	Gigabyte
Interface:	Communication with other agencies, entities, and systems to discuss subjects of common interest
POP:	Place of purchase
Smoke Test:	Test using underground smoke bombs to give visual, above-ground signs of sewer pipe leaks
Video Scanner:	A portable video camera that examines the inside surface of sewer pipe
Water Parting:	A boundary line separating the drainage districts of two streams
Watershed Area:	An area bounded by water parting and draining into a particular watercourse

Figure 7.10 Sample Glossary

Providing a glossary will help readers understand technical terms.

PROPOSAL CHECKLIST

The proposal checklist will give you the opportunity for self-assessment and peer evaluation of your writing.

Proposal Checklist

__ 1. **Title page:** Does your title page include the proposal's title, audience, author or authors, routing information (if you have written an internal proposal), and date?

__ 2. **Cover letter:** Does the cover letter state why you are writing and what you are writing about, what exactly you are providing the reader(s) within the proposal and what follow-up action is requested or suggested?

__ 3. **List of illustrations:** Does the list of illustrations include all figure and table numbers, titles, and page numbers?

__ 4. **Abstract:** Does the abstract or executive summary concisely state the problem, solution, and benefits or provide an overview of the proposal?

__ 5. **Introduction:** Does the introduction provide a simple statement of purpose and a detailed analysis of the problem?

__ 6. **Discussion:** Does the discussion explain how you will solve the readers' problem by analyzing procedures; performing research; providing specifications, timetables, materials, equipment, personnel, credentials, facilities, alternative methodologies, and costs?

__ 7. **Conclusion/Recommendations:** Does the proposal restate the benefits and recommend follow-up action?

__ 8. **Glossary:** Have you recognized your audience's level of understanding and included an alphabetized glossary to define terminology?

__ 9. **Appendix:** Have you included an appendix (optional) for additional information?

__10. **Highlighting/Page Layout:** Is your text accessible? To achieve reader-friendly ease of access, you need to use headings, boldface, italics, bullets, numbers, underlining, or graphics (tables and figures). These add interest and help your readers navigate your proposal.

Researching Information for Long Reports and Proposals

Research skills are important to help you develop a proposal or long report. Through research you can substantiate your findings or recommendations. Researched information helps you prove your contentions and shows that you have a thorough understanding of the topic.

You can research information using online catalogs, online indexes and databases, CD-ROM indexes and databases, reference books in print, online, and CD-ROM format, and by using Internet search engines and directories.

Books

All books owned by a library are listed in catalogs, usually online. Books can be searched in online catalogs in a variety of ways: by title, by author, by subject, by keyword, or by using some combination of these. No matter how you search for a book, the resulting record will look something like the following example.

Artificial Intelligence: robotics and machine evolution / David Jefferis.

Database:	DeVry University
Main Author:	Jefferis, David.
Title:	Artificial intelligence: ***robotic*** and machine evolution / David Jefferis.
Primary Material:	Book
Subjects:	***Robotics***—Juvenile literature. Artificial intelligence—Juvenile literature. Robots. Artificial intelligence.
Publisher:	New York: Crabtree Pub. Co., 1999.
Description:	32 p. : col. Ill.; 29 cm.
Series:	Megatech
Notes:	Includes index. An introduction to the past, present, and future of artificial intelligence and ***robotics,*** discussing early science fiction predictions, the dawn of AI, and today's use of robots in factories and space exploration.
Database:	DeVry University
Location:	Dallas Main Stacks
Call Number:	TJ211.1.J44 1999
Number of Items:	1
Status:	Not Changed

Periodicals

Use online, CD-ROM, or print periodical indexes to find articles on your topic. Online indexes can be searched in a variety of ways: by title, by author, by subject, by keyword, or by using a combination of these. No matter how you search, the resulting record will look something like the following example.

DH Pro Eric Carter

Bicycling; Emmaus; Mar 2001; Andrew Juskaitis.

Volume:	42
Issue:	2
Start Page:	14
ISSN:	00062073
Subject Terms:	Athletics Bicycling Organizations Bicycle Racing
Personal Names:	Carter, Eric

Abstract:
Juskaitis discusses bicycling with downhill national champion Eric Carter. Only after riding with Carter did he realize he's also at the forefront of changing the face of gravity-fed mountain **bike racing**. Carter has joined with promoter Rich Sutton to push cycling's race organizations to implement four-to-six-rider racing on highly specialized courses.

Full text: . . . (omitted here for copyright issues)

The preceding example from an online periodical database tells us that an article on the subjects Athletes, Bicycling, Organizations, and Bicycle Racing can be found in the March 2001 issue of *Bicycling* magazine, volume 42, issue 2, beginning on page 14.

The article is entitled "DH Pro Eric Carter" and was written by Andrew Juskaitis. The article contains information about a person named Eric Carter. An abstract or summary of the article is given, followed by the full text of the article itself.

Indexes to General, Popular Periodicals

Most libraries provide access to at least one of a number of indexes covering popular, nontechnical literature, and newspaper articles from a variety of subject fields. There are online, CD-ROM, and print counterparts for most of these. The online and CD-ROM indexes provide the full text of many of the articles.

- Reader's Guide to Periodical Literature
- Ebscohost
- SIRS Researcher (emphasizes social issues)
- Newsbank (emphasizes newspaper articles)

Indexes to Specialized, Scholarly, or Technical Periodicals

Many libraries provide access to one or more specialized indexes covering literature from a variety of disciplines. There are online, CD-ROM, and print counterparts for most of these. The online and CD-ROM indexes provide the full text of many of the articles.

- ABI/Inform. Covers business and management.
- Applied Science & Technology Index. Covers engineering aeronautics and space sciences, atmospheric sciences, chemistry, computer technology and applications, construction industry, energy resources and research, fire prevention, food and the food industry, geology, machinery, mathematics, metallurgy, mineralogy, oceanography, petroleum and gas, physics, plastics, the textile industry and fabrics, transportation, and other industrial and mechanical arts.
- Business Periodicals Index. Covers major U.S. publications in marketing, banking and finance, personnel, communications, computer technology, and so on.
- ERIC (Education Resources Information Center). Provides bibliography and abstracts about educational research and resources available for free through the Internet.
- General Science Index. Covers the pure sciences, such as biology and chemistry.
- MEDLINE/Medscape. Covers medical journals and allied health publications.
- NEXIS/LEXIS. Includes the full text of newspaper articles, reports, transcripts, law journals, legal reports, and other reference sources in addition to general periodical articles.
- CINAHL: Cumulative Index to Nursing and Allied Health Literature. Covers topics from a medical viewpoint.
- PsycINFO. Covers psychology and behavioral sciences.
- Social Sciences Citation Index. Covers psychology, sociology, political science, economics, and other social science topics.

The Internet

Perhaps one of the largest sources of research available today is the World Wide Web. Millions of documents from countless sources are found on the Internet. You can find material on the Internet published by government agencies, organizations, schools, businesses, or individuals (see Table 7.2, Internet Research Sources). The list of options grows daily. For example, nearly all newspapers and news organizations have online Web sites.

To find information online, use directories, search engines, or metasearch engines.

Directories

Directories, like Yahoo, AltaVista, HotBot, and Excite, let you search for information from a long list of predetermined categories, including the following:

Arts
Business
Computers
Education
Entertainment
Government
Health and Medicine
Hobbies
Money and Investing
News
Politics and Law
Recreation
Science
Sports
Society and Culture

TABLE 7.2 INTERNET RESEARCH SOURCES

Search Engines, Directories, and Metasearch Engines	Online References	Online Libraries	Online Newspapers	Online Magazines	Online Government Sites
Yahoo	Webster's Dictionary	Library of Congress	*New York Times*	*National Geographic*	United Nations
Excite	Roget's Thesaurus	New York Public Library	CNN	*HotWired*	The White House
Lycos	Britannica Online Encyclopedia	Cleveland Public Library	*USA Today*	*Atlantic*	The IRS
Infoseek	Encyclopedia Smithsonian	Guttenberg Project	*Washington Post*	*The New Republic*	U.S. Postal Service
Alta Vista	The Internet Almanac	Most city and university libraries	Most city newspapers	*U.S. News Online*	First Gov
MetaSearch	The Old Farmer's Almanac			*Time Magazine*	Most states' supreme courts, legislatures, executive offices, and local governments
MetaCrawler				*Ebony Online*	
Google					
Britannica.com					
Northern Light					
Ask Jeeves					
Dogpile					

To access any of these areas, click on the appropriate category and then "drill down," clicking on each subcategory until you get to a useful site.

Search Engines

Search engines, like Google, Northern Light, or Ask Jeeves, let you search millions of Web pages by keyword. Type a word, phrase, or name in the appropriate blank space and press the enter key. The search engine will search through documents on the World Wide Web for "hits," documents that match your criteria. One of two things will happen: either the search engine will report "no findings," or it will report that it has found thousands of sites that might contain information on your topic.

In the first instance, "no findings," you will need to rethink your search strategy. You may need to check your spelling of the keywords or find synonyms. For example, if you want to research information about online writing, you could try typing "writing online," "online writing," "electronic writing," "writing electronically," and other similar terms.

In the second instance, finding too many hits, you will need to narrow your search. For example, if you are researching ethics in business, you cannot type in "ethics." That is too broad. Instead, try "ethics+Enron," "ethics+business," "business ethics," or "ethics+business scandals," and so on.

Metasearch Engines

A metasearch engine, like Dogpile, lets you search for a keyword or phrase in a group of search engines at once, saving you the time of searching separately through each search engine.

Possible Problems with Internet Research

Researching the Internet presents at least two problems other than finding information. First, is the information you have found trustworthy? Paper-bound newspapers, journals, magazines, and books go through a lengthy publication process involving editing and review by authorities. Not all that is published on the Internet is so professional. Be wary. What you read online needs to be from a verifiable source.

Second, remember that although a book, magazine, newspaper, or journal can exist unchanged in print form for years, Web sites change constantly. A site you find today online will not necessarily be the same tomorrow. That is the nature of electronic communication.

ETHICS AND PLAGIARISM

Document your sources correctly. Your readers need to know where you found your information and from which sources you are quoting or paraphrasing. Therefore, you must document this information. Correct documentation is essential for several reasons:

- You must direct your readers to the books, periodical articles, and online reference sources that you have used in your research report or presentation. If your audience wants to find these same sources, they will depend on your documentation. If your documentation is incorrect, the audience will be confused. You want your audience to be able to rely on the correctness and validity of your research.
- Do not plagiarize. Plagiarism is the appropriation (theft) of some other person's words and ideas without giving proper credit. Communicators are often guilty of unintentional plagiarism. This occurs when you incorrectly alter part of a quotation but still give credit to the writer. Your quotation must be exactly the same as the original word, sentence, or paragraph. You cannot haphazardly change a word, a punctuation mark, or the ideas conveyed. Even if you have cited your source, an incorrectly altered quotation constitutes plagiarism.
- On the other hand, if you intentionally use another person's words and claim them as your own, omitting quotation marks and source citations, you have committed theft. This is dishonest and could raise questions about your credibility or the credibility of your research. Teachers, bosses, and colleagues will have little, if any, respect for a person who purposely takes another person's words, ideas, or visuals. It is essential, therefore, for you to cite your sources correctly.

Ethical business communicators carefully cite the source of their material. In their "Ethical Principles for Technical Communicators," the Society for Technical Communication states that "Before using another person's work, we obtain permission. We attribute authorship of material and ideas only to those who have made an original and substantive contribution" ("Code for Communicators"). The International Association of Business Communicators Code of Ethics for Professional Communicators states the following: "Professional communicators give credit for unique expressions borrowed from others and identify the sources and purposes of all information disseminated to the public" ("Code of Ethics").

Citing Sources

To document your research correctly, you must

- provide parenthetical source citations.
- supply a works cited page according to the Modern Language Association (MLA). See Appendix B: Works Cited or References.
- supply a references page according to the American Psychological Association (APA). See Appendix B: Works Cited or References.

Parenthetical Source Citations

The Modern Language Association and the American Psychological Association use a simplified form for source citations. Before 1984, footnotes and endnotes were used in research reports. In certain instances, in-text superscript numbers, footnotes, and endnotes are still correct (Chicago Style Manual). If your boss or instructor requests footnotes or endnotes, you should still use these forms. However, the most modern approach to source citations requires only that you cite the source of your information parenthetically after the quotation, paraphrase, or visual.

MLA Format

One Author. After the quotation or paraphrase, parenthetically cite the author's last name and the page number of the information.

> "Viewing the molecular activity required state-of-the-art electron microscopes" (Heinlein 193).

Note that the period follows the parenthesis, not the quotation. No comma separates the name from the page number and no lowercase *p* precedes the number.

Two Authors. After the quotation or paraphrase, parenthetically cite the authors' last names and the page number of the information.

> "Though *Gulliver's Travels* preceded Moll Flanders, few scholars consider Swift's work to be the first novel"
> (Crider and Berry 292).

Three or More Authors. Writing a series of names can be cumbersome. To avoid this, if you have a source of information written by three or more authors, parenthetically cite one author's name, followed by *et al.* (Latin for "and others") and the page number.

> "Baseball isn't just a sport; it represents man's ability to meld action with objective—the fusion of physicality and spirituality"
> (Norwood, et al. 93).

Anonymous Works. If your source has no author, parenthetically cite the shortened title and page number.

> "Robots are more accurate and less prone to errors caused by long hours of operation than humans"
> ("Useful Robots" 81).

APA Format

One Author. If you do not state the author's name or the year of the publication in the lead-in to the quotation, include the author's name, year of publication, and page number in parenthesis, after the quotation.

> "Izzy's stay in Palestine was hardly uneventful"
> (Cottrell, 1992, p. 118).

Page numbers are included for quoted material. The writer determines whether page numbers are included for source citations of summaries and paraphrases.

Two Authors. When you cite a source with two authors, always use both last names with an ampersand (&).

> "Line charts reveal relationships between sets of figures"
> (Johnson & Garcia, 1992, p. 158).

Three or More Authors. When your citation has more than two authors but fewer than six, use all the last names in the first parenthetical source citation. For subsequent citations, list the first author's last name followed by *et al.*, the year of publication, and, for a quotation, the page number.

> "Two-party politics might no longer be the country's norm next century"
> (Conners et al., 1993, p. 2).

Anonymous Works. When no author's name is listed, include in the source citation the title or part of a long title and the year. Book titles are underlined, and periodical titles are placed in quotation marks.

"Robots are more accurate and less prone to errors caused by long hours of operation than humans"
("Useful Robots," 2004, p. 81).

The Writing Process at Work

Prewriting

You can gather data and determine your objectives for writing a long report in the following ways:

- Listing/brainstorming to itemize the key components of your long report or proposal.
- Reporter's questions (who, what, when, where, why, and how) to help you gather data for any of the sections in your long report or proposal.
- Flowcharting to organize procedures or schedules.
- Mind mapping/clustering to organize your topic and long report units of discussion.
- Organizational charts for chains of command and personnel responsibilities.
- Outlining to list and prioritize your proposal components.
- Researching to find information for your long report or proposal. Research the Internet, interview vendors or clients, attend conferences, or survey employees or customers.

Writing

Review your prewriting. Double-check your listing, brainstorming, mind mapping, flowcharting, reporter's questions, interviews, research, or surveys. Do you have all of the necessary information to write the long report? If not, you might need to conduct further research, more surveys, or interview additional people.

Organize the data. Each of your long report's sections will require a different organizational pattern. Following are several possible approaches.

- Abstract: problem/solution/benefit or an overview of the proposal
- Introduction: cause/effect (The problem unit is the cause of your writing; the purpose statement represents the effect.)

- Main text: This unit will demand many different methods of organization, including
  - Analysis (cost charts, approaches, managerial chains of command, personnel biographies)
  - Chronology (procedures, scheduling)
  - Spatial (descriptions)
  - Comparison/contrast (optional approaches, personnel, products)
  - Conclusion/recommendation: analysis and importance (Organize your recommendations by importance to highlight priority or justify need.)

Rewriting

After writing a rough draft of your long report, edit and revise it for accuracy and professionalism.

Add Detail for Clarity. In addition to rereading your rough draft and adding the missing *who, what, when, where, why*, and *how*, add your graphics. Return to each section of your long report and determine where you could use any of the following:

- Tables. Your cost section lends itself to tables.
- Figures. Your introduction's problem analysis and any of the main text sections could profit from the following figures:
  - Line charts (excellent for showing upward and downward movement over a period of time. A line chart could be used to show how a company's profits have decreased, for example.)
  - Bar charts (effective for comparisons. Through a bar or grouped bar chart, you could reveal visually how one product, service, or approach is superior to another.)
  - Pie charts (excellent for showing percentages. A pie chart could help you show either the amount of time spent or amount of money allocated for an activity.)
  - Line drawings (effective for technical specifications)
  - Photographs
  - Flowcharts (a successful way to help readers understand procedures)
  - Organizational charts (excellent for giving an overview of managerial chains of command)

See Chapter 3 for a discussion of various types of graphics.

Delete Dead Words and Phrases for Conciseness. Make your text more readable through easy-to-understand language. Avoid wordiness, long sentences, and long paragraphs. Chapter 2 gives detailed explanations for achieving conciseness.

Simplify Bureaucratic Words. Use vocabulary that your readers can understand easily. Avoid words like *supersede*, writing *replace* instead. Check your text to see whether you have used words that are unnecessarily confusing.

Move Information. You might analyze according to importance, set up schedules and procedures chronologically, describe spatially, and so forth. Revise your proposal to ensure that each section maintains the appropriate organizational pattern.

Reformat for Reader-Friendly Ease of Access. To avoid wall-to-wall words, revise your proposal by reformatting.

Enhance the tone of your long report or proposal. Enhance the tone of your text by using pronouns and positive, motivational words. In addition, sell your ideas by providing persuasive information. Show reader benefit; provide facts, figures, and testimony; use logic, ethics, and emotion to sway the audience; and urge action.

SAMPLE INTERNAL PROPOSAL

The following internal proposal emphasizes a company's need for technology upgrades and new computers. See Figure 7.11 for a sample internal proposal.

Figure 7.11
Internal Proposal

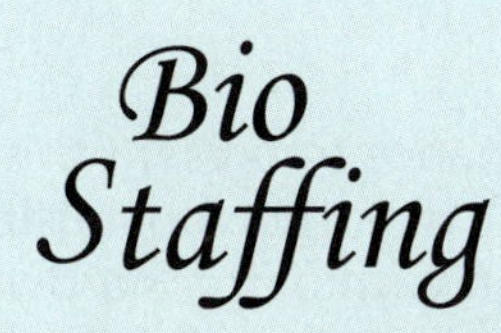

Your One-Stop Shop for Biomedical Needs

Technology Support Proposal

Submitted to
Leann Towner
Chief Financial Officer

By
Jonathan Bacon
Manager, Information Technology Department

June 12, 2006

(continued)

Figure 7.11
continued

Date: June 12, 2006
To: Leann Towner, Chief Financial Officer
From: Jonathan Bacon, Information Technology Manager
Subject: Proposal for New Corporate Technology Support

Leann, in response to your request, enclosed is a proposal to update BioStaffing's technology. The IT department has researched this topic thoroughly and is happy to recommend a revised technology policy, improved hardware and software, and a new vendor.

Among detailed analyses of our current system and proposed options, this proposal presents the following:

Thank you, Leann, for giving the IT department a chance to present this proposal. We are confident that our suggestions will maintain BioStaffing's competitive edge and increase employee satisfaction with better hardware and software. If I can answer any questions, either call me (ext. 3625) or e-mail me at jbacon@biostaff.com.

The cover memo reminds the reader why the proposal has been written and highlights key topics discussed.

Writer's Insight

When writing internal proposals, Jonathan says, "I follow a step-by-step procedure: prewriting, writing, and rewriting. First, I meet with my staff to determine what needs the company has. Often, these needs stem from e-mail messages I and my staff receive from concerned employees, frequently asked questions from the company's intranet site, or habitual problems with office equipment (desktop PCs, laptops, printers, and more). Once these corporate needs are identified, I organize my material, modeling my content according to the company's proposal boilerplate. Finally, I revise the proposal by adding visual appeal, making the proposal reader-friendly and persuasive, and checking the content for accuracy."

Figure 7.11
continued

Table of Contents

ii

(*continued*)

Figure 7.11
continued

Both Figure/Table numbers and titles are given for easy reference and clarity.

List of Illustrations

iii

An abstract often presents the problem necessitating the proposal, provides the proposed solutions, and shows the potential benefits..

1.0 Abstract

1.1 Problem

BioStaffing's current technology policies, hardware and software, and vendor agreements are inefficient. Our technology policies are not responsive to our evolving needs, our hardware and software are outdated, and our current vendor is changing ownership.

1.2 Solution

To solve these problems, the Information Technology (IT) department suggests the following:

- Upgrading our technology needs at least biannually instead of the current five-year technology changeover policy
- Purchasing new computers, printers, scanners, and digital cameras
- Hiring a new vendor to supply and repair our hardware and software

1.3 Benefits

Reviewing our technology needs frequently will help BioStaffing avoid costly repairs due to overused equipment. A biannual replacement schedule will also allow us to stay current with hardware and software advancements.

In addition, purchasing new computers, printers, scanners, and digital cameras will help our staff more effectively meet client needs.

1

Figure 7.11
continued

Hiring a new technology vendor is key to these goals. Our current vendor agreement ends June 30, 2006. By hiring a new vendor with better turnaround, pricing, maintenance, and merchandise, BioStaffing will increase customer and employee satisfaction.

2.0 Introduction

2.1 Purpose

The purpose of this proposal is to improve BioStaffing's current technology. By revising our policies, purchasing new hardware and software, and hiring a new technology vendor, we can increase both customer and worker satisfaction.

Pronouns such as "our" and "we" are appropriate for an internal proposal written to coworkers.

2.2 Problem

2.2.1 *Technology Policies*

Since 1997, we have replaced hardware and software on a five-year, rotating basis. This was an effective policy initially. In the late 1990s through the early 2000s, costs were high for new computers and software upgrades, and our technology needs were limited. Therefore, it made sense financially to change our hardware and software on an ongoing, regular basis.

However, times have changed. The current policy is not responsive to our needs, as follows:

- *Prices have gone down.* Last decade's technology prices were cost prohibitive. BioStaffing correctly chose to upgrade hardware and software only every five years. When every desktop computer cost over $2,000, for example, replacing them often did not make good business sense.

  Today, however, costs are more affordable and flexible. First, desktop computers can be purchased for approximately $1,000. Laptops can be bought for around $1,500. More importantly, handheld computers (Personal Digital Assistants [PDAs] or Pocket PCs) with wireless Internet access can be bought for under $400.

  In addition, vendors making bulk purchases will offer BioStaffing incentives. We can buy computers on an as-needed basis cost effectively by taking advantage of special sales pricing.

  In contrast, our current five-year hardware/software replacement policy is not responsive. It disregards today's changes in technology pricing, and it disallows us from taking advantage of dealer incentives.

2

(*continued*)

Figure 7.11
continued

- *Repair costs have gone up.* By replacing technology only every five years, BioStaffing has resorted to repairing and retrofitting outdated equipment. Costs for these repairs have increased more than fivefold over the last ten years, as noted in the following line graph (Figure 1).

Figure 1: Increase in Costs due to Repairs and Retrofitting Per Workstation

The line graph provides visual appeal and aids clarity by showing the trend in long-term cost increases.

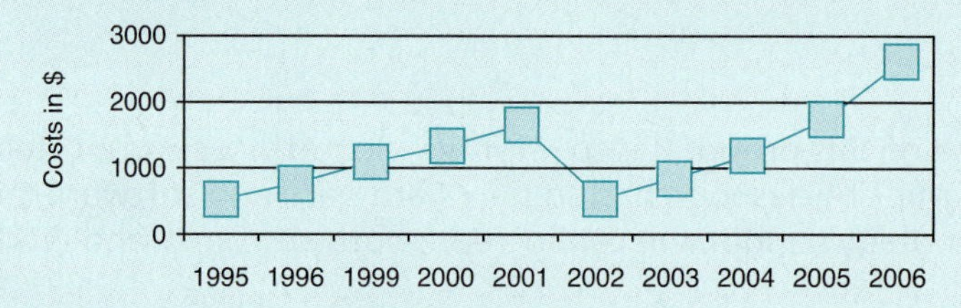

NOTE: The dip between 2001 and 2002 occurred at the five-year replacement schedule.

The talking headings, such as "BioStaffing's technology needs have expanded," summarize the content that follows.

- *BioStaffing's technology needs have expanded.* Our current five-year policy was based on our limited technology needs. In 1997, all BioStaffing needed for corporate communication purposes was one computer per department and one dot matrix printer located on every floor. No one in 1997 envisioned how technology would change communication needs.

  Today, every BioStaffing employee needs a desktop computer, Internet access, and his or her own laser printer. This is necessary due to our employees' increased correspondence responsibilities and the emergence of e-mail as a primary communication vehicle. In addition, to be productive, our employees need a fax machine and scanner in every department, at least one laptop computer per department (to be checked out for travel purposes), and/or handheld computers for work off site. Finally, to improve our user manuals, we need five digital cameras (using up-to-date jpg. images of our equipment to replace outdated line drawings).*

  We cannot wait for the completion of a five-year replacement cycle to make these purchases. Instead, technology purchases must be made as needed.

2.2.2 *Outdated Hardware and Software*

We need to purchase new technology items because our current hardware and software are outdated.

- *Hardware Limitations*
  Our current hardware is limited in several key ways:

---

* This and other terms marked with an asterisk (*) are defined in the glossary.

3

Figure 7.11
continued

1. We have insufficient memory in our computers. Most of our computers have only 2.0 GHz processors with approximately 256 MB of RAM versus the current minimal standard 3.0 GHz and 512 MB. This negatively impacts speed of document retrieval.*
2. Our computers' monitor sizes are limited. The majority of our 15-inch monitors provide only 13.8 inches of viewable screen. This is sufficient for word processing but not for the creation of our newsletters, annual reports, and Web site. Correspondence with more graphical content requires larger viewable screens. Thus, we need to upgrade to 17-inch flat panel monitors (15.9 inch viewable screens), if not 19-inch (18 inch viewable screens) plasma panels. Doing so would allow our corporate communication personnel to create more visually appealing documents.
3. Our ink jet printers are slow and produce poor quality documents. Our current black-and-white printers produce only 7 ppm with a dpi resolution of 1440 X 720.* In contrast, new black-and-white laser printers will produce up to 20 ppm at a dpi resolution of 5760 X 1440. Thus, we would gain speed for quicker turnaround time as well as improved readability (Cooper).

To clarify the importance of ppm, look at Figure 2. This bar chart shows the difference 20 ppm versus 7 ppm makes in terms of time. The figure is based on the 4,000 pages of documentation BioStaffing produces every day.

At 7 ppm (times 60 minutes per hour), BioStaffing personnel can print 420 pages per hour. When we divide the 4,000 pages per day by our current 420-page hourly capacity, you can see that printing requires **9.5 hours.**

In contrast, at 20 ppm (times 60 minutes per hour), improved printers can produce 1,200 pages per hour. Thus, at 4,000 pages divided by 1,200 pages per hour, we can accomplish the same printing task in only **3.3 hours.**

Laser printers can save our company over 6 hours of lost time each day.

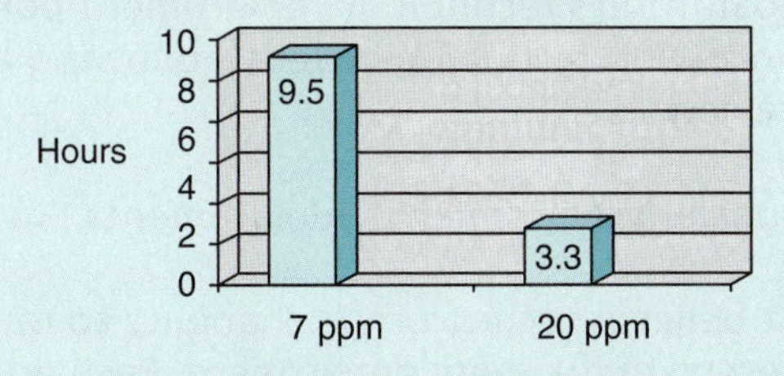

Figure 2: Time Spent Printing Based on PPM

4. We have no laptop computers, no handheld computers, no scanners, and no digital cameras. Without portable computers, our employees engaged in work-related travel cannot communicate effectively with customers or BioStaffing personnel. The lack of scanners and digital cameras is hurting our graphic designers' abilities to create quality documentation.

4

(*continued*)

Figure 7.11
continued

- *Software Limitations*

 Desktop publishing options have expanded rapidly. We could only word-process in-house in the late 1980s. We outsourced our marketing collateral (brochures, newsletters, annual reports, etc.). Now our needs have changed. Today, to create in-house proposals for clients, annual reports to our stockholders, and a Web site, we need improved software. Updated software will allow BioStaffing to import data via spreadsheets; enhance text with tables, figures, and downloadable clip art; and automatically generate tables of contents, glossaries, and indexes. To send newsletters to our employees and stakeholders, we need improved graphical and layout capabilities.

 Our current software does not allow us to create the quality documentation BioStaffing requires. We are losing our competitive edge due to poor, outdated software.

2.2.3 *Vendor Changes*

Our current technology vendor is changing ownership. After having been in business for 20 years, Business Sourcing is being sold to an offshore company, International Technologies. Our technology agreement will end as of June 30, 2006. We must renegotiate a technology contract.

 This offers us a window of opportunity. We have maintained a long-standing relationship with Business Sourcing since it also was one of our primary clients. Though they did not necessarily provide us the best pricing, we continued the arrangement for business purposes. Now, through a new vendor, we can improve quality of service as well as pricing (Norton).

A variety of highlighting techniques, such as italicized headings, textboxes, and bullets, help the reader navigate a long report.

3.0 Discussion

3.1 Revised Technology Replacement Policy

We must revise BioStaffing's technology replacement policy. It is not responsive to our growing technology needs. Following are two proposed options for your review:

3.1.1 *Option 1: Quarterly Assessments/Replacements*

The IT department believes technology is changing so rapidly that quarterly assessments are necessary for every department. Each quarter, we suggest that every department submit a Technology Needs Assessment (TNA). In this TNA, the departments would

- Prioritize their technology needs
- Explain how the technology will be used (emphasizing value added to BioStaffing)
- List the make and model of the desired hardware and software
- Provide costs

5

Figure 7.11
continued

Then, a panel (composed of representatives from all departments) will review the reports. Based on a 10-point scale (10 = essential; 1 = not essential), the panel will determine which proposals will be supported that quarter.

> **NOTE:** To ensure that all department needs are treated equally, any department that was not given technology upgrades in the prior review would receive two additional "carry over points."

Option 1 requires competition on the part of each department. In addition, it will entail added expense for BioStaffing. However, the importance of technology in our business mandates this level of responsiveness.

3.1.2 *Option 2: Biannual Rotating Replacements*

Though Option 2 is less responsive than Option 1 for technology needs, it would be more cost-effective for the company and more fair for all departments.

We propose that BioStaffing divide its 12 departments into two categories. The "Red" team would receive technology upgrades one year (even numbered years), while the "Blue" team would receive technology upgrades the next year (odd numbered years).

IT has divided the departments into teams alphabetically for fairness, as follows (Table 1):

Table 1: Biannual Rotating Technology Replacements

Red Team (*even* numbered years)	Blue Team (*odd* numbered years)
Accounting	Legal
Administration	Manufacturing
Administrative Services	Maintenance
Benefits	Personnel
Corporate Communication	Sales
Information Technology	Shipping and Receiving

3.2 Hardware Purchases

After researching department requirements, IT determined that BioStaffing needs to upgrade business application hardware as shown in Table 2 ("Allocation").

3.2.1 *Hardware Allocation Analysis*

- **Accounting.** 2 desktops, 1 laptop, 5 handhelds, and 1 laser printer
- **Administration.** 1 laptop and 4 handhelds
- **Administrative Services:** 3 desktops, 1 laptop, 1 handheld, and 1 laser printer

6

(*continued*)

Figure 7.11
continued

Table 2: Business Application Hardware Requirements	
Hardware	**Minimum Specifications (prices approximate)**
15 desktop computers	Intel® Pentium® 4 processor, 2.6 GHz* 512 MB RAM 80 GB HD CD drives 17″ color monitor (15.9″ viewable) $1,000 each
12 laptop computers	Intel® Celeron® processor, 2.2 GHz 128 MB RAM 20 GB HD ROM drive $1,200 each
50 handheld computers	32 MB RAM Hotsyncs to Outlook Wireless e-mail and Internet access Thumb-typable keyboard Compatible with Word, PowerPoint, and Excel $150 each
5 laser printers	16 MB RAM 20 ppm Letter, legal, #10 envelopes 133 MHz processor 250-sheet paper tray $400 each
3 scanners	2400 X 4800 dpi resolution Adapter for slides, negatives Auto retouch, enhancement Copy, scan, e-mail, and file $170 each
5 digital cameras	Jpg file format Auto focus 3X optical zoom 32 MB flash card $200 each

("Hardware/Software")

- **Benefits.** 1 desktop, 1 laptop, 4 handhelds, and 1 scanner
- **Corporate Communication.** 1 laptop, 7 handhelds, 1 scanner, 1 laser printer, and 2 digital cameras
- **Information Technology.** 1 laptop, 7 handhelds, 1 scanner, 1 laser printer, and 2 digital cameras
- **Legal.** 1 desktop, 1 laptop, 2 handhelds, and 1 scanner

7

Writer's Insight

Jonathan says, "To persuade my audience, I do tons of research. For this internal proposal, I knew that my audience would only be convinced by solid facts and figures. So, I researched the topic thoroughly. Notice how the proposal's focus on hardware needs and costs has been precisely detailed. Through interviews and surveys, I found out exactly how many desktops, handhelds, laptops, cameras, and scanners each department needed. Then, I itemized the costs in Table 2. Before any decisions could be made, I knew our corporate accountants would need these facts."

Figure 7.11
continued

- **Manufacturing.** 2 desktops, 1 laptop, 2 handhelds, and 1 laser printer
- **Maintenance.** 1 desktop, 1 laptop, and 2 handhelds
- **Personnel.** 2 desktops, 1 laptop, and 4 handhelds
- **Sales.** 2 desktops, 1 laptop, 10 handhelds, and 1 digital camera
- **Shipping and Receiving.** 1 desktop, 1 laptop, and 2 handhelds

The above allocations are based on department size, need, location, and recent upgrades. For example, departments located on the same floors can share equipment. Manufacturing, Shipping/Receiving and Maintenance would share a laser printer. Accounting would share a laser printer with Benefits and Legal. Administration and Administrative Services would share a laser printer.

Neither Corporate Communication nor Information Technology are requesting new desktop computers since they received upgrades last year. Only Corporate Communication, Information Technology, and Sales need digital cameras to complete their jobs. Similarly, the scanners are necessary only in Legal, Information Technology, Corporate Communication, and Benefits.

3.3 Software Purchases

To improve our business communication and create professional-looking marketing collateral, BioStaffing needs to upgrade software as shown in Table 3:

Table 3: Software Requirements for Business Communication and Marketing Collateral

Purposes	Software Required	Approximate Cost
Letters, Memos, Reports,	Microsoft Office Suite 2003 (Word, Excel, PowerPoint)	$300
Brochures, Newsletters,	Microsoft Office Suite	Listed above
Annual Reports	Quark Express	$650
	Adobe Photoshop	$250
Oral Presentations,	PowerPoint (included in	Listed above
Proposals,	Microsoft Office Suite)	
Web Design	Macromedia Suite (Dreamweaver, Flash, ColdFusion)	$950
Online Help	RoboHelp	$50

8

(*continued*)

Figure 7.11
continued

3.3.1 *Cost-Effective Purchasing*

The above software is cost-effective:

- These purchases benefit every department in BioStaffing.
- With site licenses, we can purchase only one software package and install it on all computers throughout the company.
- We are proposing the purchase of the most up-to-date, state-of-the-art software. Therefore, Information Technology anticipates that we will not need to upgrade these purchases for approximately five years.

3.4 Technology Vendor Options

Because BioStaffing's vendor agreement with Business Sourcing ends June 30, 2006, Information Technology has researched new vendor options. Our primary criteria included *pricing, quality, delivery time, maintenance, warranties,* and *customer service*. Based on these criteria, following is an overview of our findings (Table 4):

Table 4: Technology Vendor Options

Companies	Pricing	Delivery Quality	Time	Maintenance	Customer Warranties	Service
BizTech Warehouse	20% bulk discount Hardware: $18,730 Software: $1,760	Remanufac-tured name-brand hardware	Same-day delivery at no extra cost.	24-hour maintenance options at additional hourly cost—negotiable. 8:00 a.m. – 5:00 p.m. maintenance for products under warranty.	90 days. Renewable at negotiable price.	24/7 hotline. Fleet of maintenance trucks.
Technologies Today	No discount Hardware: $23,400 Software: $2,200	New name-brand hardware	Same-day delivery for home office. Overnight FedEx for other sites.	Manufacturer's maintenance (1-800 number and/or nearest dealership).	1-year manufacturer's warranty.	8:00 a.m. – 5:00 p.m. M-F office help.
***SOS* (Super-Saver Office Supplies)**	10% bulk discount Hardware: $21,000 Software: $1,980	In-house store brand hardware	Overnight FedEx.	1-800 number and/or nearest dealership.	3-year extended warranties, parts and labor.	8:00 a.m. – 5:00 p.m. M-F office help. Answering service on weekends.

9

Figure 7.11
continued

3.4.1 *Assessment of Vendor Options*

- **Pricing:** BizTech Warehouse provides the best pricing. This is an especially important factor when we consider the rapidly changing nature of technology and the constant need to upgrade.
- **Quality:** Technologies Today provides new, name-brand hardware. This is appealing. However, remanufactured hardware can be as good as new. We do not have enough data on SOS's store-brand hardware to judge its value.
- **Delivery Time:** With 25 locations nationwide, including Topeka, our home office's city, BizTech provides same-day delivery. This is an important factor considering our plans to open new office locations.
- **Maintenance:** Again, BizTech provides the most prompt maintenance service. It comes at a cost, but their vendors have assured us that we will receive a substantial discount as well as product rebates if we choose their company.
- **Warranties:** SOS's 3-year warranty is outstanding, as is Technologies Today's 1-year guarantee. However, if we plan to upgrade our hardware biannually, a 3-year warranty loses its value.
- **Customer Service:** BizTech's 24/7 accessibility is perfect for our workplace. We run our services around the clock. The other vendors do not provide the kind of customer service our business requires.

Information Technology evaluated the vendor options, giving each company points for their services, as follows:

5 = Excellent 4 = Good 3 = Average 2 = Poor 1 = Unacceptable

Based on this scale, Figure 3 shows our evaluation findings:

Figure 3: Vendor Evaluation Matrix

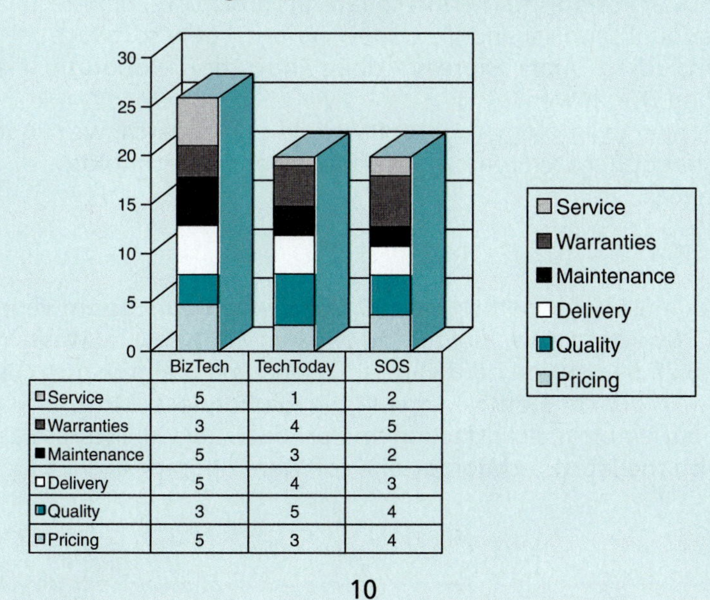

	BizTech	TechToday	SOS
Service	5	1	2
Warranties	3	4	5
Maintenance	5	3	2
Delivery	5	4	3
Quality	3	5	4
Pricing	5	3	4

10

(*continued*)

Figure 7.11
continued

To highlight the proposal's importance, the writer emphasizes critical needs.

The recommendation proposes the next course of action needed to solve the company's problems.

4.0 Conclusion

4.1 Conclusion

BioStaffing's technology needs are critical.

- Repair costs due to old hardware have risen fivefold per workstation, from as low as $500 to as high as $2,500.
- Outdated printers, producing only 7 pages per minute, require up to 9.5 hours a day to print BioStaffing's 4,000 pages of documentation.
- Insufficient software for business communication applications and marketing collateral are hurting our competitive edge.
- Expired technology vendor contracts need to be renegotiated.

Key to the above challenges is our current 5-year technology replacement policy. Technology advances demand a more responsive policy.

4.2 Recommendation

The Information Technology department, based on research and surveys, proposes the following:

- A quarterly or biannual technology replacement policy: This would allow BioStaffing to meet employee technology needs more responsively as well as benefit from vendor pricing incentives.
- New communication hardware: This would include desktop, laptop, and handheld computers; laser printers; scanners; and digital cameras.
- Upgraded software: To improve business communication needs, all computers must have access to Microsoft Office Suite. For our marketing collateral (brochures, newsletters, annual reports, and Web site), we need new software including Quark Express, Adobe Photoshop, RoboHelp, Flash, and Dreamweaver.
- A new technology vendor: Based on our research, we suggest that BizTech Warehouse will best meet our technology needs.

4.3 Benefits

These changes must occur before June 30, 2006, when our current vendor contract ends. By acting now, BioStaffing will benefit in several ways. We can save up to 20 percent on hardware, software, and maintenance costs. We will maintain our competitive edge in the marketplace. Most importantly, our employee satisfaction and productivity will increase as they work with the latest technology and software upgrades.

11

Figure 7.11
continued

5.0 Glossary

Acronym/Abbreviations	Definition
CD	Compact Disk
dpi	Dots per inch
GHz	Gigahertz
GB	Gigabyte
HD	Hard drive
jpg	Joint photographic group
MB	Megabyte
MHz	Megahertz
ppm	Pages per minute
ROM	Read only memory
RAM	Random access memory

Though all readers might not need terms defined, a glossary provides clarity to a less specialized audience.

6.0 Works Cited

"Allocation of Technology Needs." Survey. May 1, 2006.

Cooper, Jack. AAA Computers. Interview. May 15, 2006.

"Hardware/Software Specifications and Costs." *PCWorld.* 12 May 2006. http://www.pcworld.com.

Norton, Susan. *BioStaffing Annual Report.* 15 Jan. 2006. 31 May 2006. http://www.biostaff.com/annualreport.htm.

12

CASE STUDIES

1. The Corporate Communication department at BioStaffing needs to improve the company's internal communication capabilities. Currently, BioStaffing has no in-house newsletter, intranet site, or blogosphere. Without these communication channels, employees cannot easily communicate with each other or with teams. To succeed in the workplace, provide rapid communication channels, allow for give and take, and build rapport, the Corporate Communication department needs to write an internal proposal.

   As Corporate Communication manager, you have consulted with your five staff members (Jim Nguyen, Mario Lozano, Mike Thurmand, Amber Badger, and Maya Liu) to correct these problems. As a team, you have decided the company needs permission, funding, and technical support to create the following:

   - **A corporate newsletter.** This in-house document will be used to inform employees of birthdays, upcoming events (such as picnics, meetings, and training opportunities), product releases, employee changes, and revised benefits packages. The communication's team also wants this newsletter to be a fun site, complete with games, photographs, and jokes. The goal is to inform, entertain, and build rapport.
   - **Intranet site.** An electronic intranet site would provide research databases, forms, multimedia kiosks for instructional and informational purposes, policy and procedure manuals, and employee directories. An intranet would allow BioStaffing personnel access to a one-stop, internal, electronic resource center.
   - **Corporate blog.** The latest development in electronic communication for many companies is the blogosphere. Through Web logs, employees can quickly and conversationally communicate in discussion forums, participate in online polls, report on project progress, and keep up-to-date with colleagues.

Assignment

Using the criteria provided in this chapter, write an internal proposal to BioStaffing's CEO, Jim McWard. In this proposal, explain the problem, discuss the solution to this problem, and then highlight the benefits derived once the solution has been implemented. These benefits will include increased productivity, better corporate relations, increased work production, and greater employee rapport. Develop these points thoroughly and provide Mr. McWard the names of vendors for any required hardware and software. To find these vendors or technology solutions and to support your decisions, you could search the Internet, survey employees, and research periodicals.

2. RavenWood Homes, a neighborhood housing development, wants to upgrade its property. The homes' association has decided to enhance its entrance with a fountain, improve the area's landscaping, and add more street lighting for security as well as esthetics.

   RavenWood wants the entrance fountain to be made out of black granite, to stand 12 feet high, and to have the water cascade over the housing development's name, etched into the fountain's base. The name must be at least 3 feet high for visibility. Lighting should illuminate the fountain for night viewing.

The homes' association believes it needs at least 125 10-foot tall red maple trees to line both sides of RavenWood's main boulevard, Woodline Dr. This street must have 50 standard streetlights. RavenWood will also need a maintenance contract for grass cutting and tree pruning, as well as for the fountain's upkeep.

Assignment

As a landscaping architectural/engineering company, write the external proposal for RavenWood. Include pricing, a timetable, your credentials, and a list of satisfied customers. For additional information and proof to support your decisions, you could search the Internet and research periodicals.

Individual and Team Projects

1. Write an external proposal. To do so, create a product or a service and sell it through a proposal. Your product can be an improved radon detection unit, new cell phone options, services for an advertising agency, a party planning company, a temporary employment service, or in-home care for the elderly. Your service may involve dog grooming, computer maintenance, home construction (refinishing basements, building decks, or room additions), at-home occupational therapy, or telemarketing. The topic is your choice. Draw from your job experience, college coursework, or hobbies.

2. Write an internal proposal. You can select a topic from either work or school. For example, your company is considering a new venture, such as a corporate health and wellness program. Maybe your company wants to cut computer hardware or software costs, expand into a new facility, or add a new product or service. Perhaps your company needs to be more accessible to customers and co-workers with disabilities. You need to write an internal proposal to management presenting options and recommending the next course of action.

   If you choose a topic from school, you could propose a day care center, on-campus bus service, improved computer facilities, tutoring services, coed dormitories, pass/fail options, and so on. Write an internal proposal to your campus club, fraternity, or sorority. You could propose a fund-raising event, an end-of-semester party, a new method for recruiting members, or the need to outlaw prejudicial or harmful practices.

   Research your topic by reading relevant information or by interviewing/surveying students, faculty, staff, or management. Once you have gathered your data, document your findings and recommend a course of action.

3. Find a previously written proposal (at work, at school, or one already submitted by a prior student) and improve it according to the criteria presented in this chapter. Add or change any or all of the report's components, including the cover letter, table of contents, list of illustrations, abstract, introduction, main text, conclusion/recommendation, or glossary.

4. Find a previously written proposal (at work, at school, or one already submitted by a prior student) and improve it by adding graphics where needed. You can use tables and figures to clarify a point and to make the report more cosmetically appealing.

Problem Solving Think Pieces

Stinson, Heinlein, and Brown Accounting, LLC, employs over 2,000 workers, including accountants, computer information specialists, a legal staff, paralegals, and office managers. The company requires a great deal of written and oral communication with customers, vendors, governmental agencies, and coworkers. For example, a sample of their workplace communication includes the following:

- Written reports to judges and lawyers.
- Letters and reports to customers.
- E-mail and memos to coworkers.
- Oral communication in face-to-face meetings, videoconferences, and sales presentations.

Unfortunately, not all employees communicate effectively. The writing companywide is uneven. Discrepancies in style, grammar, content, and format hurt the company's professionalism. The same problems occur with oral communication.

George Hunt, a mid-level manager, plans to write an internal proposal to the company's principal owners, highlighting the problems and suggesting solutions.

Assignment

What must Mr. Hunt include in his proposal—beyond the obvious proposal components (a title page, table of contents, abstract, introduction, and so forth)—to persuade the owners to accept his suggestions? Suggest ways in which the problem can be solved.

Web Workshop

1. By typing "RFP," "proposal," "online proposal," or "online RFP" in an Internet search engine, you can find tips for writing proposals and requests for proposals (RFPs), software products offered to automatically generate e-proposals and winning RFPs, articles on how to write proposals, samples of RFPs and proposals, and online RFP and proposal forms.

   Read these articles, research the tips you find, or review the samples. Report your findings, either in an oral presentation or in writing (e-mail, memo, letter, or report).

2. To perform a more limited search, type in phrases like "biomedical equipment RFP," "computer maintenance RFP," "human resource RFP," "web design RFP," and many more topics. You will find examples of both proposals and RFPs from businesses, school systems, city governments, and various industries.

   To enhance your understanding of business and industry's focus on proposal writing, search the Web for information on RFPs and proposals. Using the criteria in this chapter and your knowledge of effective workplace communication techniques, analyze your findings.

   - How do the online proposals or RFPs compare to those discussed in this textbook, in terms of content, tone, layout, and persuasiveness?
   - What information provided in this textbook is missing in the online discussions?
   - What are some of the industries that are requesting proposals, and what types of products or services are these industries interested in?
   - Report your findings, either in an oral presentation or in writing (e-mail, memo, letter, or report).

Technical Applications

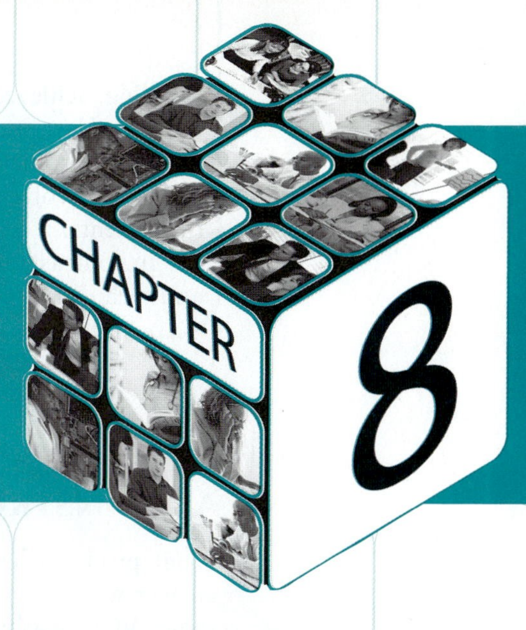

OBJECTIVES

When you complete this chapter, you will be able to do the following:

1. Understand the purpose of technical descriptions and instructions
2. Apply the criteria for writing technical descriptions
3. Apply the criteria for writing instructions
4. Use graphics effectively in technical descriptions and instructions
5. Use the writing process—prewriting, writing, and rewriting—to write effective technical descriptions and instructions
6. Test your knowledge of technical descriptions and instructions through end-of-chapter activities:
   - Case Studies
   - Individual and Team Projects
   - Problem Solving Think Pieces
   - Web Workshop

COMMUNICATION AT WORK

The PhlebotomyDR scenario shows the importance of instructions and technical descriptions in many industries, including biotechnology.

PhlebotomyDR is a medical consulting firm that trains medical technicians to perform blood collections. PhlebotomyDR facilitates training workshops, led by John Staples, to teach venipuncture standards and venipuncture procedures.

PhlebotomyDR creates the following venipuncture instructions, complete with technical descriptions of the equipment:

- Proper equipment selection, sterilization, use, and cleaning
- Proper labeling procedures
- Order of phlebotomy draw
- Patient care before, during, and following venipuncture
- Safety and infection control procedures
- Procedures to follow when meeting quality assurance regulations

Each of the above instructions requires numerous steps, complete with technical descriptions and visual aids.

PhlebotomyDR's outstanding staff realizes that trained technicians make an enormous difference.

Training, achieved through instructional manuals, ensures the health and safety of patients. Untrained technicians make errors that cost us all. Medical errors create insurance problems, the need to redo procedures, increased medical bills, the potential involvement of regulators and legislators, and dangerous repercussions for patients.

In contrast, effective communication, achieved through successful instructions and technical descriptions, saves money, time, and lives.

Purpose of Technical Descriptions and Instructions

What products have you purchased that came packaged with instructions? A safe answer would be "everything." Your iPod came with instructions. So did your clock radio, VCR, washing machine, printer, computer, and automobile. Manufacturers include instructions in the packaging of a mechanism, tool, or piece of equipment. The instructions help the end user construct, install, operate, and service the equipment. In addition, instructions often include technical descriptions. Technical descriptions provide the end user with information about the mechanism's features or capabilities. A technical description helps the reader visualize the mechanism and may tell the user which components are enclosed in the shipping package, clarify the quality of these components, specify what function these components serve in the mechanism, or allow the user to reorder any missing or flawed components.

Criteria for Writing Technical Descriptions

When writing your technical description, include the following:

Title

Preface your text with a title precisely stating the topic of your description. This could be the name and serial number of the mechanism, tool, or piece of equipment you are writing about.

Overall Organization

In the introduction, you could specify and define what topic you are describing, explain the mechanism's functions or capabilities, and list its major components. In the discussion, you could use highlighting techniques (call-outs, itemization, headings, tables, white space) to describe the mechanism's components.

Your conclusion depends on your purpose in describing the topic. Some options are as follows:

- Sales. "Implementation of this product will provide you and your company . . . "
- Uses. "After implementation, you will be able to use this XYZ to . . . "
- Guarantees. "The XYZ carries a 15-year warranty against parts and labor."
- Testimony. "Parishioners swear by the XYZ. Our satisfied customers include . . . "

- **Comparison/contrast.** "Compared to our largest competitor, the XYZ has sold three times more . . . "
- **Reiteration of introductory comments.** "Thus, the XYZ is composed of five parts: belts, pulleys, gears, nuts, and bolts."

Use graphics in your technical descriptions to help your reader visualize the mechanism or piece of equipment. You can use line drawings, photographs, clip art, exploded views, or sectional cutaway views of your topic, each accompanied by callouts (labels identifying key components of the mechanism). We discuss these and other aspects of graphics in Chapter 3, Visual Communication: Page Layout and Graphics.

Internal Organization

When describing your topic in the discussion portion of the technical description, itemize the topic's components in some logical sequence. Components of a piece of equipment, tool, or product can be organized by importance, spatial organization, or comparison.

When a topic is spatially organized, you describe the components as they are seen either from left to right, from right to left, from top to bottom, from bottom to top, from inside to outside, or from outside to inside. A description organized by comparison lets the reader see optional components.

Development

To describe your topic clearly and accurately, you might need to include the following:

Weight
Size (dimensions)
Color
Shape
Density
Materials (composition)
Identifying numbers
Make/model
Texture
Capacity

Word Usage

Your word usage, either photographic or impressionistic, depends on your purpose. For factual, objective technical descriptions, use photographic words. For subjective, sales-oriented descriptions, use impressionistic words. Photographic words are denotative, quantifiable, and specific. Impressionistic words are vague and connotative. Table 8.1 shows the difference.

TABLE 8.1 PHOTOGRAPHIC VERSUS IMPRESSIONISTIC WORD USAGE

Photographic (Denotative)	Impressionistic (Connotative)
6′9″	tall
350 lb.	heavy
gold	precious metal
6,000 shares of United Can	major holdings
700 lumens	bright
0.030 mm	thin
1966 XKE Jaguar	impressive car

USES OF TECHNICAL DESCRIPTIONS

User Manuals

Instructions and user manuals can include technical descriptions written in paragraph form, as seen in the following example:

> The Modern Electronics Tone Test Tracer, Model 77A
>
> Housed in a yellow, high-impact plastic case, the Tone Test Tracer measures $1\frac{1}{4}$ inch × 2 inch × $2\frac{1}{4}$ inch, weighs 4 ounces, and is powered by a 1604 battery. Red and black test leads are provided. The 77A has a standard four-conductor cord, a three-position toggle switch, and an LED for line polarity testing. A tone selector switch located inside the test set provides either solid tone or dual alternating tone. The Tracer is compatible with the EXX, Setup, and Crossbow models.

Product Demand Specifications

Sometimes a company needs a piece of equipment that does not currently exist. To acquire this equipment, the company writes a product demand specifying its exact needs, as shown in the following memo:

> Subject: Requested Pricing for EDM Microdrills
>
> Please provide us with pricing information for the construction of 50 EDM Microdrills capable of meeting the following specifications:
>
> - Designed for high-speed, deep-hole drilling
> - Capable of drilling to depths of 100 times the diameter using 0.012-inch to 0.030-inch diameter electrodes
> - Able to produce a hole through a 1.000-inch-thick piece of AISI D2 electrode
>
> We need your response by May 31, 2006.

Study Report

Companies often hire consulting engineering firms to study a problem and provide a descriptive analysis. The resulting study report is used as the basis for a product demand specification requesting a solution to the problem. One firm, when asked to study crumbling cement walkways, provided the following technical description in its study report:

The slab construction consists of a wearing slab over a $\frac{1}{2}$-inch-thick waterproofing membrane. The wearing slab ranges in thickness from $3\frac{1}{2}$ inches to $8\frac{1}{2}$ inches, and several sections have been patched and replaced repeatedly. The structural slab varies in thickness from $5\frac{1}{2}$ inches to 9 inches with as little as 2 inches over the top of the steel beams. The removable slab section, which has been replaced since original construction, is badly deteriorated and should be replaced. Refer to Appendix A, Photo 9, and Appendix C for shoring installed to support the framing prior to replacement.

Sales Literature

Companies want to make money. One way to market equipment or services is to describe the product. Such descriptions are common in sales letters, proposals, and on Web sites. Figure 8.1 is a technical description from Hewlett-Packard.

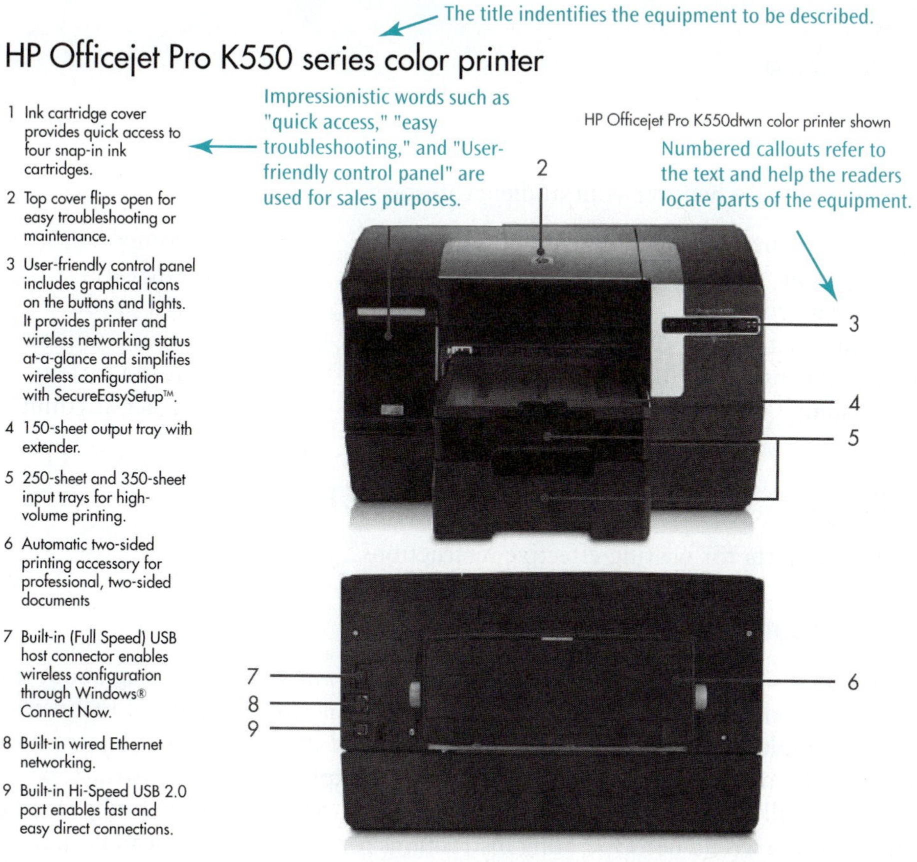

The title indentifies the equipment to be described.

HP Officejet Pro K550 series color printer

Impressionistic words such as "quick access," "easy troubleshooting," and "User-friendly control panel" are used for sales purposes.

HP Officejet Pro K550dtwn color printer shown

Numbered callouts refer to the text and help the readers locate parts of the equipment.

1 Ink cartridge cover provides quick access to four snap-in ink cartridges.

2 Top cover flips open for easy troubleshooting or maintenance.

3 User-friendly control panel includes graphical icons on the buttons and lights. It provides printer and wireless networking status at-a-glance and simplifies wireless configuration with SecureEasySetup™.

4 150-sheet output tray with extender.

5 250-sheet and 350-sheet input trays for high-volume printing.

6 Automatic two-sided printing accessory for professional, two-sided documents

7 Built-in (Full Speed) USB host connector enables wireless configuration through Windows® Connect Now.

8 Built-in wired Ethernet networking.

9 Built-in Hi-Speed USB 2.0 port enables fast and easy direct connections.

2

3

4

5

6

7

8

9

Figure 8.1 Technical Description

Source: Courtesy of Hewlett-Packard

Technical Description Checklist

The technical description checklist will give you the opportunity for self-assessment and peer evaluation of your writing.

Technical Description Checklist

__ 1. Does the technical description have a title, noting your topic's name and any identifying numbers?

__ 2. Does the technical description's introduction (a) state the topic, (b) mention its functions or the purpose of the mechanism, and (c) list the components?

__ 3. Does the technical description's discussion use headings to itemize the components for reader-friendly ease of access?

__ 4. Do you need to define the mechanism and its main parts?

__ 5. Is the detail within the technical description's discussion photographically precise? That is, does the discussion portion of the description specify the following:

Colors	Capacities
Sizes	Textures
Materials	Identifying numbers
Shapes	Weight
Density	Make/model

__ 6. Are all of the dimensions and measurements correct?

__ 7. Do you sum up your discussion using any of the optional conclusions discussed in this chapter?

__ 8. Does your technical description provide graphics that are correctly labeled, appropriately placed, neatly drawn or reproduced, and appropriately sized?

__ 9. Do you write using an effective technical style (low fog index) and a personalized tone?

__10. Have you avoided grammatical errors?

Reasons for Writing Instructions

Include instructions whenever your audience needs to know how to

Operate a mechanism	Restore a product	Test components
Install equipment	Correct a problem	Use software
Manufacture a product	Service equipment	Set up a product
Package a product	Troubleshoot a system	Maintain equipment
Unpack equipment	Clean a product	Monitor a system
Repair a system	Assemble a product	Construct anything

Criteria for Writing Instructions

Follow these criteria for writing effective instructions.

Title Your Instructions

Preface your text with a title that explains two things: *what* you are writing about (the name of your product or service) and *why* you are writing (the purpose of the instructions). For example, to title your instructions "Overhead Projector" would be uninformative. This title names the product, but it does not explain why the instructions are being written. Will the text discuss maintenance, setup, packing, or operating instructions? A better title would be "Operating Instructions for the XYZ Overhead Projector."

Organize Your Instructions

Well-organized instructions help readers follow your directions. To organize your instructions effectively, include the following:

Introduction. Begin your instructions by telling your readers two things: what *topic* you will be discussing and your *reasons* for writing. The topic names the product or service. Your reason for writing either explains the purpose of the instructions ("maintaining the machine will increase its longevity") or comments about the product's capabilities or ease of use.

Required Tools or Equipment. In addition to an introductory overview focusing on the topic and reasons for writing, you might want to tell your readers what tools or equipment they will need to perform the procedures. You can provide this information through a list or graphically.

Hazard Notations. Decide whether you should preface your instructions with dangers, warnings, cautions, or notes. This is an essential consideration to avoid costly lawsuits and to avoid potentially harming an individual or damaging equipment.

When including hazard alerts, consider the following:

- Placement. You can place hazard alerts before the text, in the text (next to appropriate steps), or after the text.
- Access. Make the caution, warning, or danger notice obvious. To do so, vary your typeface and type size, use white space to separate the warning or caution from the surrounding text, or box the warning or caution.
- Definitions. What does *caution* mean? How does it differ from *warning*, *danger*, or *note*?

  Follow the hierarchy of definitions, which clarifies the degree of hazard:
  1. *Danger*. The potential for death.
  2. *Warning*. The potential for serious personal injury.
  3. *Caution*. The potential for damage or destruction of equipment.
  4. *Note*. Important information, necessary to perform a task effectively or to avoid loss of data or inconvenience.
- Colors. Another way to emphasize your hazard message is through a colored window or text box around the word. *Note* is printed in blue or black, *Caution* in yellow, *Warning* in orange, and *Danger* in red.
- Text. To further clarify your terminology, provide the readers text to accompany your hazard alert. Your text should have the following three parts:
  1. *A one- or two-word identification alerting the reader.* Words such as "High Voltage," "Hot Equipment," "Sharp Objects," or "Magnetic Parts," for example, will warn your reader of potential dangers, warnings, or cautions.
  2. *The consequences of the hazards, in three to five words.* Phrases like "Electrocution can kill," "Can cause burns," "Cuts can occur," or "Can lead to data loss," for example, will tell your readers the results stemming from the dangers, warnings, or cautions.
  3. *Avoidance steps.* In three to five words, tell the readers how to avoid the consequences noted: "Wear rubber shoes," "Don't touch until cool," "Wear protective gloves," or "Keep disks away."
- Icons. Equipment is manufactured and sold globally; people speak different languages. Your hazard alert should contain an icon—a picture of the potential consequence—to help everyone understand the caution, warning, or danger.

Discussion. Itemize and thoroughly discuss the steps in your instructions. Organize them chronologically—as a step-by-step sequence. To operate machinery, monitor a system, or construct equipment, your readers must follow a chronological sequence.

Conclusion. As with a technical description, you can conclude your instructions in various ways. You can end your instructions with a (a) comment about warranties; (b) sales pitch highlighting the product's ease of use; (c) reiteration of the product's applications; (d) summary of the company's credentials; (e) troubleshooting guide; (f) frequently asked questions (FAQs); and (g) corporate contact information.

Use Graphics to Highlight Steps

As with technical descriptions, clarify your points graphically. What the reader has difficulty understanding, your graphic can help explain pictorially.

Techniques for Writing Effective Instructions

To write easy-to-understand instructions, follow these criteria for effective style:

1. *Number your steps.* Do not use bullets or the alphabet. Numbers, which you can never run out of, help your readers refer to a specific step. In contrast, if you use bullets, your readers would have to count to locate the steps—seven bullets for step 7, and so on. If you used the alphabet, you would be in trouble when you reached step 27.
2. *Use highlighting techniques.* Boldface, different font sizes and styles, emphatic warning words, color, and italics call attention to special concerns. A danger, caution, warning, or a specially required technique must be evident to your reader. If this special concern is buried in a block of unappealing text, it will not be read. This could be dangerous to your reader or costly to you and your company. To avoid lawsuits or to help your readers see what is important, call it out through formatting.
3. *Limit the information within each step.* Don't overload your reader by writing lengthy steps.

before

Overloaded Steps

Start the engine and run it to idling speed while opening the radiator cap and inserting the measuring gauge until the red ball within the glass tube floats either to the acceptable green range or to the dangerous red line.

after

Separated Steps

1. Start the engine.
2. Run the engine to idling speed.
3. Open the radiator cap.
4. Insert the measuring gauge.
5. Note whether the red ball within the glass tube floats to the acceptable green range or up to the dangerous red line.

4. *Develop your points thoroughly.* Clarify your content by providing precise details.

before

After rotating the discs correctly, grease each with an approved lubricant.

after

1. Rotate the discs clockwise so that the tabs on the outside edges align.
2. Lubricate the discs with 2 oz of XYZ grease.

5. *Use short words and phrases.* Conciseness in workplace communication helps to create easy-to-understand documents. See Chapter 2 for tips on achieving conciseness.
6. *Begin your steps with verbs—the imperative mood.* Note that each of the numbered steps in the following example begins with a verb.

Verbs Begin Steps

1. *Number* your steps.
2. *Use* highlighting techniques.
3. *Limit* the information within each step.
4. *Develop* your points thoroughly.
5. *Use* short words and phrases.
6. *Begin* your steps with a verb.

7. *Personalize your text.* Involve your readers in instructions by using pronouns ("you," "your," "our," etc.).
8. *Do not omit articles (a, an, the).*

before

1. Press right arrow button to scroll through list of programs.
2. Select program to scan.
3. Place item to scan face down on scanner glass in upper left corner.

after

1. Press the right arrow button to scroll through a list of programs.
2. Select the program you want to scan.
3. Place an item to scan face down on the scanner glass in the upper left corner.

Sample Instructions

See Figure 8.2 for a sample instruction.

Figure 8.2 Dry Mounting Instructions

Title clarifying intent

INSTRUCTIONS FOR DRY MOUNTING PHOTOGRAPHS

Dry mounting photographs keeps them safe for your family and generations to come. To successfully dry mount your photographs, follow the seven steps we've provided.

Dry mounting requires the following materials:

Materials

Required tools and equipment listed →

1. Press
2. Tacking iron
3. Mat knife
4. Mat board
5. Dry-mount tissue
6. Ruler
7. Print (photographed image)

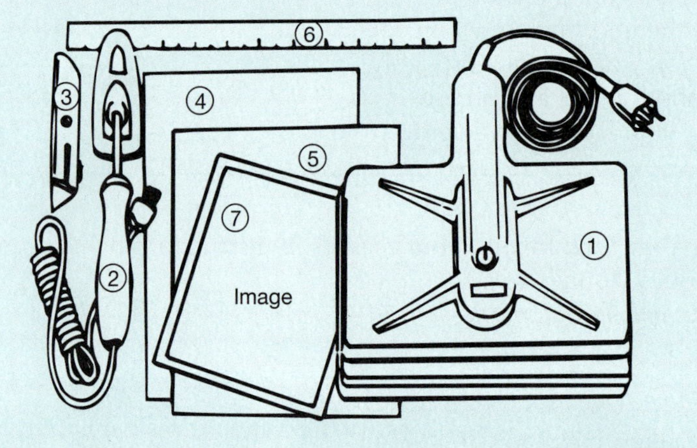

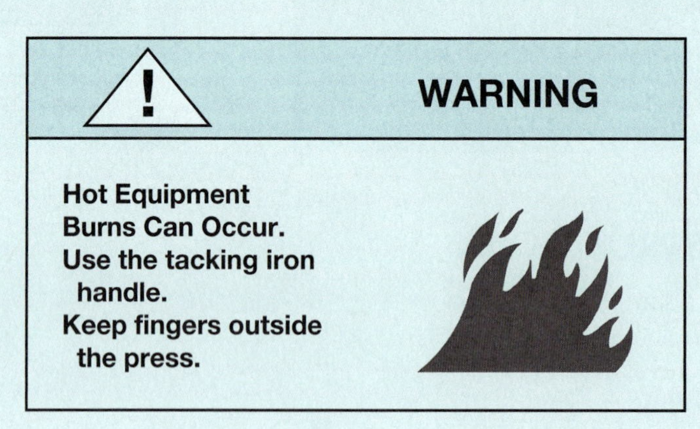

WARNING
Hot Equipment
Burns Can Occur.
Use the tacking iron handle.
Keep fingers outside the press.

Figure 8.2
continued

1.0 DRYING THE MATERIAL

1.1 Plug in the press and turn it on. Wait to see the red light.
1.2 Heat the press to 250 degrees by adjusting the dial on the lid.
1.3 Predry the mount board and the print.
1.4 Place them in the press and hold down the lid for 30 seconds.

The Institute of Electrical and Electronics Engineers (IEEE) created a numbering system using decimal points. This system allows you to organize the steps.

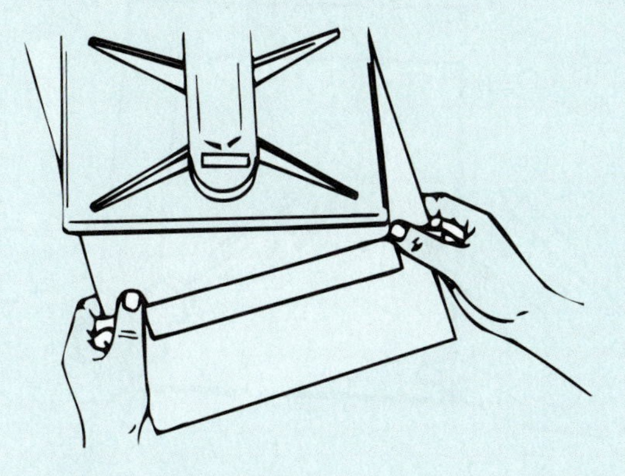

2.0 WIPING THE MATERIALS CLEAN

2.1 After the materials have been baked, let them cool for two to three minutes.
2.2 Wipe away any loose dust or dirt with a clean cloth.

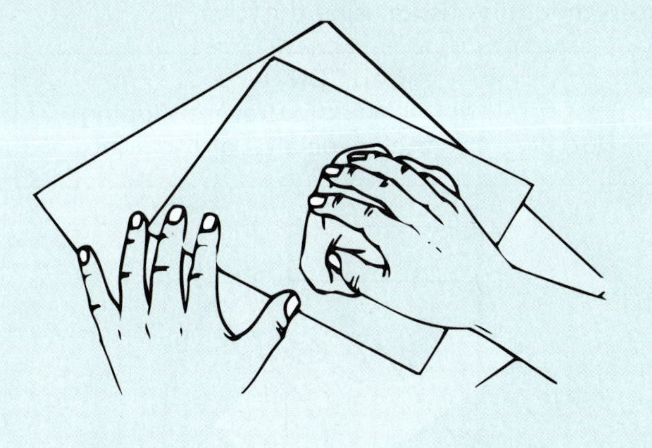

Visual aids enhance the instructions.

(*continued*)

Figure 8.2
continued

3.0 TACKING THE MOUNTING TISSUE TO PRINT

3.1 Plug in the tacking iron and turn it on.
3.2 With the print face down and the mounting tissue on top, tack the tissue to the center of the print.

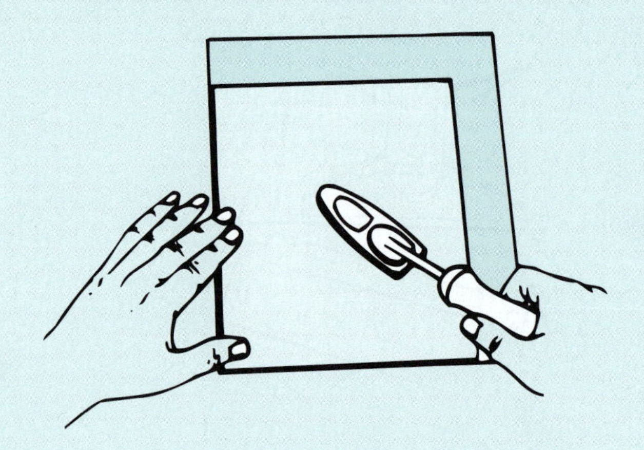

4.0 TRIMMING THE MOUNTING TISSUE

4.1 Place a piece of smooth cardboard under the print so that you won't cut your table.
4.2 Trim off excess mounting tissue using the ruler.

CAUTION
Press firmly on the ruler to prevent slipping!
Don't cut into the image area of the print!

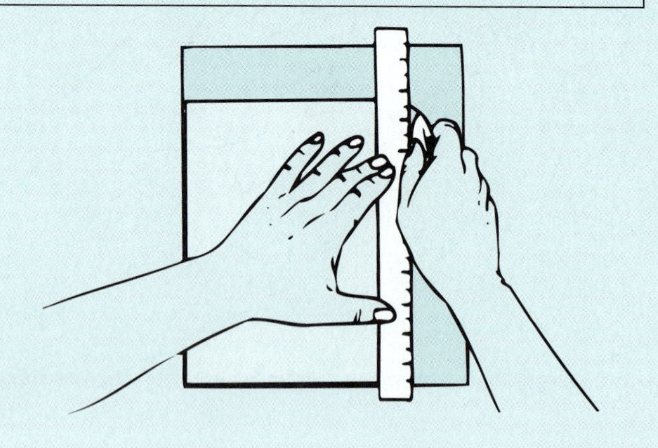

Figure 8.2 *continued*

5.0 ATTACHING THE MOUNTING TISSUE AND PRINT TO THE BOARD

> CAUTION
> The tissue must lie completely flat on the board to prevent wrinkles under the print!

5.1 Position the print and the tissue, face up, on the top of the mount board.
5.2 Slightly raise one corner of the print and touch the tacking iron to the tissue on top of the board.
5.3 Press down the corner lightly, taking care to avoid wrinkles.

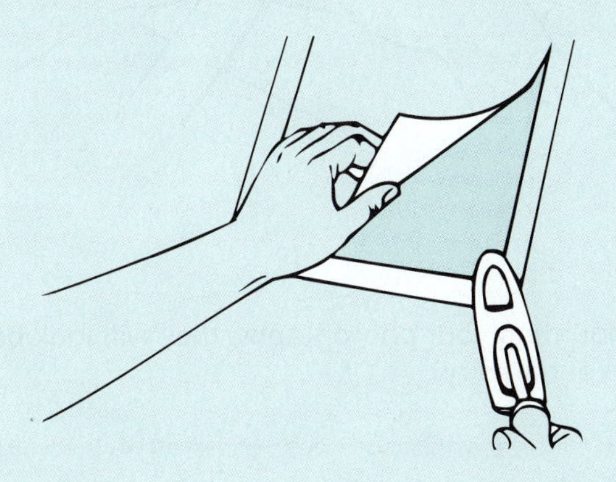

6.0 MOUNTING THE PRINT

6.1 Put the board, tissue, and print (with cover sheet on top) into the press.
6.2 Hold down the press lid firmly for 30 seconds.

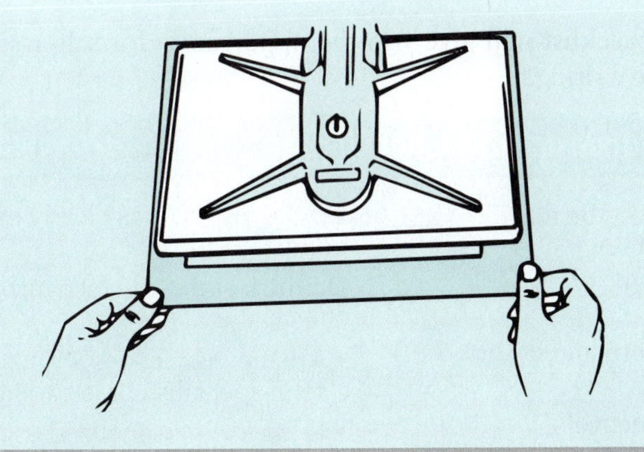

(continued)

Figure 8.2
continued

7.0 TRIMMING THE MOUNTED PRINT

7.1 Pressing down firmly on the ruler to avoid slipping, trim the edges of the mounted print with a sharp mat knife.

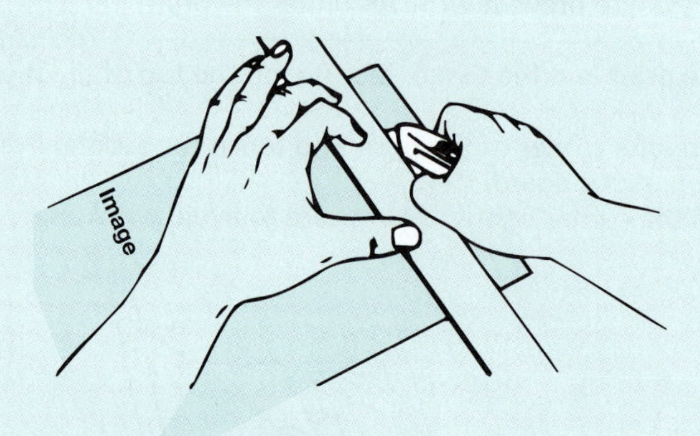

Conclusion to sum up the instruction →

By carefully dry mounting your photographs, they will look beautiful and last longer. Enjoy your artistry.

Instructions Checklist

The instructions checklist will give you the opportunity for self-assessment and peer evaluation of your writing.

Instructions Checklist

__ 1. Do the instructions have an effective title that clarifies the topic, and an introduction mentioning the topic and the reasons for performing the instructions?

__ 2. Do the instructions include a list of required tools or equipment?

__ 3. Are hazard alert messages used effectively?
- Are the hazard alert messages placed correctly?
- Is the correct term used (*Danger, Warning, Caution,* or *Note*)?

__ 4. Are the instructions organized chronologically?

__ 5. Does each step begin with a verb, avoid overloading by presenting one clearly defined action, and include articles?

__ 6. Do the instructions clearly explain each point?

__ 7. Do the instructions recognize the audience effectively?
- Are the instructions written for a specialist, semi-specialist, or lay audience?
- Is the text personalized using pronouns?

__ 8. Does the instructions' document design use highlighting techniques effectively, such as graphics and headings?

__ 9. Have you used short words, short sentences, and short paragraphs?

__10. Have you corrected grammatical errors?

THE WRITING PROCESS AT WORK

To clarify the importance of the writing process, look at how an employee at PhlebotomyDR used prewriting, writing, and rewriting to write instructions.

Prewriting

To write instructions, use reporter's questions to gather data and to determine objectives.

- *Who* is my reader (specialists, semi-specialists, or lay readers)?
- *What* detailed information is needed for my audience? For example, a specialist understands "inflate sufficiently," whereas an instruction geared toward semi-specialists or lay readers needs to be more specific: "inflate to 25 pounds per square inch (psi)."
- *How* must the instructions be organized? For instructions, the answer is chronological.
- *When* should the procedure occur?
  - Daily, weekly, monthly, quarterly, biannually, or annually?
  - As needed?
- *Why* should the instructions be carried out? Will the instructions be used to help readers repair, maintain, install, or operate?

Once you have gathered this information, you can sketch a list of the steps required in the instructions and the detail needed for clarity.

Writing

Once you have listed the steps in prewriting, write the rough draft. In Figure 8.3, the PhlebotomyDR employee wrote a rough draft of instructions.

Rewriting

Based on John Staple's revision suggestions and additional peer assistance, the PhlebotomyDR employee rewrote the rough draft. To do so, he did the following:

- *Add* missing detail for clarity.
- *Delete* dead words and phrases for conciseness.
- *Simplify* unnecessarily complex words and phrases to allow for easier understanding.
- *Move* around information (cut and paste) to ensure that the most important ideas were emphasized.
- *Reformat* (using highlighting techniques) to ensure reader-friendly ease of access.
- *Enhance* the tone and style of the text.
- *Correct* any errors to ensure accurate grammar and content.

Figure 8.4 shows the finished instructions.

Figure 8.3 Rough Draft Instructions

Venipuncture procedures

You must strictly adhere to venipuncture procedures to obtain accurate blood test results. Venipuncture should be performed in designated areas such as venipuncture stations with venipuncture chairs.

- Wash hands before and after procedure. Use hand sanitizer as needed.

Procedure:

- Check the physician's order to determine which blood tests required. Assemble required materials.
- Refer to the venipuncture manuals for the current appropriate amounts of blood required for specimen collection.
- Gloves shall be worn by all personnel.
- Identify the patient utilizing at least two patient identifiers, tell the patient what procedure is being performed and why this is necessary.
- Apply tourniquet.
- Cleanse the blood draw site with alcohol. Dry the site.
- Insert the needle into the vein.
- Fill each test tube with the desired amount of blood and remove the needle. You always want to cover the needle site with a Band-Aid. Apply pressure on the site for three or so minutes.

Writer's Insight

John Staples, the PhlebotomyDR training coordinator, read the rough draft and suggested these revisions: "This is a great start for the instructions. For our finished copy, you'll need to number the steps and start each step with a verb. Notice that 'Gloves shall be worn' needs to be rewritten in the active voice: 'Wear gloves.'

In addition, couldn't we make that first sentence less demanding? How about beginning with a 'please'? I think I see why you bulleted that first point 'Wash hands,' but I'd add it to the list of numbered steps, even putting it first. Also, write 'Wash *your* hands.' Without the pronoun, the sentence sounds sort of strange. I'm not an English teacher, but it doesn't sound grammatically correct.

Finally, these instructions need graphics to help the readers follow the steps more easily. And to make sure that everyone is aware of how dangerous blood draws can be, highlight the warnings/cautions/dangers—with graphics or larger fonts or boxes. Let me see your next version, please."

Figure 8.4 Revised Instructions

VENIPUNCTURE PROCEDURES

To obtain accurate blood test results, please adhere strictly to the following venipuncture procedures.

Danger Incorrect venipuncture can lead to serious patient health problems, including misdiagnosis.

Necessary Equipment

Syringe and collection tubes	Tourniquet	Surgical gloves	Cotton balls	Rubbing alcohol
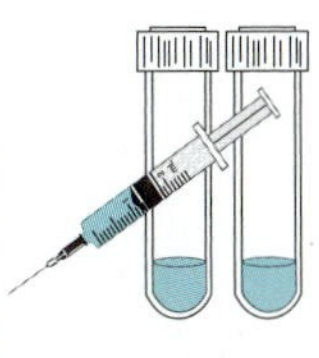	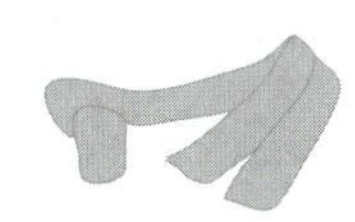			

Procedure

1. Check the physician's order to determine which blood tests are required.

2. Carefully read the venipuncture manual to learn what amounts of blood are required for the specific specimen collection.

3. Wash your hands before and after each blood draw.

Caution To ensure cleanliness, use the hand sanitizer located on the wall next to the venipuncture station.

4. Wear surgical gloves for your protection and the safety of your patient.

5. Gather the required syringe(s) and blood collection tubes.

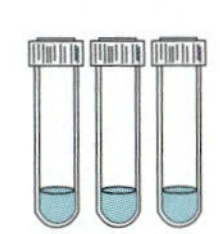

(continued)

Figure 8.4
continued

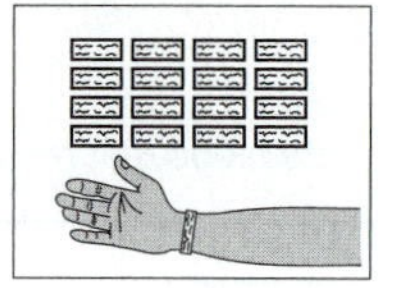

6. Identify the patient using at least two patient identifiers (name and patient ID number, for example).

7. Tell the patient what procedure is being performed and why it is necessary.

8. Apply the tourniquet above the puncture site (upper arm in most instances).

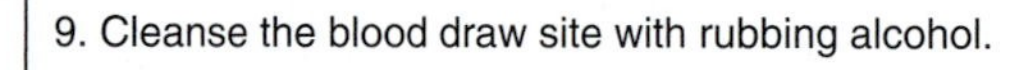

9. Cleanse the blood draw site with rubbing alcohol.

10. Allow the alcohol site to air dry.

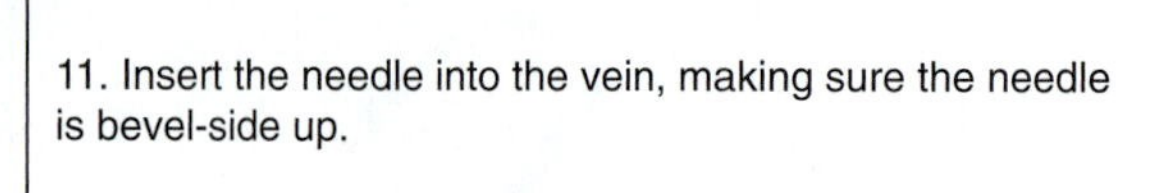

11. Insert the needle into the vein, making sure the needle is bevel-side up.

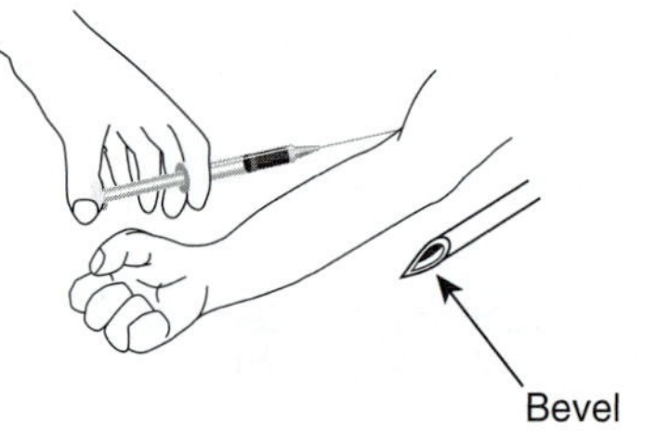

12. Fill each test tube with the required amount of blood, as specified in the venipuncture manual.

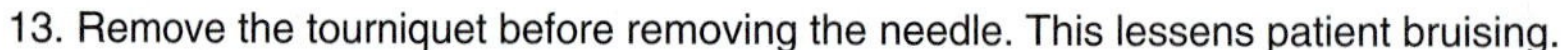

13. Remove the tourniquet before removing the needle. This lessens patient bruising.

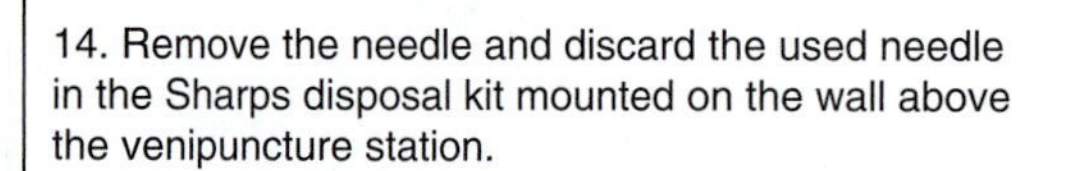

14. Remove the needle and discard the used needle in the Sharps disposal kit mounted on the wall above the venipuncture station.

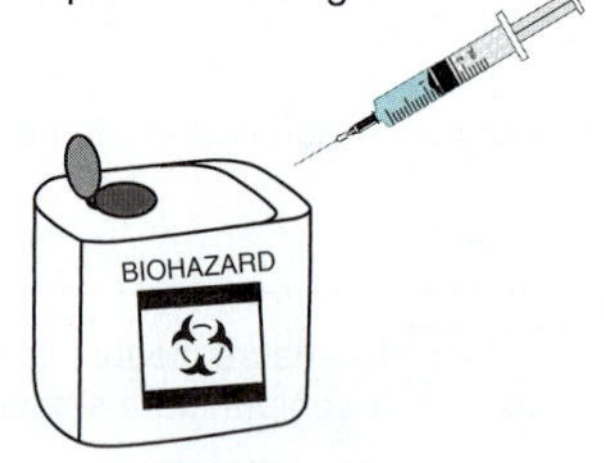

15. Cover the puncture site with a bandage.

If you have any questions or concerns during the blood draw, ask the attending nurse or supervisor for help. For additional assistance, call **1-800-BLOODDR.**

Developing Workplace Skills

Case Study

PhlebotomyDR needs to provide instructions for their staff (nurses, doctors, and technicians) regarding the correct procedures to ensure cleanliness and employee protection. These include hand washing procedures and the correct use of sterile equipment, such as gloves, masks, aprons, and shoe coverings.

Following is a rough draft of one set of instructions for personnel safety.

Assignment

Revise the rough draft according to the criteria provided in this chapter. Correct the order of information, the grammar, and the instruction's content. In addition, improve the instruction by including appropriate graphics.

Hand washing

- Lather hands to cover all surfaces of hands and wrists.
- Wet hands with water.
- Rub hands together to cover all surfaces of hands and fingers. Pay special attention to areas around nails and fingers. Lather for at least 15 seconds.
- Dry thoroughly.
- Rinse well with running warm water.
- Avoid using hot water. Repeated exposure to hot water can lead to dermatitis.
- Use paper towels to turn off faucet.

Gloves

- Replace damaged gloves as soon as patient safety permits
- Don gloves immediately prior to task.
- Remove and discard gloves after each use involving any bodily fluids

Masks

Wear masks and eye protection devices (goggles or eye shields) to avoid droplets, spray, or splashes and to prevent exposure to mucous substances. Masks are also worn to protect nurses, doctors, and technicians from infectious elements during close contact with patients.

Aprons and Other Protective Clothing

- Wear aprons or gowns to avoid contact with body substances during patient care procedures
- Remove and dispatch aprons and other protective clothing before leaving work area.
- Some work areas might require additional protective clothing such as surgical caps and shoe covers or boots.

INDIVIDUAL AND TEAM PROJECTS

1. Write a set of instructions. To do so, first select a topic. You can write instructions telling how to monitor, repair, test, package, plant, clean, operate, manage, open, shut, set up, maintain, troubleshoot, install, use software, and so on. Choose a topic from your field of expertise or one that interests you. Follow the writing process techniques to complete your instructions. Prewrite, write a draft, and rewrite to perfect your text.

2. Find examples of instructions for consumer products. These can include instructions for assembling children's toys, refinishing furniture, insulating attics or windows, setting up stereo systems, flushing out a radiator, installing ceiling fans, and so on. Once you find some examples, bring them to class. Then, applying the criteria for good instructions presented in this chapter, determine the success of the examples. If they are successful, explain why and how. If they fail, show where the problems are and rewrite the instructions to improve them.

3. Microsoft Word's Help program provides instructional steps for hundreds of word processing operations. Access the Word Help Index to find instructions. Open 3 to 5 of these instructions and accomplish the following:

   - Discuss how they are similar to or different from the instructions discussed in this chapter.
   - Select one of the instructions and rewrite it, adding graphics (screen captures), cautions or notes, and a glossary of terms.

4. Write a technical description, either individually or as a team. To do so, first select a topic. You can describe any tool, mechanism, or piece of equipment. However, don't choose a topic too broad to describe accurately. To provide a thorough and precise description, you will need to be exact and minutely detailed. A broad topic, such as a computer, an oscilloscope, a respirator, or a Boeing airliner, would be too demanding for a two- to four-page description. On the other hand, do not choose a topic that is too small, such as a paper clip, a nail, or a shoestring. Choose a topic that provides you with a challenge but that is manageable. Consider describing any of the following topics:

USB Flash Drive	Computer disk	VCR remote control
Wrench	Computer mouse	Mechanical pencil
Screwdriver	Lightbulb	Ballpoint pen
Pliers	Calculator	Computer monitor
Wall outlet	Automobile tire	Cell phone

   Once you or your team has chosen a topic, prewrite (listing the topic's components), write a draft (following the criteria provided in this chapter), and rewrite (revising your draft).

5. Find examples of professionally written technical descriptions in technical books and textbooks, professional magazines and journals, or on the Internet. Bring these examples to class and discuss whether they are successful according to the criteria presented in this chapter. If the descriptions are good, specify how and why. If they are flawed, state where, suggest ways to improve them, and rewrite the flawed descriptions.

Problem Solving Think Pieces

1. Good writing demands revision. Following are flawed instructions. Rewrite the text, following the criteria for instructions and the rewriting techniques included in this chapter.

> Date: November 1, 2006
> To: Maintenance Technicians
> From: Second Shift Supervisor
> Subject: Oven Cleaning
>
> The convection ovens in kiln room 33 need extensive cleaning. This would consist of vacuuming and wiping all walls, doors, roofs, and floors. All vents and dampers need to be removed and a tack cloth used to remove loose dust and dirt. Also, all filters need replacing. I am requesting this because when wet parts are placed in the ovens to cure the paint, loose particles of dust and dirt are blown onto the parts, which causes extensive rework. I would like this done twice a week to ensure cleanliness of product.

2. Read the following instructional steps. Are they in the correct chronological order? How would you reorder these steps to make the instruction more effective?

> **Changing Oil in Your Car**
>
> Run the car's engine for approximately 10 minutes and then drain the old oil.
>
> Park the car on a level surface, set the parking brake, and turn off the car's engine.
>
> Gather all of the necessary tools and materials you might need.
>
> Open the hood.
>
> Jack up and support the car securely.
>
> Place the funnel in the opening and pour in the new oil.
>
> Replace the cap when you have finished pouring in new oil.
>
> Locate the oil filter cap on top of the engine and remove the cap.
>
> Tighten the plug or oil filter if you find leakage.
>
> Run the engine for a minute, then check the dipstick. Add more oil if necessary.
>
> Pour the used oil into a plastic container and dispose of it safely and legally.

Web Workshop

1. Review any of the following Web site's online instructions. Based on the criteria provided in this chapter, are the instructions successful or not?

   - If the answer is yes, explain why and how the instructions succeed.
   - If the answer is no, explain why the instructions fail.
   - Rewrite any of the flawed instructions to improve them.

Web Sites	Topics
http://www.hometips.com/diy.html	Electrical systems, plumbing, kitchen appliances, walls, windows, roofing, and siding
http://www.hammerzone.com	Kitchen projects, tubs, sinks, toilets, showers, and water heaters
http://dmoz.org/Home/Home_Improvement/	Links to step-by-step procedures for painting, welding, soldering, plumbing, walls, windows, and door repair and installation
http://directory.google.com/Top/Home/Home_Improvement/	Links to sites for instructions on decorating, electrical, flooring, furniture, lighting, painting, plumbing, welding, windows, and doors
http://www.quakerstate.com/pages/carcare/oilchange.asp	Instructions for changing oil
http://www.csaa.com/yourcar/takingcareofyourcar	AAA instructions for car care, including general maintenance and checking fluids, hoses, drive belts, electrical systems, and tires
http://www.gateway.com/index.shtml	Access the FAQs for upgrading systems or correcting problems with printers, drivers, monitors, memory, and more

2. You are ready to purchase a product. This could include printers, monitors, digital cameras, scanners, PCs, laptops, speakers, cables, adapters, automotive engine hoists, generators, battery chargers, jacks, power tools, truck boxes, screws, bolts, nuts, rivets, hand tools, and more. A great place to shop is online. By going to an online search engine, you can find not only prices for your products but also technical descriptions, technical specifications, or data sheets. These will help you determine if the product has the size, shape, materials, and capacity you are looking for. Notice how the Hewlett-Packard printer includes a data sheet describing the product's size, shape, and capabilities, found by clicking on the "learn more about" link.

(Courtesy of Hewlett-Packard)

Go online to search for a product of your choice and review the technical description or specifications provided. Using the criteria in this chapter and your knowledge of effective technical writing techniques, analyze your findings.

- How do the online technical descriptions compare to those discussed in this textbook?
- Are graphics used to help you visualize the product?
- Are call-outs used to help you identify parts of the product?
- Are high-tech terms defined?
- Is the use of the product explained?

The Job Search

OBJECTIVES

When you complete this chapter, you will be able to do the following:

1. Search for job openings applicable to your interests, education, and experience.
2. Compose effective letters of application that gain attention and are persuasive.
3. Choose either to write a functional or reverse chronological resume.
4. Write effective resumes consisting of your objectives, summary of qualifications, work experience, education, and professional skills.
5. Decide on the correct method of delivery of your resume, either through the mail, as an e-mail attachment, or scannable.
6. Understand effective interview techniques that demonstrate your professionalism.
7. Write appropriate follow-up correspondence to restate how you can benefit the company.
8. Test your knowledge of the job search through end-of-chapter activities:
   - Case Studies
   - Individual and Team Projects
   - Problem Solving Think Pieces
   - Web Workshop

COMMUNICATION AT WORK

In this scenario, a business owner interviews potential job applicants.

The job search involves at least two people—the applicant and the individual making the hiring decision. Usually, more people than two are involved, however, since companies typically hire based on a committee's decision. That's the case at DiskServe. This St. Louis-based company is hoping to hire a customer service representative for their computer technology department. DiskServe asked applicants to apply using e-mail. The applicants submitted an e-mail letter of application and an attached scannable resume.

DiskServe advertised this opening in the career placement centers at local colleges, in the city newspaper, and online at their Web site: *www.DiskServe.com*.

In addition to DiskServe's CEO, Harold Irving, the hiring committee will consist of two managers from other DiskServe departments, the former employee whose job is being filled, and two coworkers in the computer technology department.

Ten candidates were considered for the position. All ten first had teleconference interviews. Harold called the ten, while the other hiring committee members listened on a speaker phone. After the telephone interviews, four candidates were invited to DiskServe's work site for personal interviews—Macy Heart, Aaron Brown, Rosemary Lopez, and Robin Scott.

Each job candidate was asked a series of questions. These included the following: "What is your greatest strength, and give me an example of how this reveals itself on the job?" "What did you like most about your last job?" "How have you handled customer complaints in the past?" "Where do you see yourself in five years?" Then, each candidate was taken to the computer repair lab and confronted with an actual hardware or software problem. The candidates were asked to solve the problem, and their work was timed. Finally, the applicants were allowed to ask questions about DiskServe and their job responsibilities.

Harold Irving takes the hiring process seriously. He wants to hire the best person because he hopes that employee will stay with his company a long time. Hiring well is a good corporate investment.

How to Find Job Openings

When it is time to look for a job, how will you begin your search? Approach the job search systematically. Try the following techniques.

Visit Your College or University Job Placement Center

- Your school's job placement service will have job counselors to counsel you regarding your skills and job options.
- Your job placement center can give you helpful hints on preparing resumes, letters of application, and follow-up letters.
- The center will post job openings.
- The center will be able to tell you when companies will visit campus for job recruiting.
- The service can keep on file your letters of recommendation or portfolio. The job placement center will send these out to interested companies upon your request.

Attend a Job Fair

Many colleges and universities host job fairs. A job fair will allow you to research job openings, make contacts for internships, or submit resumes for job openings. If you attend a job fair, treat it like an interview. Dress professionally and take copies of your resume and letters of recommendation.

Talk to Your Instructors

Whether in your major field or not, instructors can be excellent job sources. They will have contacts in business, industry, and education. They may know of job openings or people who may be helpful in your job search. Instructors might know which types of jobs or work environments best suit you.

Network with Friends and Past Employers

A July 2003 *Smart Money* magazine article reports that 62 percent of job searchers find employment through "face-to-face networking" (Bloch 12). A study performed by Drake Beam Morin confirms this, stating that "64 percent of . . . almost 7,500 people surveyed said they found their jobs through socializing and meeting people" (Drakeley 5). Tell friends and acquaintances that you are looking for a job. They might know someone for you to call.

Check Your Professional Affiliations and Publications

If you belong to a professional organization, this could be a source of employment in two ways: Sponsors or board members might be aware of job openings; your organization might publish a listing of job openings. In addition, you can gain professional experience by serving on boards or being an officer in an organization. These experiences can be included in your resume under professional skills, work experience, and affiliations.

Read the Want Ads

Check the classified sections in newspapers. These want ads list job openings, requirements, and salary ranges.

Get Involved in Your Community

Consider volunteering for a community committee, pursuing religious affiliations, joining community clubs, or participating in fund-raising events. Take classes in accounting, HTML, RoboHelp, or Flash at your local community college. This is a great way to network and acquire additional expertise in your field.

Read Newspaper or Business Journal Articles to Find Growing Businesses/Business Sectors

Which companies are receiving grants, building new sites, winning awards, or creating new service or product lines? Which companies have just gained new clients or received expanded contracts? Newspapers and journals report this kind of news, and a growing company or business sector might be good news for you. If a company is expanding, this means more job opportunities for you to pursue.

Take a "Temp" Job

Temporary jobs, accessed through staffing agencies, pay you while you look for a job, help you acquire new skills, allow you to network, and can lead to full-time employment.

Get an Internship

Internships provide you outstanding job preparedness skills, help you meet new people for networking, and improve your resume. An unpaid internship in your preferred work area might lead to full-time employment. An internship "gives you the opportunity to show your skills, work ethic, positive attitude, and passion for your work." By interning, you can prove that you should be "the next employee the company hires" (Drakeley 7). In fact, the National Association of Colleges and Employers lists internships as one of the top ten skills employers want ("Planning Job Choices").

Job Shadow

Job shadowing allows you to visit a work site and follow employees through work activities for a few hours or days. This allows you to learn about job responsibilities in a certain field or work environment. In addition, job shadowing also helps you find out if a company or industry is hiring, allows you to make new contacts, and places you in a favorable position for future employment at that company.

Set Up an Informational Interview

In an informational interview, you talk with people currently working in a field, asking them questions about career opportunities and contacts. Informational interviews allow you to

- explore careers and clarify your career goal.
- expand your professional network.
- build confidence for your job interviews.
- access the most up-to-date career information.
- identify your professional strengths and weaknesses.

Research the Internet

In the mid-to-late 90s, the Internet was the preferred means by which job seekers found employment. That has changed drastically. *Newsweek* magazine calls the Internet "a time waster." Quoting the head of a Chicago outplacement firm, *Newsweek* writes that some of the Internet's popular job sites are "big black holes"—your resume goes in, but you never hear from anyone again (Stern 67).

Others are equally pessimistic about the Internet's value as a source for jobs. Only 10 percent of technical or computer-related jobs are found from electronic job searches, "about 13 percent of interviews for managerial-level jobs result from responding to an online posting," and a mere 4 percent of jobs overall are found through the Internet (Bloch 12).

Nonetheless, you should make the Internet part of your job search strategy. If it is not the best place to find a job, the Internet still provides numerous benefits. Internet job search engines provide excellent job search resources. Research the Internet sites for job openings (Table 9.1).

Criteria for Effective Resumes

Resumes are usually the first impression you make on a prospective employer. Your resume should present an objective, easily accessible, detailed biographical sketch. Do not try to include your entire history. The primary goal of your resume, together with your letter of application, is to get you an interview. Use your interview to explain in more detail information that does not appear on your resume.

Table 9.1 Sites for Internet Job Searches

- **www.Monster.com** lets you post resumes, search for jobs, and access career advice.
- **www.CareerJournal.com** is the *Wall Street Journal*'s career search site, provides salary and hiring information, a resume database, and job hunting advice.
- **www.FlipDog.com** provides national job listings in diverse fields by job title, location, and date of listing.
- **www.HotJobs.yahoo.com** lets you search for jobs by keyword, city, and state.
- **www.WantedJobs.com** lists jobs in the United States and Canada.
- **www.WetFeet.com** gives you company and industry profiles, resume help, city profiles, and international job sites.
- **www.CareerLab.com** provides a cover letter library.
- **www.CareerCity.com** offers detailed discussions and samples of functional versus chronological resumes.
- **www.JobOptions.com** provides a jobs search for accounting, customer service, engineering, human resources, sales, and technology.
- **www.CareerMag.com** allows for online job searches by keyword.
- **www.JobLocator.com** offers access to many other Internet job search engines.
- **www.CareerShop.com** helps you post and edit resumes, offers career advice, and lists hiring employers.
- **www.CareerBuilder.com** lets you search for jobs by company, industry, and job type.

When writing your resume, you have two optional approaches. You can write either a reverse chronological resume or a functional resume.

Reverse Chronological Resume

Write a reverse chronological resume if you

- are a traditional job applicant (recent high school or college graduate, age 18 to 25).
- hope to enter the profession in which you have received college training or certification.
- have made steady progress in one profession (promotions or salary increases).
- plan to stay in your present profession.
- are calling attention to a stable work history.

Functional Resume

Write a functional resume if you

- are entering the job market for the first time or pursuing an entry-level position and lack applicable work history for a reverse chronological resume.
- are making a career change.
- are emphasizing the skills relevant to the future rather than past job responsibilities.

- are a nontraditional job applicant (returning to the workforce after a lengthy absence, older, not a recent high school or college graduate).
- plan to enter a profession in which you have not received formal college training or certification.
- have changed jobs frequently.

Key Resume Components

Whether you write a reverse chronological or functional resume, include the following.

Identification and Contact Information

Begin your resume with the following: Your name, address, and telephone number. You can also include your e-mail address, Web site address, and fax number. Be sure that your e-mail address is professional sounding. An e-mail address, such as "ILuvDaBears," "Hotrodder," or "HeavyMetalDude," is not likely to inspire a company to interview you.

Career Objectives

A career objective line focuses on a specific job and shows how you will benefit the company. Too often, career objectives are so generic that their vagueness does more harm than good. In addition, bad career objectives suggest that you are looking for a job to benefit yourself rather than the company.

before

Career Objective Seeking employment in a business environment offering an opportunity for professional growth

This "Objective" is too generic. It neither states the writer's goals nor shows how the writer will add value to the company.

after

Career Objective Provide financial counseling to ensure positive client relations and build corporate growth

This improved career objective not only specifies which job the applicant is seeking but also how he or she will benefit the company.

Summary of Qualifications

As an option to beginning your resume with career objectives, you might want to consider starting with a summary of qualifications.

A summary of qualifications should include the following:

- An overview of your skills, abilities, accomplishments, and attributes
- Your strengths in relation to the position you are applying for

Summary of Qualifications

- Over four years combined experience in marketing and business
- Developed winning bid package for promotional brochures
- Promoted to shift manager in less than two years
- Maintained database of customers, special-ordered merchandise, and tracked inventory

Employment

The employment section lists the jobs you have held. This information must be presented in reverse chronological order (your current job listed first, previous jobs listed next). In the employment section, include the following:

- Your job title (if you have or had one)
- The name of the company you worked for
- The location of this company (either street address, city, and state or just the city and state)
- The time period during which you worked at this job
- Your job duties, responsibilities, and achievements (see Table 9.2 for a list of verbs to highlight achievements)

Assistant Manager

McConnel Oil Change, Beauxdroit, LA 2005 to present

- Tracked and maintained over $25,000 inventory
- Trained a minimum of four new employees quarterly
- Achieved 10 percent growth in customer car count for three consecutive years
- Developed a written manual for hazardous waste disposal, earning a "Citizen's Recognition Award" from the Beauxdroit City Council

Listing your job title, company name, location, and dates of employment merely shows where you were in a given period of time. To prove your contributions to the company, provide specific details highlighting achievements.

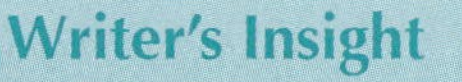

Writer's Insight

DiskServe's CEO, Harold Irving, reads resumes carefully. He says, "I look for two strengths, above all others in the resume. First, I want to see evidence of skills and credentials applicable to the job. Next, I need proof. Before I hire someone and start paying them wages and benefits, I need evidence of success. In addition to skills, I want a concrete example that prove the applicant's accomplishments. For example, I want a resume to read something like, 'reduced turnover by a specific percent within such and such a time.'

Having to review dozens of resumes for every job opening, I also want the resumes to be brief, with bulleted points instead of paragraphs. I do not want to read resumes that include personal information, hobbies, and pictures of the applicant. Why would I care if a job applicant plays the guitar or reads books for pleasure? And I can't hire or fire on someone's looks—that's illegal. I don't need extraneous information," says Harold. "I need specific facts that will help me hire the best person for the job."

Education

Document your educational experiences in reverse chronological order (most recent education first; previous schools, colleges, universities, military courses, or training seminars next). When listing your education, provide the following information:

- **Degree.** If you have not yet received your degree, you can write "Anticipated date of graduation June 2006" or "Degree expected in 2006."
- **Area of specialization.**
- **School attended.** Do not abbreviate. Although you might assume that everyone knows what UT means, your readers will not understand this abbreviation. Is UT the University of Texas, the University of Tennessee, the University of Tulsa, or the University of Toledo?
- **Location.** Include the city and state.
- **Year of graduation or years attended.**
- **Educational accomplishments,** such as grade point average; academic club memberships and leadership offices held; special class projects; academic honors, scholarships, or awards; fraternity/sorority leadership offices held; internships; and volunteerism.
- **Include your grade point average** if it is higher than 3.0 on a 4.0 scale.
- **Include your high school graduation** if you are a recent high school graduate.

Professional Skills

If you are changing professions or reentering the workforce after a long absence, you should write a functional resume. Many functional resumes begin with a professional skills section and then list work and education. Professional skills include the following:

- Computer expertise (hardware and software knowledge)
- Machinery you can operate
- Procedures you can perform
- Special accomplishments and awards
- On-the-job training you have received
- Training you have provided
- New techniques you have created or implemented
- Number of people you have managed
- Languages you can speak or read
- Certifications
- Publications

Professional Skills

- Proficient in Microsoft Word, Excel, PowerPoint, and FrontPage
- Knowledge of HTML, Java, Visual Basic, and C++
- Certified OSHA Hazardous Management Safety Trainer
- Fluent in Spanish and English
- Completed Second Shift Administration Certificate

Highlight your professional skills that will set you apart from other potential employees.

Military Experience

Military experience can be listed under your work experience heading or education (if you have received training while in the military). If you served for several years in the military, you might want to describe this service in a separate section. You would include the following:

- Rank
- Service branch
- Location (city, state, country, ship)
- Years in service
- Discharge status
- Special clearances
- Achievements and professional skills
- Training seminars attended or education received

Professional Affiliations

If you belong to regional, national, or international clubs or have professional affiliations, you might want to mention these.

Portfolios

As an optional component for your job search, consider using a portfolio. If you are in fashion merchandising, heating/ventilation/air conditioning (HVAC), engineering, drafting, architecture, or graphic design, for example, you might want to provide samples of your work. These samples could include schematics, drawings, photographs, or certifications.

However, avoid sending an unsolicited portfolio electronically. Many companies are reluctant to open unknown attachments. Make hard copies of your work or a CD-ROM. When you have an interview with a prospective employer, you can give the employer the hard copy or CD-ROM.

Resume Style

To make your resume accessible at a glance, use the following techniques:

Choose Appropriate Font Types and Sizes. As with most workplace communication, the best font types are Times New Roman and Arial. These are readable and professional looking. Avoid "designer fonts," such as Comic Sans, and cursive fonts, such as Shelley Vollante. In addition, use a 10- to 12-point font for your text. Smaller font sizes are hard to read; larger font sizes look unprofessional. Headings can be boldface and 12- to 14-point font size.

Avoid Sentences. Sentences create four problems in a resume. First, if you use sentences, the majority of them will begin with the first-person pronoun "I." You will write, "I have . . . ," "I graduated . . . ," or "I worked. . . . " Such sentences are repetitious and egocentric. Second, if you choose to use sentences, you will risk committing grammatical errors: run-ons, dangling modifiers, agreement errors, and so forth. Next, sentences take up room in your resume, making it longer than necessary. Finally,

Table 9.2 Active Verbs to Highlight Achievements

accomplished	designed	initiated	planned
achieved	developed	installed	prepared
analyzed	diagnosed	led	presented
awarded	directed	made	programmed
built	earned	maintained	reduced
completed	established	managed	resolved
conducted	expanded	manufactured	reviewed
coordinated	gained	negotiated	sold
created	implemented	ordered	supervised
customized	improved	organized	trained

sentences will lead to paragraphing. Paragraphs create large blocks of text that are not easily accessible at a glance.

Itemize Your Achievements. Use bulleted lists to highlight your accomplishments, awards, unique skills, and so on.

Begin Your Lists with Verbs. To convey a positive, assertive tone, use verbs when describing your achievements. See Table 9.2 for a list of verbs you might use.

Quantify Your Achievements. Your resume should not tell your readers how great you are. It should prove your worth. To do so, quantify by specifically explaining your achievements.

before

Maintained positive customer relations with numerous clients.

Improved field representative efficiency through effective training.

Achieved production goals.

after

Maintained positive customer relations with 5,000 retail and 90 wholesale clients.

Improved field representative efficiency by writing corporate manuals for policies and procedures.

Achieved 95 percent production, surpassing the company's desired goal of 90 percent.

Format Your Resume for Reader-Friendly Ease of Access. Make your resume more accessible. To do so, capitalize and boldface headings, indent subheadings to create white space, underline subheadings and italicize major achievements, use bullets to create lists, and use different font sizes to highlight headings, subheadings, and achievements.

Make Your Resume Perfect. You cannot afford to make an error in your resume. Remember, your resume is the first impression you will make on your prospective employer. Errors in your resume will make a poor first impression.

Methods of Delivery

When writing either a reverse chronological or a functional resume, you can deliver your document several ways.

Mail Version

The traditional way to deliver a resume is to insert it into an envelope and mail your document. This resume can be highly designed, using bullets, boldface, horizontal rules, indentations, and different font sizes. Since this document will be a hard copy, what the reader sees will be exactly what you mail. See Figures 9.1 and 9.2 for traditional mail versions of a reverse chronological and functional resume.

E-Mail Resume

When using e-mail, you can send your resume either as an attachment or as part of the actual e-mail message. If you send the resume as an attachment, write a brief e-mail cover message (discussed later in the chapter, under letters of application). In addition, you must clarify what software you have used for this attachment—Word, Works, WordPerfect, or Rich Text Formats (rtf) for example.

Scannable Resume

Many companies use computers to screen resumes with a technique called *electronic applicant tracking*. The company's computer program scans resumes as raster (or bitmap) images, using optical character recognition (OCR) software. This software uses artificial intelligence to read the text, scanning for keywords. To accommodate resume scanning software, you want to create a "bare bones" resume, one without typical document design enhancements like boldface, designer fonts, computer-generated bullets, indentations, and different font sizes.

Figure 9.3 is an example of a scannable resume.

Criteria for Effective Letters of Application

Your resume, whether paper or online, will be prefaced by a letter of application (or cover e-mail message if you submit the resume as an e-mail attachment). Criteria for an effective letter of application include the following.

Essential Components

Letters contain certain mandatory components (see Chapter 4 for more information about letter formats, spacing, and font selection):

- Writer's address
- Date
- Reader's address
- Salutation
- Text
- Complimentary close
- Signed name
- Typed name
- Enclosure notation

Sharon J. Barenblatt

1901 Rosebud Avenue
Boston, MA 12987
Cell phone: 202-555-2121
E-mail: sharonbb@juno.com

OBJECTIVE:

Employment as an account manager in public relations, using my education, work experience, and interpersonal communication skills to generate business.

EDUCATION:

BS, Business. Boston College. Boston, MA 2005
- 3.2 GPA
- Social Justice Chair, Sigma Delta Tau, 2004
- Study Abroad Program, Milan, Italy, 2003
- Internship, Ace Public Relations, Boston, MA 2002

Frederick Douglas High School. Newcastle, MA 2001
- 3.5 GPA
- Member, Honor Society
- Captain, Frederick Douglas High School tennis team

WORK EXPERIENCE:

Salesperson/assistant department manager. Jessica McClintock Clothing Store. Boston, MA 2004 to present.
- Prepare nightly deposits, input daily receipts
- Open and close the store
- Provide customer service
- Trained six new employees

Salesperson. GAP Clothing. Newcastle, MA 1999 to 2001
- Assisted customers
- Stocked shelves

PROFESSIONAL SKILLS:
- Made oral presentations to the Pan-Hellenic Council to advertise sorority philanthropic activities, 2004
- Helped plan community-wide "Paul Revere's Ride Day," 2002
- Created advertising brochures and fliers, 2002

Figure 9.1 Reverse Chronological Resume

You could replace the "Objective" line with a "Summary of Qualifications," as follows:

Summary of Qualifications
- Over five years of customer service experience
- Bachelor's degree in business
- Proven record of written and interpersonal communication skills
- Fluent in Italian

List your education and work experience in reverse chronological order.

Do not only list where you worked and when you worked there. Also include your job responsibilities.

If you are submitting an electronic resume along with an online letter of application, you will not need these letter essentials. Type the application letter as part of your online resume or send the letter as an e-mail attachment (discussed later in this chapter).

Figure 9.2
Functional Resume

In a functional resume, emphasize skills you have acquired that relate to the advertised position. Also quantify your accomplishments.

In a functional resume, you still should list education and work experience in reverse chronological order.

A functional resume is organized by importance. Begin with the skills or accomplishments that will get you the job. Place less important information lower in the resume.

JODY R. SEACREST
1944 W. 112th Street
Salem, OR 64925
(513) 451-4978 jseacrest12@hotmail.com

PROFESSIONAL SKILLS

- Operated sporting goods/sportswear mail-order house. Business began as home-based but experienced 125 percent growth and was purchased by a national retail sporting goods chain.
- Managed a retail design studio producing over $500,000 annually.
- Hired, trained, and supervised an administrative staff of 15 employees for a financial planning institution.
- Sold copiers through on-site demonstrations. Exceeded corporate sales goals by 10 percent annually.
- Provided purchaser training for office equipment, reducing labor costs by 25 percent.
- Acquired modern management skills through continuing education courses.

WORK EXPERIENCE

Office manager, Simcoe Designs, Salem, OR 2004 to present.
Sales representative, Hi-Tech Office Systems, Salem, OR 2000 to 2004.
Office manager, Lueck Finances, Portland, OR 1998 to 2000.
President, Good Sports, Inc., Portland, OR 1996 to 1998.

COMPUTER PROFICIENCY

Microsoft Office XP, Visual Basic 6, C++, Oracle, Microsoft SQL Server, Network Administration.

MILITARY EXPERIENCE

Corporal, U.S. Army, Fort Lewis, WA 1992–1996. Honorably Discharged.

- Served as Company network administrator.
- Planned and budgeted all IT purchases.

EDUCATION

BA, Communication Studies, Portland State University, Portland, OR 1992.

Introduction

In your introductory paragraph, include the following: tell where you discovered the job opening, state which specific job you are applying for, and sum up your best credentials.

Discussion

In the discussion paragraph(s), sell your skills: focus on your assets uniquely applicable to the advertised position, and quantify your abilities.

Rochelle J. Kroft
1101 Ave. L
Tuscaloosa, AL 89403
Home: (313) 690-4530
Cell: (313) 900-6767
E-mail: rkroft90@aol.com

OBJECTIVES
==
To use HAZARDOUS WASTE MANAGEMENT experience and knowledge to ensure company compliance and employee safety.

PROFESSIONAL ACCOMPLISHMENTS
==
* HAZARDOUS WASTE MANAGEMENT with skills in teamwork, end-user support, quality assurance, problem solving, and written documentation.
* Five years' experience working with international and national businesses and regulatory agencies.
* Skilled in assessing environmental needs and implementing hazardous waste improvement projects.
* Able to communicate effectively with multinational, cross-cultural teams, consisting of clients, vendors, coworkers, and local and regional stakeholders.
* Excellent customer service, using strong problem-solving techniques.
* Effective project management skills, able to multitask.

COMPUTER PROFICIENCY
==
Microsoft Windows XP, FrontPage, PowerPoint, C++, Visual Basic, Java, CAD/CAM

EXPERIENCE
==
Hazardous Waste Manager
Shallenberger Industries, Tuscaloosa, AL (2000 to present)
* Assess client needs for root cause analysis and recommend strategic actions.
* Oversee waste management improvements, using project management skills.
* Conduct and document follow-up quality assurance testing on all newly developed applications to ensure compliance.
* Develop training manuals to ensure team and stakeholder safety. Shallenberger has had NO injuries throughout my management.
* Manage a staff of 25 employees.
* Achieved "Citizen's Recognition" Award from Tuscaloosa City Council for safety compliance record.

Hazardous Waste Technician
CleanAir, Montgomery, AL (1998–2000)
* Developed innovative solutions to improve community safety.
* Created new procedure manuals to ensure regulatory compliance.

EDUCATION
==
B.S. Biological Sciences, University of Alabama, Tuscaloosa, AL (1997)
* Biotechnology Honor Society, President (1996)
* Golden Key National Honor Society

AFFILIATIONS
==
Member, Hazardous Waste Society International

Figure 9.3
Scannable Resume

Place your name at the top of a scannable resume and avoid centering text.

Use key words to summarize your accomplishments.

Use equal signs for design elements, such as line dividers, and asterisks (*) for bullets.

Type your scannable resume in Courier, Arial, Verdana, or Helvetica. Avoid "designer" fonts like Comic Sans, Lucida, or Corsiva.

Write past jobs in past tense, present jobs in present tense.

Conclusion

Your final paragraph should ask for an interview and provide contact information. You could say, "I will call you next week to discuss the possibility of an interview" or "I am looking forward to discussing my application with you in greater detail." You should tell your reader how to get in touch with you: "I can be reached at (913) 469-8500." In addition to these suggestions, you should mention your enclosed or attached resume. See Figure 9.4 for a sample letter of application.

E-Mail Cover Message

If you submit the resume electronically, attach it to a brief e-mail cover message. This e-mail message will serve the same purpose as a hard-copy letter of application. Include an informative subject line, an introductory paragraph, a body, and a conclusion.

- **Subject line.** Corporate employees receive approximately 50 e-mail messages a day. Many of these messages are spam and could contain viruses. Thus, employees fear opening unknown e-mail messages and will delete them quickly. An effective subject line will help you avoid this problem. In the subject line of an e-mail cover message, announce your intentions or the contents of the e-mail clearly. Try subject lines as follows: "Resume—Vanessa Diaz" or "Response to Accountant Job Opening."
- **Introduction.** Tell the reader which job you are applying for and where you learned of this position.
- **Body.** State that you have attached a resume. Since the resume will not be a hard-copy printout, you need to ensure that the audience can open your file. Tell the reader which software you have used to write your resume.

> "I have attached a resume for your review. To open this document, you will need Microsoft Word."
>
> "I have attached a resume, saved as an RTF (rich text format) file."

  Briefly explain why you are the right person for the job. You can do this in a short paragraph or by briefly listing three to five key assets.

- **Conclusion.** Sum up your e-mail message pleasantly. Tell your reader that you look forward to an interview.

Online Application Etiquette

If you send your resume as an attachment to an e-mail message, be sure to follow online etiquette:

- Do not use your current employer's e-mail system.
- Avoid unprofessional e-mail addresses, such as "Mustang65@aol.com," "Hangglider@yahoo.com," or "IluvDaBears@juno.com." These will create a poor and unprofessional first impression.
- Send one e-mail at a time to one prospective employer.

Figure 9.5 shows an effective cover e-mail prefacing an attached resume.

Figure 9.4 Letter of Application

11944 West 112th Street
St. Louis, MO 66221

December 11, 2006

Mr. Harold Irving
CEO
DiskServe
9659 W. 157th St.
St. Louis, MO 78580

Dear Mr. Irving:

I am responding to your advertisement in the November 24, 2006, issue of *The St. Louis Courier* for a job in your computer technology department. Because of my two years of experience in computer information systems, I believe I have the skills you require.

The introduction mentions where the writer found the job listing, which job he is applying for, and key qualifications for the position.

Although I have enclosed a resume, let me elaborate on my achievements. While working as computer technician for Radio Shack, I troubleshot motherboards, worked the help desk, and made service calls to businesses and residential customers. Your advertisement listed the importance of customer service and technical skills. I have expertise in both areas.

The letter's body quantifies the writer's abilities. The body also shows how the writer meets the job requirements.

Your job description also mentioned the importance of working in a team environment. At Radio Shack, I frequently made service calls with a team of technicians. Together, we provided quality service.

I would like to meet with you and discuss employment possibilities at your company. Please call me at (913) 469-8500 so that we can set up an interview at your convenience.I appreciate your consideration.

The conclusion ends positively, using words and phrases like "convenience," "Please," and "appreciate." It also provides follow-up information.

Sincerely,

Macy G. Heart

Enclosure (Resume)

Writer's Insight

Macy says, "I started looking for a job six months before graduation. I attended my college's job fair, sent out over two dozen applications, tried online job searches, and networked with all of my professors. First, I applied to every job opening I found even remotely related to my degree. After going on six interviews and getting no offers, however, I limited my focus. I was more selective, looking only for jobs that I knew fit my interests and degree exactly. Finding a job that perfectly suited my credentials took time, but the patience and effort were worth the wait. After I wrote this letter to DiskServe, I got an interview and a job offer. I start my job in two weeks!"

Figure 9.5 E-mail Cover Message

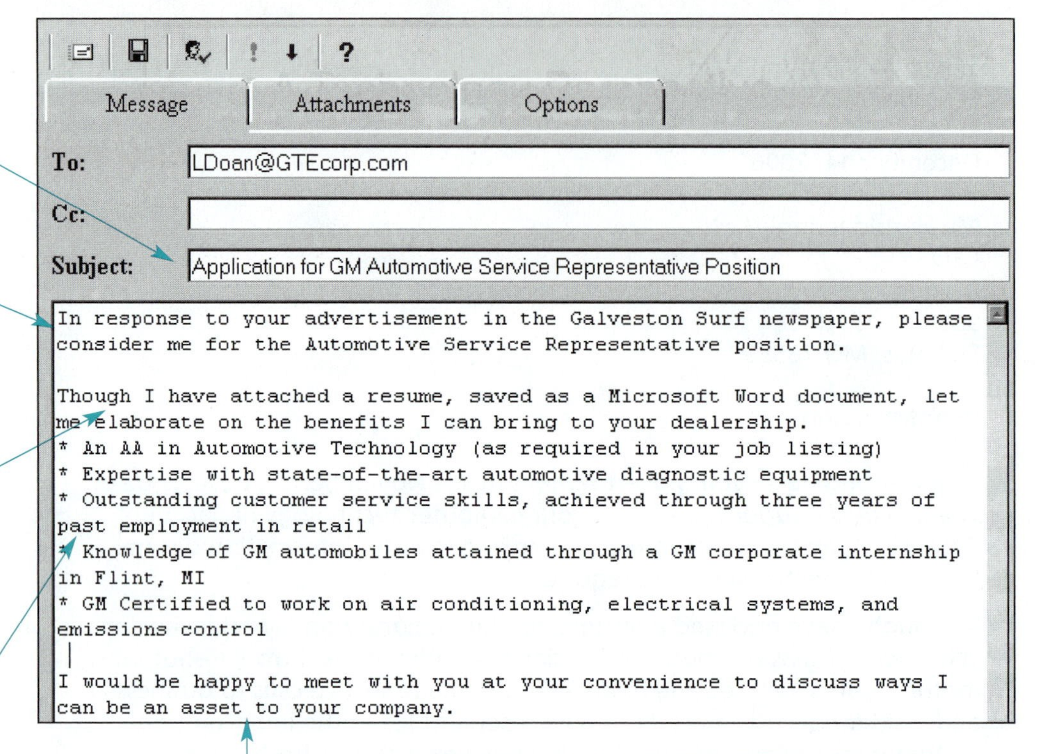

Message | Attachments | Options

To: LDoan@GTEcorp.com

Cc:

Subject: Application for GM Automotive Service Representative Position

```
In response to your advertisement in the Galveston Surf newspaper, please
consider me for the Automotive Service Representative position.

Though I have attached a resume, saved as a Microsoft Word document, let
me elaborate on the benefits I can bring to your dealership.
* An AA in Automotive Technology (as required in your job listing)
* Expertise with state-of-the-art automotive diagnostic equipment
* Outstanding customer service skills, achieved through three years of
past employment in retail
* Knowledge of GM automobiles attained through a GM corporate internship
in Flint, MI
* GM Certified to work on air conditioning, electrical systems, and
emissions control

I would be happy to meet with you at your convenience to discuss ways I
can be an asset to your company.
```

Subject line clarifying the purpose of the e-mail.

Introduction stating where you found the job listing and what job you are applying for.

Lead-in to body stating what software has been used for the attachment.

A brief, itemized list of your assets, specifically related to the job opening.

A conclusion ending upbeat and positive.

NOTE: Standard keystrokes, such as asterisks for bullets and paragraph spacing, are used for highlighting. This ensures that any e-mail package can read the text.

Techniques for Interviewing Effectively

Some sources suggest that the interview is *the most important stage of your job search*. The Society for Human Resource Management states that 95 percent of respondents to a 2003 survey ranked "interview performance" as "very influential" when deciding to hire an employee. "Interview performance [is] more influential than 17 other criteria, including years of relevant work experience, resume quality, education levels, test scores or references" (Stafford L1).

To interview successfully, use the following techniques:

Look Professional. Professionalism starts with your appearance. The key to successful dressing is to wear clean, conservative clothing. No one expects you to spend money on high-fashion, stylish clothes, but everyone expects you to look neat and acceptable. Business suits are still best for both men and women.

Be on Time. Plan to arrive at your interview at least 20 to 30 minutes ahead of schedule.

Watch Your Body Language. To make the best impression, don't slouch, chew your fingernails, play with your hair or jewelry, or check your watch. These actions will make you look edgy and impatient. Sit straight in your chair, even leaning forward a little to show your enthusiasm and energy. Look your interviewer in the eye. Smile and shake your interviewer's hand firmly.

Don't Chew Gum, Smoke, or Drink Beverages During the Interview. The gum might distort your speech; the cigarette will probably offend the interviewer, particularly if he or she is a non-smoker; and you might spill the beverage.

Turn off Your Cell Phone. Today, cell phones are commonplace. However, the interview room is one place where cell phones must be avoided. Taking a call while you are being interviewed is rude and will ensure that you will not be hired.

Watch What You Say and How You Say It. Speak slowly, focus on the conversation, and don't ramble. Once you have answered the questions satisfactorily, stop.

Bring Supporting Documents to the Interview. Supporting documents can include extra copies of your resume, a list of references, letters of recommendation, employer performance appraisals, transcripts, and a portfolio of your work.

Research the Company. Show the interviewer that you are sincerely interested in and knowledgeable about the company. Dr. Judith Evans, vice president of Right Management Consultants of New York, says that the most successful job candidates show interviewers that they "know the company inside and out" (Kallick D1).

Be Familiar With Typical Interview Questions. You want to anticipate questions you will be asked and be ready with answers. Some typical questions include the following:

- What are your strengths?
- What are your weaknesses?
- Why do you want to work for this company?
- Why are you leaving your present employment?
- What did you like most about your last job?
- What did you like least about your last job?
- What computer hardware are you familiar with, and what computer languages do you know?
- What machines can you use?
- What special techniques do you know, or what special skills do you have?
- How do you get along with colleagues and with management?
- Can you travel?
- Will you relocate?
- What starting salary would you expect?
- What do you want to be doing in five years, ten years?
- What about this job appealed to you?
- How would you handle this (hypothetical) situation?
- What was your biggest accomplishment in your last job or while in college?
- What questions do you have for us?

Writer's Insight

DiskServe's CEO, Harold Irving, takes interviewing seriously. To hire the best candidate, he asks what he calls "behavioral questions" to learn more about a prospective hire's evidence of success on the job. "For example, I'll ask the potential job candidate to 'tell me about a situation in which you had to deal with an upset customer.' Then, I look for an articulate answer that clearly and specifically proves the applicant has solved problems in the past. People tend to repeat behavior, so they will probably respond on the job just as they have in similar past situations," Harold says. "If the applicant responds, 'I tried to help but the customer kept on yelling, so I just quit trying to help and walked away,' that's a red flag.

I also gauge the applicant's preparation. If they arrive at the interview with extra copies of resumes, names and telephone numbers of references, or transcripts, then I assume they are professional and more likely to be prepared on the job. In contrast, if they have poor posture, speak in a monotone, don't make eye contact, and dress inappropriately, then their general day-to-day job enthusiasm and preparedness might be lacking as well.

Trust me—a job interview tells me more about the applicant than the resume or cover letter."

Criteria for Effective Follow-Up Correspondence

Once you have interviewed, do not sit back and wait, hoping that you will be offered the job. Write a follow-up letter or e-mail message. This follow-up correspondence

For an e-mail follow-up, you would include your reader's e-mail address and a subject line, such as "Thank You for the July 8 Interview" or "Follow-up to July 8 Interview."

For a hard-copy follow-up letter, you would include all letter components: writer's address, date, reader's address, salutation, complimentary close, and signature.

Thank you for allowing me to interview with Acme Corporation on July 8. I enjoyed meeting you and the other members of the team to discuss the position of Account Representative.

You stated in the interview that Acme is planning to expand into international marketing. With my Spanish speaking ability and my study-abroad experience, I would welcome the opportunity to become involved in this exciting expansion.

Again, thank you for your time and consideration. I look forward to hearing from you. Please e-mail me at gfiefer21@aol.com.

Figure 9.6 Follow-up Correspondence

accomplishes three goals. It thanks your interviewers for their time, keeps your name fresh in their memories, and gives you an opportunity to introduce new reasons for hiring you.

A brief follow-up letter or e-mail message contains the following:

- **Introduction.** Tell the reader(s) how much you appreciated meeting them. Be sure to state the date on which you met and the job for which you applied.
- **Discussion.** Emphasize or add important information concerning your suitability for the job. Add details that you forgot to mention during the interview, clarify details that you covered insufficiently, or highlight your skills that match the job requirements.
- **Conclusion.** Thank the reader(s) for their consideration, and remind them how they can get in touch with you for further information.

See Figure 9.6 for a sample follow-up correspondence.

CHECKLISTS

The following checklists will give you the opportunity for self-assessment and peer evaluation of your writing.

Job Search Checklist

__ 1. Did you visit your college or university job placement center?
__ 2. Did you talk to your professors about job openings?
__ 3. Have you networked with friends or past employers?
__ 4. Have you checked with your professional affiliations or looked for job openings in trade journals?
__ 5. Did you read the want ads in the newspapers?
__ 6. Did you search the Internet for job openings?
__ 7. Did you consider taking a "temp" job or applying for an internship?
__ 8. Have you had any informational interviews?

Resume Checklist

__ 1. Have you decided to write a reverse chronological resume or a functional resume?

__ 2. Have you chosen which method of delivery to use (mail, e-mail, or scannable)?

__ 3. Have you included contact information (name, address, phone numbers, e-mail address, Web site URL, or fax number)?

__ 4. Have you provided an Objective line that is specific and focuses on company benefits?

__ 5. In the Work Experience section, have you included your job title, company name, city, state, and date of employment?

__ 6. In the Education section of the resume, have you included your degree, school, city, state, and date of graduation?

__ 7. Have you itemized specific achievements at work and school, beginning with verbs?

__ 8. Have you avoided using sentences, passive voice, and the word "I"?

__ 9. Does your resume use highlighting techniques to make it reader-friendly?

__ 10. Have you proofread your resume to find grammatical and mechanical errors?

Letter of Application Checklist

Letter of Application

__ 1. Have you included all of the letter essentials?

__ 2. Does your introductory paragraph state where you learned of the job, which job you are applying for, and your interest in the position?

__ 3. Does your letter's discussion unit pinpoint the ways in which you will benefit the company?

__ 4. Does your letter's concluding paragraph end cordially and explain what you will do next or what you hope your reader will do next?

__ 5. Is your letter free of all errors?

E-mail Cover Message

__ 1. Have you avoided an unprofessional e-mail address?

__ 2. Do you have a precise subject line?

__ 3. Did you include an introduction, body, and conclusion?

__ 4. Did you tell the reader which software you have used for your attached resume?

__ 5. Have you kept your e-mail message short (20–25 lines)?

__ 6. Have you considered online e-mail application etiquette?

Interview Checklist

__ 1. Will you dress appropriately?

__ 2. Will you arrive ahead of time?

__ 3. Will you avoid gum, cigarettes, and beverages?

__ 4. Have you practiced answering potential questions?

__ 5. Have you researched the company so you can ask informed questions?

__ 6. Will you bring to the interview additional examples of your work or copies of your resume?

Follow-Up Correspondence Checklist

__ 1. Have you included all the letter essentials, such as writer's address, date, reader's address, etc.?

__ 2. Does your introductory paragraph remind the readers when you interviewed and what position you interviewed for?

__ 3. Does the discussion unit highlight additional ways in which you might benefit the company?

__ 4. Does the concluding paragraph thank the readers for their time and consideration?

__ 5. Does your letter avoid all errors?

Case Studies

1. Macy G. Heart is looking for a job. He plans to submit the resume provided on page 277. However, the resume has flaws.

Assignment

Rewrite the resume to improve it as follows:

- Revise the stylistic and organizational errors.
- Create three different types of resumes—a chronological resume, a functional resume, and a scannable resume.

2. DiskServe, a St. Louis-based company, is hoping to hire a customer service representative for their computer technology department. In addition to DiskServe's CEO, Harold Irving, the hiring committee will consist of two managers from other DiskServe departments and two coworkers in the computer technology department.

   The position requires a bachelor's degree in information technology (or a comparable degree) and/or four years' experience working with computer technology. Candidates must have knowledge of C++, Visual Basic (VB), SQL, Oracle, and Microsoft Office applications. In addition, customer service skills are mandatory.

   Four candidates were invited to DiskServe's work site for personal interviews: Macy Heart, Aaron Brown, Rosemary Lopez, and Robin Scott.

   Macy has a bachelor's degree in computer information systems. He has worked two years part-time in his college's technology lab helping faculty and students with computer hardware and software applications, including Microsoft Office and Visual Basic. He worked for two years at a computer hardware/software store as a salesperson. His supervisor considers Macy to be an outstanding young man who works hard to please his supervisors and to meet customer needs. According to the supervisor, Macy's greatest strength is customer service, since Macy is patient, knowledgeable, and respectful.

   Aaron has an information technology certificate from Microsoft, where he has worked for five years. Aaron began his career at Microsoft as a temporary office support assistant, but progressed to a full-time salesperson. When asked where he saw himself in five years, Aaron stated, "The sky's the limit." References proved Aaron's lofty goals by calling him a self-starter, very motivated, hardworking, and someone with excellent customer service skills. He is taking programming courses at night from the local community college, focusing on C++, Visual Basic, and SQL.

   Rosemary has an associate's degree in information technology. She has five years of experience as the supervisor of Oracle application. Prior to that, Rosemary worked with C++, VB, and SQL. She also has extensive knowledge of Microsoft Office. Rosemary was asked how she handled customer complaints in the past. She responded, "I rarely handle customer complaints. In my past job, I assigned that work to my subordinates."

1890 Arrowhead Dr.
Utica, MO 51246
710-235-9999

Resume of Macy G. Heart

Objective: Seeking a position in Computer Technology Customer Service where I can use my many technology and people-person skills. I want to help troubleshoot software and hardware problems and work in a progressive company which will give me an opportunity for advancement and personal growth.

Work Experience

Jan. 2004 to now Aramco.net St. Louis, Missouri Tech Support Specialists
Primarily I provide customer support for customer problems with C++, Visual Basic, Java, Networking, and Databases (Access, Oracle, SQL). I provide solutions to software and hardware problems and respond to e-mail queries in a timely manner.

Oct. 2002 to Dec. 22, 2003 DocuHelp Chesterfield, MO Computer Consultant
Provide technical support for PC's and Mac's. I also trained new PC and Mac users in hardware applications. When business was slow, I repaired computer problems, using my many technology skills.

May 2002 to Oct. 2002 Ram-on-the-Run East St. Louis, Illinois Computer Salesman
Sold laptops, PCs, printers, and other computer accessories to men and women. Answered customer questions. Won "Salesman of the Month Award" three months in a row due to exceeding sales quotas.

Jan. 2002 to May 2002 Carbondale High School Carbondale, IL Lab Tech
Worked in the school's computer lab, helping Mr. Jones with computer-related classwork. This included fixing computer problems and tutoring new students having trouble with assignments.

Education

Aug. 2004 Bachelor's Degree, Information Technology, Carbondale Institute of Technology, Carbondale, IL
Concentration: Database/Programming Applications
Relevant classes: Business Information Systems, Hardware Maintenance, Database Management, Visual Basic, Systems Analysis and Design, C++, Web Design

May 2000 Graduate Carbondale High School
Member of the Computer Technology Club
Member of FFA and DECA
Principal's Honor Roll, senior year

Computer Expertise

Cisco Certified, knowledge of C++, VB, Microsoft Office Suite XP, HTML, Java, SML

References

Mr. Oscar Jones, Computer Applications Teacher, Carbondale High School
Mr. Renaldo Gomez, Manager, Aramco.net
Mr. Ted Harriot, Technical Support Supervisor, DocuHelp

Additional Information

A good team player, who works well with others
Made all A's in my college major classes
Built my own computer from scratch in high school
Starting football player in high school, Junior Varsity tight end

Robin has a bachelor's degree in information technology. To complete her degree, Robin took courses in C++, VB, Oracle, and SQL. She is very familiar with Microsoft Office. Since Robin just graduated from college, she has no full-time experience in the computer industry. However, she worked in various retail jobs (food services, clothing stores, and bookstores) during high school, summers, and in her senior year. She excelled in customer service, winning the "Red Dragon Employee of the Month" award from her last job as a server in a Chinese restaurant.

Assignment

Who would you hire? Give an oral presentation or write an e-mail or memo to Harold Irving, DiskServe's CEO, explaining which of the candidates he should hire. Explain your choice.

INDIVIDUAL AND TEAM PROJECTS

1. Practice a job search. To do so, find examples of five to ten job openings in your degree area. Perform this job search by looking at the following: newspapers, professional journals, your college or university's career placement service, and online sources. Once you have found examples of job listings, complete the following:
   - Determine which job listings are most precise in their requirements.
   - Draft a letter of application to the top three job listings.
2. Researching your newspaper's classified section, your school's career planning and placement office, your work site's personnel office, or a trade journal, find a job advertisement to which you could apply (based on your education, work experience, or professional skills). Then write a hard-copy letter of application according to the suggestions provided in this chapter.
3. Write a functional or reverse chronological resume focusing on your work experience, education, and professional skills. To do so, follow the suggestions provided in this chapter. Once you have constructed this resume, bring it to class for peer review. In small groups, discuss each resume's successes and areas needing improvement. Then rewrite the resume based on your team's suggested revisions.
4. Practice a job interview in small groups (three to five students) as follows:
   - Ask each student to find a job opening to which he or she could apply (based on the student's education, work experience, and skills).
   - Have the students bring the hard-copy job listing to class.
   - Allow each student to be the interviewee while other students act as an interview panel.
   - Let the interview panel ask the applicant 10 sample interview questions provided in this chapter or other questions found online.
   - Rotate each student through this process, asking 10 different questions to each interviewee.

   After all interviews have concluded, write a memo or e-mail to your instructor explaining which student (other than yourself) should be hired. Base your decision on the criteria for successful interviews provided in this chapter.

Problem Solving Think Pieces

1. You need to submit a resume for a job opening. However, you have problems with your work history, such as the following:

   - You have had no jobs.
   - You have only worked as a babysitter, or you have cut grass in your neighborhood.
   - You have been fired from a job.
   - You have had five (or more) jobs in one year.

   Consider how you would meet the challenges of your job history.

2. You have found a job that you want to apply for. The job requires a bachelor's degree in a specific field. Though you had been enrolled in that specific degree program for three years, you never completed the degree.

   What should you say in your resume and letter of application to apply for this position, even though you do not meet the degree requirement? Which type of resume (reverse chronological or functional) should you write? Explain your answer.

3. You are in the middle of an interview. Though the interview had been scheduled for 1:00 p.m. to 1:45 p.m., it is running late. You had planned to pick up your son from day care at 2:00 p.m. How do you handle this problem?

4. You have just completed the interview process for a job in your field. You have impeccable credentials, meeting and exceeding all of the requirements. Your interview went very well.

   At the close of the interview, one of the interviewers says, "Thank you for interviewing with us. You did a great job answering our questions, and your credentials are truly excellent. However, I think you would find the job unchallenging, maybe even boring. You are overqualified."

   How should you handle this situation?

Web Workshop

1. Using a search engine, type in *http://www.resumedoctor.com/Resource Center.htm.* to access The ResumeDoctor.com Resource Center. This Web site provides articles about employment advice, resume, interviews, and job seeking. For example, you can scroll to click on links about tax tips for the unemployed, recruiter resume pet peeves, innovative job seeking tips, and seasonal employment tips.

   Select an article that interests you, read it, and report your findings either in writing (memo, letter, e-mail, or report) or give an oral presentation to your class.

2. Using an Internet search engine, find job openings in your area of interest. Which companies are hiring, and what skills do they want from prospective employees? Report your findings either to your instructor by writing a memo or e-mail or give an oral presentation to your class about the job market in your field.

Oral Communication

OBJECTIVES

When you complete this chapter, you will be able to do the following:

1. Communicate orally with customers, vendors, and coworkers.
2. Use the telephone to communicate successfully.
3. Use voice mail effectively.
4. Deliver effective informal oral presentations on the job.
5. Participate in teleconferences.
6. Communicate effectively in a videoconference or Web conference.
7. Deliver formal oral presentations.
8. Use a variety of visual aids to enhance your oral presentations, including techniques for effective PowerPoint presentations.
9. Use the writing process—prewriting, writing, and rewriting—for informal and formal oral presentations.
10. Test your knowledge of oral communication through end-of-chapter activities:
    - Case Studies
    - Individual and Team Projects
    - Problem Solving Think Pieces
    - Web Workshop

COMMUNICATION AT WORK

In the TechStop scenario, a customer, Carolyn Jensen, complains about malfunctioning equipment. To do so, she uses a variety of oral communication channels, including face-to-face conversation, voice-mail messages, and telephone calls.

TechStop sells DVDs, VCRs, TVs, audio components, computers, and computer peripherals. Shuan Wang is the customer service representative. Carolyn Jensen, a longtime customer at TechStop, walks up to Mr. Wang's desk to discuss a malfunctioning printer. During their conversation, Mr. Wang learns the following:

Ms. Jensen's printed copy looks different from the text on the screen, she cannot print envelopes, and her graphics are not printing in color though her color ink cartridge is full. She purchased the printer 12 months and two weeks ago. The printer's in-store warranty was for 12 months, guaranteeing full replacement of parts and labor coverage.

Ms. Jensen realizes that the warranty expired two weeks ago. However, she contends that several facts should negate this deadline. The deadline expired over a Labor Day weekend, she was out of town on business, and a severe thunderstorm delayed her return flight. Perhaps most importantly, her company has done business with TechStop for six years, purchasing over 1,000 pieces of equipment, including computers, printers, paper items, cell phones, pagers, and radios for a fleet of trucks.

Shuan Wang empathizes with her dilemma, but he is powerless at the moment. Shuan tells her that he has to research the situation, talk with his supervisor, and discuss the issue with the printer manufacturer. He says he will contact her by telephone when he gets more information.

One week later, Ms. Jensen has still not received the promised phone call. She calls the store's customer service department to talk with Shuan. He is out of the office, so she leaves a voice-mail message. In the message, she gives her name, phone number, the problem ticket number for her printer, and the date of their conversation. Carolyn asks Mr. Wang to return her call with a progress update. She waits three days but still does not hear from Shuan.

When Carolyn next calls TechStop, she asks for the customer service supervisor. She speaks to the supervisor, Hal Lang, telling him her name, company name, telephone number, the situation with her printer, and her recent contact with Mr. Wang. Hal has no record of Ms. Jensen's complaint.

Two days later, Hal Lang returns Carolyn's call and does not reach her but leaves a voice-mail message. In the message, he tells Carolyn that TechStop will honor her request for extended warranty coverage. Hal also tells Carolyn that he is implementing a training program to improve customer service at TechStop.

Types of Oral Communication

Many people, even the seemingly most confident, are afraid to speak in front of other people. This chapter offers techniques to make your oral communication experience rewarding rather than frightening.

You may have to communicate orally with your peers, your subordinates, your supervisors, and the public. Oral communication is an important component of your business success because you will be required to speak

- On an *everyday basis* to colleagues, customers, and vendors
- *Informally* to coworkers and clients
- *Formally* to large and small groups

Everyday Oral Communication

"Hi. My name is Bill. How may I help you?" Think about how often you have spoken to someone today or this week at your job. You constantly speak to customers, vendors, and coworkers face to face, on the telephone, or by leaving messages on voice mail.

- If you work an 800-hotline, your primary job responsibility is oral communication.
- If you are an account executive, every time you answer the telephone, your oral communication skills are on display.
- As a customer service representative, your first line of communication might be oral.
- When you return the dozens of calls you receive or leave voice-mail messages, each instance reveals your communication abilities.

The goal of effective oral communication is to ensure that your verbal skills make a good impression and communicate your messages effectively.

Telephone and Voice Mail

You speak on the telephone dozens of times each week. When speaking on a telephone, make sure that you do not waste either your time or your listener's time.

Ten Tips for Telephone and Voice-Mail Etiquette

1. Know what you are going to say before you call.
2. Always speak clearly. Enunciate each syllable.
3. Avoid rambling conversationally.
4. Avoid lengthy pauses.
5. Leave brief messages.
6. Avoid communicating bad news.
7. Repeat your phone number twice, including the area code.
8. Offer your e-mail address as an option.
9. Sound pleasant, friendly, and polite.
10. If a return call is unnecessary, say so.

Informal Oral Presentations

As a team member, manager, supervisor, employee, or job applicant, you often will speak to a coworker, a group of colleagues, or a hiring committee. You will need to communicate orally in an informal setting for several reasons:

- Your boss needs your help preparing a presentation. You conduct research, interview appropriate sources, and prepare reports. When you have concluded your research, you might be asked to share your findings with your boss in a brief, informal oral presentation.
- Your company is planning corporate changes (staff layoffs, mergers, relocations, increases in personnel, etc.). As a supervisor, you want to provide your input in an oral briefing to a corporate decision maker.
- In a team meeting, you participate in oral discussions regarding agenda items.
- Your company is involved in a project with coworkers, contractors, and customers from distant sites. To communicate with these individuals, you participate in a video- or teleconference, orally sharing your ideas.
- You are applying for a job. Your interview, though not a formal, rehearsed presentation, requires that you speak effectively before a hiring committee.

Many informal oral communication instances could be accomplished best through video- and teleconferences.

Video- and Teleconferences

If you are communicating with a group and you want to hear what everyone else is saying simultaneously, video- or teleconferences are an answer. Consider using a video- or teleconference when three or more people at separate locations need to talk.

In a video- or teleconference, you want all participants to feel as if they are in the same room facing each other. With expensive technology in place, such as cameras, audio components, coder/decoders, display monitors, and user interfaces, you want to avoid wasting time and money with poor communication.

Ten Tips for Video- and Teleconferences

1. Inform participants of the conference date, time, time zone, and expected duration.
2. Make sure participants have printed materials before the teleconference.
3. Ensure that equipment has good audio quality.
4. Choose your room location carefully for quiet and privacy.
5. Consider arrangements for hearing impaired participants. You might need a text telephone (TTY) system or simultaneous transcription in a chat room.
6. Introduce all participants.
7. Direct questions and comments to specific individuals.
8. Do not talk too loudly, too softly, or too rapidly.
9. Turn off cell phones and pagers.
10. Limit side conversations.

Web Conferencing

Due to rising costs of airline tickets, time-consuming travel, and the need to complete projects or communicate information quickly, many companies and organizations are using Web conferencing. Web conferencing, sometimes called webinars or virtual meetings, include web-based seminars, lectures, presentations, or workshops transmitted over the Internet (Coyner; Murray).

Ten Tips for Web Conferencing

1. Limit Web conferences to 60 to 90 minutes.
2. Limit a Web conference's focus to three or four important ideas.
3. Limit introductory comments.
4. Keep it simple. Instead of using too many Internet tools, limit yourself to simple and important features like polling and messaging.
5. Plan ahead. Make sure that all Web-conference participants have the correct Internet hardware and software requirements; know the correct date, time, agenda, and Web login information for your Web conference; and have the correct Web URL or password.
6. Before the Web conference, test your equipment, hypertext links, and PowerPoint slide controls.
7. Use both presenter and participant views. One way to ensure that all links and slides work is by setting up two computer stations. Have a computer set on the presenter's view and another computer logged in as a guest. This will allow you to view accurately what the audience sees and how long displays take to load.
8. Involve the audience interactively through questions and/or text messaging.
9. Personalize the presentation. Introduce yourself, other individuals involved in the presentation, and audience members.
10. Archive the presentation. (Coyner; Murray)

FORMAL ORAL PRESENTATIONS

You might need to make a formal presentation for the following reasons:

- Your company asks you to visit a civic club meeting and to provide an oral presentation to maintain good corporate or community relations.
- Your company asks you to represent it at a city council meeting. You will give an oral presentation explaining your company's desired course of action or justifying activities already performed.
- Your company asks you to represent it at a local, regional, national, or international conference by giving a speech.
- A customer has requested a proposal. In addition to writing this long report, you and several coworkers also need to make an oral presentation promoting your service or product to the potential customer.

See Figure 10.1 for the oral presentation process.

Parts of a Formal Oral Presentation

A formal oral presentation consists of an introduction, a discussion (or body for development), and a conclusion.

Introduction. The introduction should arouse and capture your audience's attention and interest. This is the point in the presentation where you are drawing in your listener, hoping to create enthusiasm and a positive impression. You can use a variety of

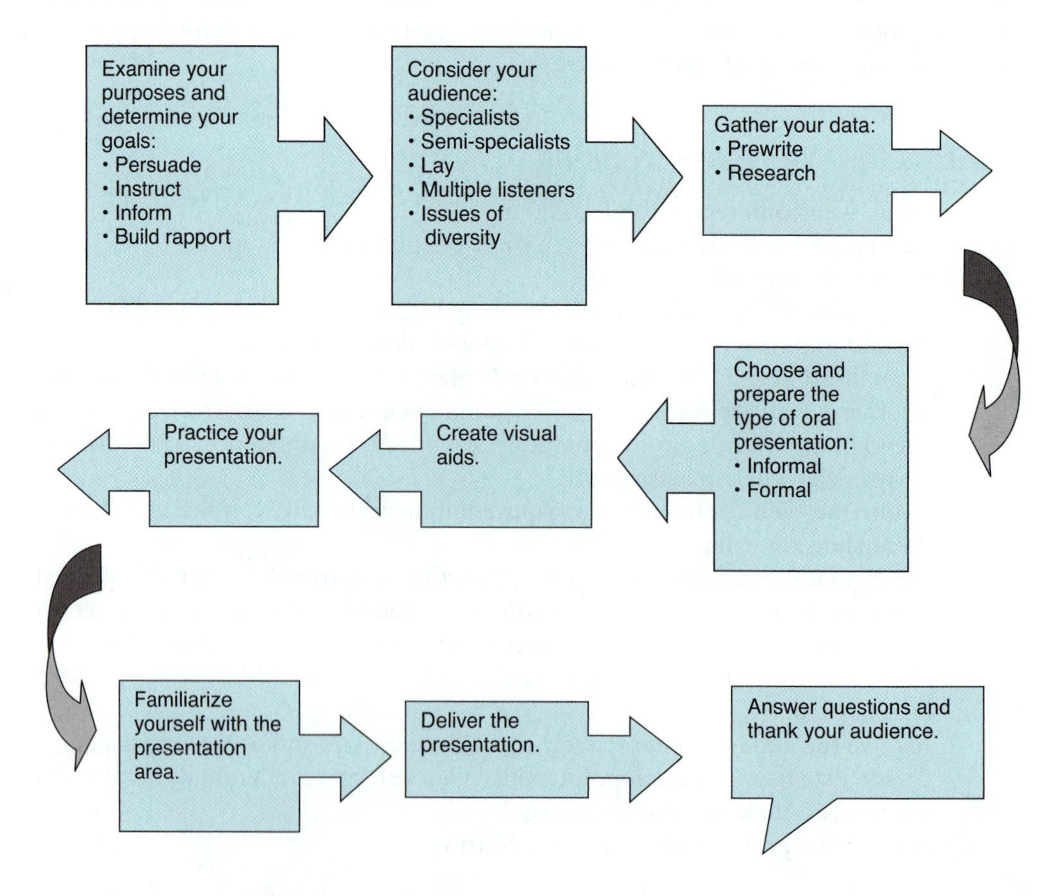

Figure 10.1 Oral Presentation Process

techniques, such as anecdotes, a question or a series of questions, a quotation from a famous person, or facts and figures to lead into your presentation.

Lead-ins to Arouse Reader Interest.

- **An Anecdote.** Anecdotes are short, interesting, and relevant stories. For example, at a workplace communication workshop for engineers, the training facilitator began as follows:

An Anecdote

"Recently I met an engineer who told me this story. Bob had been hired for his engineering expertise. After all, that was what he had trained to do; that was his educational background. On his first month on the job, he needed to write a progress report. He wrote it and turned it in to his boss. A few days later, the boss called him in to his office and said, 'That was an excellent report! It was clear, concise, and easy to read. Thanks for a great job.' Little by little, things started changing for Bob. Other engineers came to him for help, his boss asked him to write important reports, and Bob was promoted rapidly. Bob is now a manager, having moved up the ladder faster than some of his colleagues for whom written communication was more of a challenge. When I asked Bob to explain his success, he simply replied, 'I had added value. In addition to my engineering talents, I could write well.'"

- **A Question or a Series of Questions.** Asking questions involves the audience immediately. The question or questions must be relevant to the topic. The training facilitator in the preceding example could have begun the workshop as follows:

Questions to Involve Your Audience

"How many reports do you write each week or month? How often do you receive and send e-mail messages to customers and colleagues? How much time do you spend on the telephone? Face it—workplace communication is a larger part of your engineering job than you ever imagined."

- **A Quotation from a Famous Person.** Warren Buffet is a famous investor and businessperson. A quote from Mr. Buffet carries weight. If he says something, you can trust that his comments have value. Using the training facilitator as an example, note how beginning a formal oral presentation with a quote can involve the audience and clarify the topic's importance.

A Quote to Arouse Audience Interest

"How important is effective workplace communication? Just listen to what Warren Buffet has to say on the topic:

'For more than forty years, I've studied the documents that public companies file. Too often, I've been unable to decipher just what is being said or, worse yet, had to conclude that nothing was being said. . . . Perhaps the most common problem . . . is that a well-intentioned and informed writer simply fails to get the message across to an intelligent, interested reader. In that case, stilted jargon and complex constructions are usually the villain. . . . When writing Berkshire Hathaway annual reports, I pretend that I'm talking to my sisters. I have no trouble picturing them: Though highly intelligent, they are not experts in accounting or finance. They will understand plain English, but jargon may puzzle them. My goal is simply to give them the information I would wish them to supply me if our positions were reversed. To succeed, I don't need to be Shakespeare; I must, though, have a sincere desire to inform.'

That's what I want to impart to you today: good writing is communication that is easy to understand. If simple language is good enough for Mr. Buffet, then that should be your goal."

(Buffet 1–2)

- **Facts and Figures.** Not everyone wants to hear stories, answer questions, or listen to quotes. Facts are a good way to involve your audience objectively and with quantifiable impact. Notice how the training facilitator begins the speech this time.

Facts

"How important is oral and written communication in your work? If you are like most people, you are spending lots of time writing and speaking on the job. A Pitney Bowes study tells us the following:

- You talk on the phone and listen to voice-mail messages **75** times each day.
- Between pagers, faxes, and e-mail, you are involved in electronic communication **68** times a day.
- You write and read **33** hard-copy memos and letters daily.
- You read and write **30** sticky notes each day.

That's 206 pieces of workplace communication a day. And you thought you went to school to learn to become an engineer!"

Thesis (Overview of Key Points). After you have captured your audiences' attention, you need to clarify the topic of your presentation. To do so, provide a *thesis statement* or a clear *overview of your key points*. With this statement (one or two sentences), you let your audience know exactly what you plan to talk about.

For example, look at how the workshop facilitator could have begun the formal presentation, after arousing the audience's attention:

Thesis Statement for Formal Oral Presentation

"What we are going to talk about, and in this order, is the importance of clarity, conciseness, ease of access, audience, organization, and accuracy. These are key parts of effective workplace communication."

Discussion (or Body). After you have aroused your listeners' attention and clarified your goals, you have to prove your assertions. In the *discussion* section of your formal oral presentation, provide details to support your thesis statement.

Comparison/Contrast. In your presentation, you could compare different makes of office equipment, employees you are considering for promotion, different locations for a new office site, vendors to supply and maintain your computers, different employee benefit providers, and so forth. Comparison/contrast is a great way to make value judgments and provide your audience options.

Problem/Solution. You might develop your formal oral presentation by using a problem-to-solution analysis. For example, you might need to explain to your audience why your division needs to downsize. Your division has faced problems with unhappy customers, increased insurance premiums, decreased revenues, and several early retirements of top producers. In your speech, you can then suggest ways to solve these problems ("We need to downsize to lower outgo and ultimately increase morale"; "Let's create a 24-hour, 1-800-hotline to answer customer concerns"; "We should compare and contrast new employee benefits packages to find creative ways to lower our insurance costs.")

Argument/Persuasion. Almost every oral presentation has an element of argument/persuasion to it—as does much written workplace communication. You will usually be persuading your audience to do something based on the information you share with them in the presentation.

Importance. Prioritizing the information you present from least to most important (or most important to least) will help your listeners follow your reasoning more easily. To ensure the audience understands that you are prioritizing, provide verbal cues. These include simple words like "first," "next," "more important," and "most important."

Chronology. A chronological oral presentation can outline for your audience the order of the actions they need to follow or the order in which events have occurred. For example, you might need to make an oral presentation to employees, instructing them how to prepare a yearly sales report. Provide your audience with target deadlines and with the specific steps they must follow in their reports each quarter.

Maintaining Coherence. To maintain coherence, guide your audience through your speech as follows:

- **Use clear topic sentences.** Let your listeners know when you are beginning a new, key point: "Next, let's talk about the importance of *conciseness* in your workplace communication."

- **Restate your topic often.** Constant restating of the topic is required because listeners have difficulty retaining spoken ideas. A reader can refer to a previously discussed point by turning back a page or two, scrolling to prior e-mail messages, or retrieving bookmarked information online. Listeners do not have this option.

  Furthermore, a listener is easily distracted from a speech by noises, room temperature, uncomfortable chairs, or movement inside and outside the meeting site. Restating your topic helps your listener maintain focus.
- **Use transitional words and phrases.** Help your listeners follow your speech. Some good transitional words to consider using frequently in your presentation include *first*, *second*, *third*, *therefore*, *moreover*, *furthermore*, *for example*, *another idea is*, and *in conclusion*.

Conclusion. Conclude your speech by restating the main points, by recommending a future course of action, or by asking for questions or comments.

Restating Main Points

"Therefore, as I have mentioned throughout this presentation, the key to successful workplace communication is clarity, conciseness, ease of access, audience, organization, and accuracy. If you can accomplish these goals, your communication will succeed."

Recommending Future Action

"Now that we have discussed important aspects of successful workplace communication, what should you do next? Here are some helpful tips:

1. Let others read your text or listen to your speech before a formal presentation.
2. Use your word processing spell checkers and grammar checkers (but don't trust them!).
3. Put yourself in your audience's shoes. Will they understand the acronyms you have used, for example?
4. Memorize the phone number to your local college's grammar hotline.
5. Above all, do not fear communication. The techniques I have shared with you today will help you succeed."

Asking for Questions or Comments

"Thank you so much for your time and involvement. You have been a great group. Do you have any questions or comments?"

USING VISUAL AIDS IN ORAL PRESENTATIONS

Most speakers find that visual aids enhance their oral communication. Although PowerPoint, slide shows, graphs, tables, flip charts, and overhead transparencies are powerful means of communication, you must be the judge of whether visual aids will enhance your presentation. Avoid using them if you think they will distract from your presentation or if you lack confidence in your ability to create them and integrate them effectively. However, with practice, you probably will find that visuals add to the success of most presentations.

Advantages and Disadvantages of Visual Aids

Table 10.1 lists the advantages, disadvantages, and helpful hints for using visual aids. For all types of visual aids, practice using them before you actually make your presentation. When you practice your speech, incorporate the visual aids you plan to share with the audience.

TABLE 10.1 VISUAL AIDS: ADVANTAGES AND DISADVANTAGES

Type	Advantages	Disadvantages	Helpful Hints
Chalkboards	Are inexpensive Help audiences take notes Allow you to emphasize a point Allow audiences to focus on a statement Help you be spontaneous Break up monotonous speeches	Can be messy Can be noisy Make you turn your back to the audience Can be hard to see from a distance Can be hard to read if your handwriting is poor	Clean the board well Have extra chalk Stand to the side as you write Print in large letters Write slowly Avoid talking with your back to the audience Don't erase too soon
Chalkless White Boards	Same as above	Are expensive Require unique, erasable pens Can stain clothing Some pens can be hard to erase if left on the board too long Pens that run low on ink create light, unreadable impressions	Use blue, black, or red ink Cap pens to avoid drying out Use pens made especially for these boards Erase soon after use

(continued)

Table 10.1 Continued

Type	Advantages	Disadvantages	Helpful Hints
Flip Charts	Can be prepared in advance Are neat and clean Can be reused Are inexpensive Are portable Help you avoid a non-stop presentation Allow for spontaneity Help audiences take notes Allow you to emphasize key points Encourage audience participation Allow easy reference by turning back to prior pages Allow highlighting with different colors	Are limited by small size Require an easel Require neat handwriting Won't work well with large groups You can run out of paper Markers can run out of ink	Have two pads Have numerous markers Use different colors for effect Print in large letters Turn pages when through with an idea so audience will not be distracted Don't write on the back of pages where print bleeds through
Overhead Transparencies	Are inexpensive Can be used with lights on Can be prepared in advance Can be reused Can be used for large audiences Allow you to return to a prior point Allow you to face audiences	Require an overhead projector Can be hard to focus Require an electrical outlet and cords Can become scratched and smudged Can be too small for viewing Bulbs burn out	Use larger print Protect the transparencies with separating sheets of paper Frame transparencies for better handling Turn the overhead off to avoid distractions Focus the overhead before beginning your speech Keep spare bulbs Don't write on transparencies Face the audience

Table 10.1 Continued			
Videotapes	Allow instant replay Can be freeze-framed for emphasis Can be economically duplicated Can be rented or leased inexpensively Are entertaining	Require costly equipment (monitor and recorder) Are bulky and difficult to move Can malfunction Require dark rooms Require compatible equipment Deny easy note taking Deny speaker-audience interaction	Practice operating the equipment Avoid long tapes
Films	Are easy to use Are entertaining Have many to choose from Can be used for large groups	Require dark rooms Require equipment and outlets Make note taking difficult Deny speaker-audience interaction Can malfunction and become dated	Use up-to-date films Avoid long films Provide discussion time Use to supplement the speech, not replace it Practice with the equipment
PowerPoint Presentations	Are entertaining Offer flexibility, allowing you to move from topic with a mouse click Can be customized and updated Can be used for large groups Allow for speaker-audience interaction Can be supplemented with handouts easily generated by PPT Can be prepared in advance Can be reused Allow you to return to a prior slide Can include animation and hyperlinks	Require computers, screens, and outlets Work better with dark rooms Computers can malfunction Can be too small for viewing Can distance the speaker from the audience	Practice with the equipment Bring spare computer cables Be prepared with a backup plan if the system crashes Have the correct computer equipment (cables, monitors, screens, etc.) Make backup transparencies Practice your presentation

PowerPoint Presentations

One of the most powerful oral communication tools is visual—PowerPoint (PPT). Whether you are giving an informal or formal oral presentation, your communication might be enhanced by PowerPoint slides.

Today, you will attend very few meetings where the speakers do not use PowerPoint slides. "PowerPoint, the public speaking application included in the Microsoft Office software package, is one of the most pervasive and ubiquitous technological tools ever concocted. In less than a decade, it has revolutionized the worlds of business, education, science and communications" (Keller). According to Microsoft, "300 million PowerPoint presentations take place every day: 1.25 million every hour" (Mahin).

Benefits of PowerPoint

When you become familiar with PowerPoint, you will be able to achieve the following benefits:

- Choose from many different presentation auto layouts and designs. PowerPoint auto layouts let you vary your presentation slides. You can include graphics, tables, columns, and bulleted points to enhance your text.
- Create your own designs by choosing colors and color schemes from preselected designs.
- Add, delete, or rearrange slides as needed. By left-clicking on any slide, you can copy, paste, or delete it. Also, by left-clicking between any of the slides, you can add a new blank slide for additional information.
- Insert art from the Web, add images, or create your own drawings.
- Incorporate sound or video.
- Add hyperlinks either to slides within your PowerPoint presentation or to external Web links. When you right-click on your PowerPoint text, a drop-down box lets you add a hypertext link.

Tips for Using PowerPoint

Creating the Slides

- Create optimal contrast. Use dark backgrounds for light text or light backgrounds for dark text.
- Choose an easy-to-read font size and style. Use common fonts, such as Times New Roman, Courier, or Arial. Use at least a 24-point font size for text and 36-point font for headings.
- Use few words per screen to ensure readability. Limit yourself to no more than 50 or so words per screen.
- Limit the text to six or seven lines per slide and six or seven words per line. Think 6 × 6 (or 7 × 7).
- Use headings for readability. To create a hierarchy of headings, use larger fonts for a first-level heading and smaller fonts for second-level headings.
- Create variety in your screens. Use figures, graphs, pictures from the Web, or other line drawings.
- Use emphasis techniques. To call attention to a word, phrase, or idea, use color (sparingly), boldface, all caps, or arrows. Use a layout that includes white space.
- Develop a coherent flow of text from screen to screen. This can be accomplished through headings, consistent layout design, or transitional words or phrases.
- End with an *obvious* concluding screen.
- Prepare handouts.

During the Presentation

- Avoid reading your screens to your audience.
- Refer to each screen.
- Elaborate on each screen.
- Leave enough time for questions and comments.

Sample PowerPoint Slides

Figure 10.2 and 10.3 are samples of successful PowerPoint slides.

Figure 10.2 PowerPoint Slide with Line Graph

Heading in 36 pt. Arial font for readability.

Graphics to customize PPT slide and add visual interest.

TechStop

Increased Customer Complaints

# of complaints

70
60
50
40
30
20
10
0

Jan-Mar Apr-Jun Jul-Sep Oct-Dec

22 31 45 64

Line graph to show increased trend in customer complaints. The color line and data point maintain color coordination with slide design.

Figure 10.3 (Six Lines of Text/Six Words per Line) PowerPoint Slide

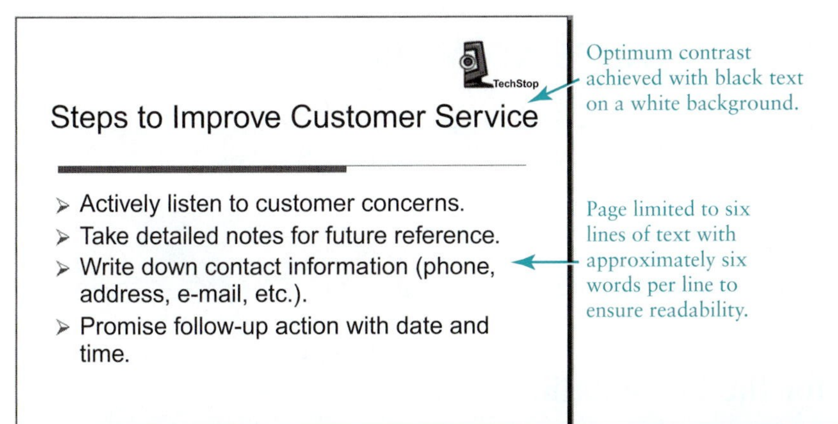

TechStop
Steps to Improve Customer Service
Actively listen to customer concerns.
Take detailed notes for future reference.
Write down contact information (phone, address, e-mail, etc.).
Promise follow-up action with date and time.

Optimum contrast achieved with black text on a white background.

Page limited to six lines of text with approximately six words per line to ensure readability.

Writer's Insight

Hal Lang, the customer service manager at TechStop, says, "Oral communication is a large part of my job. Sometimes I make formal presentations like a training session. More commonly I just talk to customers or sales personnel, either in person or on the phone.

When I make a formal presentation to my staff, I use PowerPoint. Doing so allows me to highlight important points visually. I also always like to make handouts of the slides so my employees can review the content at a later date.

I learned a long time ago that I should never just stand in front of a group and, with my back to the audience, read the PowerPoint slides. Instead, I limit the text of each slide and use the words like I would a note card. The text on the slides reminds me what to say and in what order to cover each point. I then elaborate on the text with additional comments, examples, anecdotes, or steps to follow.

I don't want the text to replace my authority in the room; the slides should only help the audience follow my oral presentation more easily."

PowerPoint Checklist

The PowerPoint checklist will give you the opportunity for self-assessment and peer evaluation of your presentation slides.

PowerPoint Checklist

__ 1. Does the presentation include headings for each slide?
__ 2. Have you used an appropriate font size for readability?
__ 3. Did you choose an appropriate font type for readability?
__ 4. Did you limit yourself to no more than three different font sizes per screen?
__ 5. Has color been used effectively for readability and emphasis, including font color and slide background?
__ 6. Did you use special effects effectively vs. overusing them?
__ 7. Have you limited text on each screen (remembering the 6 x 6 rule)?
__ 8. Did you size your graphics correctly for readability, avoiding those that are too small and/or too complex?
__ 9. Have you used highlighting techniques (arrows, color, white space) to emphasize key points?
__10. Have you edited for spelling and grammatical errors?

The Writing Process at Work

Prewriting for the Presentation

Prewriting gets you started with your presentation. Similar to prewriting for written communication, when you plan a speech, you should accomplish the following:

Consider the Purpose. Why are you making an oral presentation? Ask yourself questions like the following:

- Are you selling a product or service to a client?
- Do you want to inform your audience of the features in your newly created software?
- Has your boss asked for your help in preparing a presentation? After you research the content, will you have to present your information orally?
- Are you a supervisor justifying workforce cuts to your division?
- Did a customer request information on solutions to a problem?
- Are you representing your company at a conference by giving a speech?
- At the division meeting, are you reporting orally on the work you and your team have completed and the future activities you plan for the project?

Inform. When your speech is to *inform*, you want to update your listeners. You might tell an audience about new software platforms at work, changes in company policies, altered work schedules, or promotions.

Instruct. You might speak to *instruct*. In an instruction, you will teach an audience how to follow procedures.

Persuade. You might give a *persuasive* oral presentation about the need to hold more regular and constructive meetings. You might tell your audience that teamwork will enhance productivity. Maybe you are giving an oral presentation about the value of quality controls to enhance product development. In each instance, you want your audience to leave the speech ready and inspired to act on your suggestions.

Build Trust. You might give an oral presentation to *build trust*. Let's say you are speaking at an annual meeting. Your goal might be to inform the audience of your company's status, and to instill the audience with a sense of confidence about the company's practices. You could explain that the company is acting with the audience's best interests in mind. Similarly, in a departmental meeting, you might speak to build rapport. As a supervisor, you will want all employees to feel empowered and valued. Speaking to build trust will accomplish this goal.

Consider Your Audience. When you plan your oral presentation, consider your audience. Ask yourself questions such as the following:

- Are you speaking to specialists, semi-specialists, or lay audience?
- Are you speaking up to supervisors?
- Are you speaking down to subordinates?
- Are you speaking laterally to peers?
- Are you speaking to the public?
- Are you addressing multiple audience types (supervisors, subordinates, and peers)?
- Is your audience friendly and receptive, hostile, or neutral?
- Are you speaking to a captive audience (one required to attend your presentation)?
- Is your audience diverse in terms of culture, gender, or age?

To prewrite for your oral presentation, use a presentation plan. This will help you consider purpose, audience, and content (see Figure 10.4).

Gather Information. The best delivery by the most polished professional speaker will lack credibility if the speaker has little of value to communicate. As you prewrite for your presentation, you must study and research your topic thoroughly before you write a rough draft. We discuss research thoroughly in Chapter 6 and provide documentation examples in Appendix B.

You can rely on numerous sources when you research a topic for an oral presentation. For example, you could use any of the following sources:

- Interviews
- Questionnaires and surveys
- Visits to job sites
- Conversations in meetings or on the telephone
- Company reports
- Internet research

Figure 10.4
Presentation Plan for Oral Communication

Presentation Plan

Topic: ______________________________

Objectives:
- What do you want your audience to believe or do as a result of your presentation?
- Are you trying to persuade, instruct, inform, build trust, or combinations of the above?

Development: What main points are you going to develop in your presentations?

1.
2.
3.

Organization: Will you organize your presentation using *analysis, comparison/contrast, chronology, importance,* or *problem/solution*?

Visuals: Which visual aids will you use?

- Library research including periodicals and books
- Market research

Figure 10.5 is a sample questionnaire used by a team. The team was researching the feasibility of adding a day care center to its university campus for use by students and staff.

Writing the Presentation

After you obtain your information, your next step is to write a draft and consider visual aids for the presentation. The writing step in the communication process lets you use the research you gathered during prewriting. When you organize your information, you will determine whether or not additional material is needed or if you can delete some of the material you have gathered.

Avoid writing out the complete text of the presentation. Use an outline or note cards to present your speech.

Outline. A skeleton speech outline (Figure 10.6) provides you with a template for your presentation.

Note Cards. If you decide that presenting your speech from the outline will not work for you, consider writing highlights on 3″ × 5″ note cards. Avoid writing complete sentences or filling in the cards from side to side. Write short notes (keywords or short phrases) that will aid your memory when you make your presentation.

Figure 10.5 Sample Questionnaire

Student and Staff Questionnaire for Proposed Day Care Center

1. Are you male or female?
2. Are you in a single- or double-income family?
3. Are you a student or staff?
4. How many children do you have?
5. What are the ages?
6. Would you be interested in having a day care center on campus?
7. How much would you be willing to pay per hour for childcare at this center?
8. Do you think a day care center would increase enrollment at this university?
9. How many hours per week would you enroll your child/children?
10. What hours of operation should the day care center cover?
11. What credentials should the day care providers have?
12. What should be the number of children per classroom?

If student:

13. Are you enrolled full time or part time at the university?
14. Do you attend mornings, afternoons, evenings, weekends, or a combination of the above (please specify)?
15. Would you be willing to work in the day care center part-time?

If staff:

16. What hours do you work at the university?
17. What hours would you need to use for day care at the center?

Additional comments (if any):

Thank you for your assistance.

Rewriting the Presentation

In the rewriting step of the writing process, consider all aspects of style, delivery, appearance, body language, and gestures. Then, most importantly, practice. Even if you have excellent visual aids and well-organized content, if you fail to deliver effectively, your audience could miss your intended message.

Style. As with good writing, effective oral communication demands clarity and conciseness. To achieve clarity, stick to the point. Remember to speak so that your audience can understand you and your level of vocabulary. You should speak to communicate rather than to impress your listeners.

Delivery. Effective oral communicators interact with and establish a dynamic relationship with their audiences by using a variety of delivery techniques, as follows:

Eye Contact. Avoid keeping your eyes glued to your notes. Look at your audience. Try looking into different peoples' eyes as you move through your presentation (or look slightly above their heads if that makes you more comfortable).

Figure 10.6
Skeleton Outline

Skeleton Outline

Title:
Purpose:
- I. Introduction
  - A. Attention getter:
  - B. Focus statement:
- II. Body
  - A. First main point:
    - 1. Documentation/subpoint:
      - a. Documentation/subpoint:
      - b. Documentation/subpoint:
    - 2. Documentation/subpoint:
    - 3. Documentation/subpoint:
  - B. Second main point:
    - 1. Documentation/subpoint:
    - 2. Documentation/subpoint:
      - a. Documentation/subpoint:
      - b. Documentation/subpoint:
    - 3. Documentation/subpoint:
  - C. Third main point:
    - 1. Documentation/subpoint:
    - 2. Documentation/subpoint:
    - 3. Documentation/subpoint:
      - a. Documentation/subpoint:
      - b. Documentation/subpoint:
- III. Conclusion
  - A. Summary of main points:
  - B. Recommended future course of action:

Rate. Since your audience wants to listen and learn, you need to speak at a rate slow enough to achieve those two goals. Consider how you normally speak, and slow down.

Enunciation. Speak each syllable of every word clearly and distinctly.

Pitch. When you speak, your voice creates high and low sounds. That's *pitch*. In your presentation, capitalize on this fact. Vary your pitch by using even more high and low sounds than you do in your normal, day-to-day conversations. Modulate to stress certain keywords or major points in your oral presentations.

Pauses. One way to achieve a successful pace is to pause within the oral presentation. Pause to ask for and to answer questions, to allow ideas to sink in, to use visual aids, or to give the audience handouts.

Emphasis. You will not be able to underline or boldface comments you make in oral presentations. However, just as in written communication, you will want to emphasize key ideas. Your body language, pitch, gestures, and enunciation will enable you to highlight words, phrases, or even entire sentences.

Interaction with Listeners. You might need your audience to be active participants at some point in your oral presentation, so you will want to encourage this response. Your attitude and the tone of your delivery are key elements contributing to an encouraging atmosphere. To achieve audience involvement, consider these options:

- Ask questions.
- Call on individuals by name.
- Use an audience member's experiences as an example.

Conflict Resolution. You might be confronted with a hostile listener who either disagrees with you or does not want to be in attendance. You need to be prepared to deal with such a person. If someone disagrees with you or takes issue with a comment you make, try these responses:

- "That is an interesting perspective."
- "Thanks for your input."
- "Let me think about that some more and get back to you."
- "I have got several more ideas to share. We could talk about that point later, during a break."

Practice. Practice your speech including manipulation of your visual aids so you use them at appropriate times and places during the presentation. If you can, videotape yourself practicing the speech. As you practice, you will grow more comfortable and less dependent on your note cards or outline. Use the "Oral Presentation Checklist" to determine if you are sufficiently prepared for your oral presentation.

Oral Presentation Checklist

The oral presentation checklist will give you the opportunity for self-assessment and peer evaluation of your speech.

Oral Presentation Checklist

__ 1. Does your speech have an introduction that
- Arouses the audience's attention?
- Clearly states the topic of the presentation?

__ 2. Does your speech have a body that
- Explains exactly what you want to say?
- Develops your points thoroughly?

__ 3. Does your speech have a conclusion that
- Suggests what is next?
- Explains when (due date) a follow-up should occur?
- States why that date is important?

__ 4. Does your presentation provide visual aids to help you make and explain your points?

__ 5. Do you modulate your pace and pitch?

__ 6. Do you enunciate clearly so the audience will understand you?

__ 7. Have you used body language effectively by
- Maintaining eye contact?
- Using hand gestures?
- Moving appropriately?
- Avoiding fidgeting with your hair or clothing?

__ 8. Have you prepared for possible conflicts?

__ 9. Do you speak slowly and remember to pause so the audience can think?

__10. Have you practiced with any equipment you might use?

Developing Workplace Skills

Case Studies

1. Read the TechStop scenario that begins this chapter. In this scenario, a customer, a customer service representative, and a supervisor discuss problems with a printer and its warranty. The communication entails a face-to-face meeting, a phone conversation, and two voice-mail messages. The oral communication between the customer and the service representative is ineffective.

Assignment

Assess the communication skills of the speakers in this scenario. In a brief oral presentation, explain your conclusions based on the guidelines provided in this chapter and suggest ways in which the oral communication could have been improved.

2. Jessica Studin is the information technology manager for the City of Gullwing, Texas. Every month she updates the city's decision makers and constituents about Gullwing's technology needs. To do so, she makes an oral presentation at the city's board meeting. Her immediate audience includes the mayor, other city officials, invited guests, and any city residents in attendance. In addition, Jessica's oral presentation is televised locally to all of Gullwing's citizens.

   To make her oral presentations, Jessica works from a presentation plan, speaks from an outline, and uses PowerPoint slides to enhance her speech.

   This month, Ms. Studin must focus on financial and technology needs confronting the city. Specifically, Gullwing City Hall has old and unreliable computers, limited printers, software that is outdated, and a Web site that needs revisions. Her challenge is to solve these problems in a fiscally responsible manner.

   Some possible options include the following:

   - Purchasing new laser printers that can be shared by several offices.
   - Finding new hardware and software providers that can offer government discounts.
   - Hiring a Web designer.

Assignment

Using the guidelines presented in this chapter, help Jessica make her oral presentation. To do so,

- Organize the above information according to a logical order. This could include problem/solution, cause/effect, or importance.
- Create an outline for Jessica's speech. You might need to research the above topics further to find more detailed information.
- Create a PowerPoint presentation, using the above information and any additional information you consider necessary.

Individual and Team Projects

Evaluating an Oral Presentation

Listen to a speech. You could do so on television; at a student union; at a church, synagogue, or mosque; at a civic event, city hall meeting, or community organization (Lions Club, Rotary Club, Boy Scouts, etc.); at a company activity; in your classroom; etc. Then, answer the following questions:

1. What was the speaker's goal? Did he or she try to persuade, instruct, inform, or build trust? Explain your answer by giving examples.
2. What type of introduction did the speaker use to arouse listener interest (anecdote, question, quote, facts, startling comments, etc.)? Give examples to support your decision.
3. Were the introductions successful? If so, why and how? If not, what type of introductory method would have been better?
4. What visual aids were used in the presentation? Were these visual aids effective? Explain your answers.
5. Was the speaker's delivery effective? Use the following Presentation Delivery Checklist to assess the speaker's performance.

Presentation Delivery Checklist			
Delivery Techniques	Good	Bad	Explanation
Eye Contact			
Rate of Speech			
Enunciation			
Pitch of Voice			
Use of Pauses			
Emphasis			
Interaction with Listeners			
Conflict Resolution			

Giving an Oral Presentation

1. Find an advertised job opening in your field of expertise or degree program. You could look either online, in the newspaper, at your school's career placement center, or at a company's human resource office. Then, perform a mock interview for this position. To do so, follow this procedure:

   - Designate one person as the applicant.
   - He or she should prepare for the interview by making a list of potential questions (ones that the applicant believes he or she will be asked, as well as questions to ask the search committee).
   - Designate others in the class to represent the search committee.
   - The mock search committee should prepare a list of questions to ask the applicant.

2. Research a topic in your field of expertise or degree program. The topic could include a legal issue, a governmental regulation, a news item, an innovation in the field, a published article in a professional journal or public magazine, etc. Make an oral presentation about your findings.
3. Research job opportunities in your degree program. Are jobs decreasing or increasing? What is the pay scale? Which companies are hiring in your field? What skills are required for the job? Make an oral presentation about your findings.
4. Interview employees at your job site or at any company of your choice to find out the following:
    - The estimated amount of time they spend in oral communication on the job.
    - Whether their oral communication is everyday, informal, or formal.
    - If their oral communication goals are to persuade, instruct, inform, or build trust.
    - The types of visual aids they use.
    - Whether they have participated in teleconferences and/or videoconferences.

    After concluding your interviews, share your findings with your class in an oral presentation.

Creating a PowerPoint Slide Presentation

For assignments 2–4 above (Giving an Oral Presentation), create PowerPoint slides to enhance your oral communication. To do so, follow the guidelines provided in this chapter.

Assessing PowerPoint Slides

After reviewing the following PowerPoint slides, determine which are successful, which are unsuccessful, and explain your answers, based on the guidelines provided in this chapter.

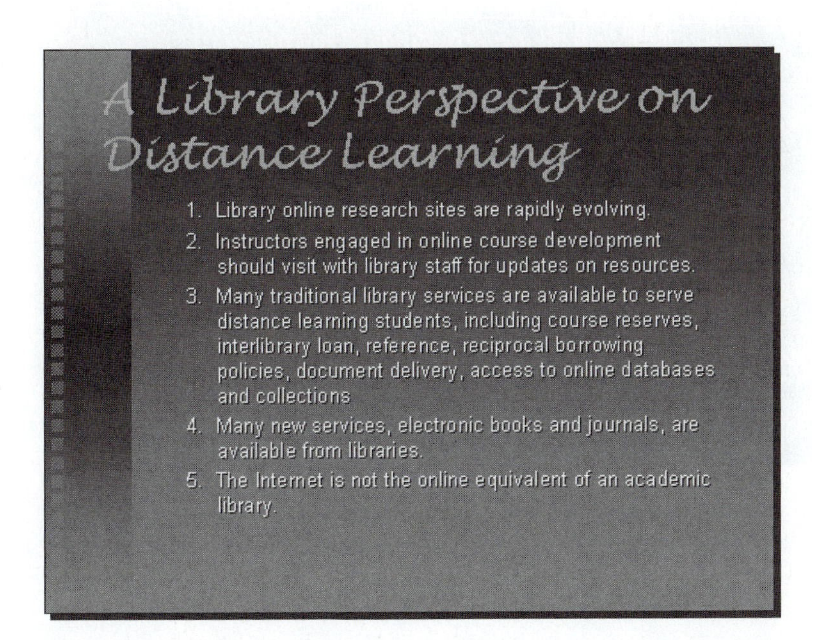

A Technology Perspective on Online Education

Students - 24-hour Call Centers answer student hardware/software needs

Faculty–The Tech Center helps faculty with course creations and tech resources

As DISTANCE LEARNING classes have risen in popularity, many *diverse areas* of interest across campuses have converged.

These include counseling, library services, technology, faculty, administration, continuing education, etc.

To ensure student success, Distance Learning Coordinating Committees need to draw from these multiple disciplines and collaborate on solutions to arising technology challenges.

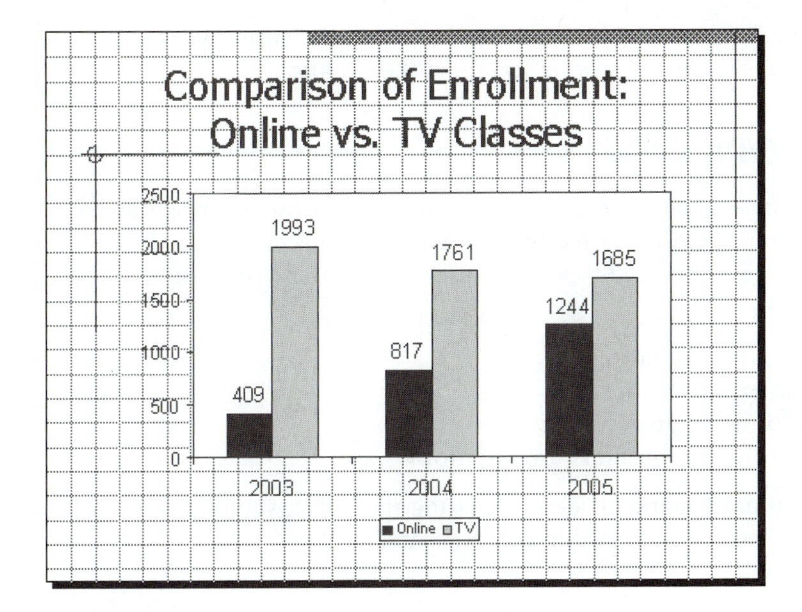

PROBLEM SOLVING THINK PIECES

After reading the following scenarios, answer these questions in an informal classroom discussion:

- Would your oral presentation be *everyday, informal*, or *formal*?
- Would you use a *videoconference, teleconference, web conference*, or *face-to-face meeting*?
- Which *visual aids* would work best for your presentation?
- Is your oral presentation goal to *persuade, instruct, inform*, or *build trust*?

1. You work at FlashCom Electric. Your company has created a new interface for modems. Your boss has asked you to make a presentation to sales representatives from 20 potential vendors in the city. The oral presentation will explain to the vendors the benefits of your product, the sales breaks you will offer, and how the vendors can increase their sales.

2. After working for Friendly's, a major discount computer hardware and software retailer, for over a year, you have created a new organizational plan for their vast inventory. In an oral presentation, you plan to show the CEO and Board of Directors why your plan is cost-effective and efficient.

3. You are the manager of an automotive parts supply company, Plugs, Lugs, and More. Your staff of ten in-store employees lacks knowledge of the store's new merchandise, has not been meeting sales goals, and does not always treat customers with respect and care. It's time to address these concerns.

4. As CFO of your engineering/architectural firm (Levin, Lisk, and Lamb), you must downsize. Business is decreasing, and costs are rising. To ensure third-quarter profitability, 10 percent of the staff must be laid off. That will amount to a layoff of over 50 employees. You must make an oral

presentation to your stockholders and the entire workforce at your company's annual meeting to report the situation.

WEB WORKSHOP

PowerPoint has proven to be a useful tool in business presentations. However, PowerPoint must be used correctly, as discussed in this chapter, to be effective. Access an online search engine and type in phrases such as these (include any additional ones you create):

- using powerpoint in business
- effective powerpoint use
- tips for using powerpoint
- powerpoint + business presentations

Once you have found articles discussing this topic, summarize your findings and make an oral presentation (using PowerPoint slides).

Grammar Handbook: Grammar, Punctuation, and Numbers

GRAMMAR RULES

Agreement Between Pronoun and Antecedent (Referent)

A pronoun has to agree in gender and number with its antecedent.

Examples:

Susan went on *her* vacation yesterday.
The *people* who quit said that *they* deserved raises.

1. Problems often arise when a singular indefinite pronoun is the antecedent. The following antecedents require singular pronouns: anybody, each, everybody, everyone, somebody, and someone.

   incorrect

   Anyone can pick up *their* applications at the job placement center.

   correct

   Anyone can pick up *his or her* applications at the job placement center.

2. Problems also arise when the antecedent is separated from the pronoun by numerous words.

   incorrect

   Even when the best *employee* is considered for a raise, *they* often do not receive it.

   correct

   Even when the best *employee* is considered for a raise, *she* often does not receive it.

Agreement Between Subject and Verb

Subjects and verbs must agree. If a subject is plural, it needs a plural verb; if a subject is singular, it needs a singular verb. Problems occur for the following reasons:

1. Writers sometimes create disagreement between subjects and verbs, especially if other words separate the subject from the verb. To ensure agreement, ignore the words that come between the subject and verb.

   incorrect

   Her *boss* undoubtedly *think* that all the employees want promotions.

   correct

   Her *boss* undoubtedly *thinks* that all the employees want promotions.

   incorrect

   The *employees* who sell the most equipment *is* going to Hawaii for a week.

   correct

   The *employees* who sell the most equipment *are* going to Hawaii for a week.

2. If a sentence contains two subjects (a compound subject) connected by *and*, use a plural verb.

   incorrect

   Joe and Alicia *was* both selected employee of the year.

   correct

   Joe and Alicia *were* both selected employee of the year.

   incorrect

   The bench workers and their supervisor *is* going to work closely to complete this project.

   correct

   The bench workers and their supervisor *are* going to work closely to complete this project.

Add a final *s* or *es* to create most plural subjects or singular verbs, as follows:

Plural Subjects	**Singular Verbs**
boss*es* hire	a boss hire*s*
employee*s* demand	an employee demand*s*
experiment*s* work	an experiment work*s*
attitude*s* change	the attitude change*s*

3. If a sentence has two subjects connected by *either . . . or*, *neither . . . nor*, or *not only . . . but also*, the verb should agree with the closest subject.

   - *Either* the salespeople *or* the warehouse worker deserves raises.
   - *Not only* the warehouse worker *but also* the salespeople deserve raises.
   - *Neither* the salespeople *nor* the warehouse worker deserves raises.

4. If an indefinite pronoun acts as the subject of your sentence, use singular verbs.

   Indefinite Pronouns

Another	Everything
Anybody	Neither
Anyone	Nobody
Anything	No one
Each	Nothing
Either	Somebody
Everybody	Someone
Everyone	Something

   Anyone who works in factories *is* guaranteed maternity leave.
   Everybody wants the company to declare a profit this quarter.

5. If you use collective nouns as the subject of your sentence, use singular verbs.

   Collective Nouns

Class	Organization
Corporation	Platoon
Department	Staff
Group	Team

   The *staff is* sending the boss a bouquet of roses.

Run-on Sentences (Comma Splices and Fused Sentences)

Run-on sentences occur for two reasons:

1. Two independent clauses are joined by a comma.

   incorrect

   Sue was an excellent employee, she got a promotion.
   [Note: this is called a comma splice.]

2. Two independent clauses are joined by no punctuation.

   incorrect

   Sue was an excellent employee she got a promotion.
   [Note: this is called a fused sentence.]

Both of these errors can be corrected in the following ways:

- Separate the independent clauses with a period.
- Separate the independent clauses with a semicolon.
- Separate the independent clauses with a comma preceding a coordinate conjunction: *and, or, but, nor, for, so, yet*

- Separate the two independent clauses with a semicolon, a conjunctive adverb, and a comma. Conjunctive adverbs include words such as *consequently, furthermore, however,* and *therefore.*
- Use a subordinate conjunction to make one of the independent clauses into a dependent clause. Subordinate conjunctions include words such as *after, although, because, before, once, since, unless,* and *when.*

Fragments

A fragment occurs when a group of words is *not an independent* clause. Remember, an independent clause must contain a subject and a predicate, must express a complete thought, and must be able to stand alone as a sentence.

incorrect

Working with computers.

Though this begins with a capital "W" and ends with a period, it's not an independent clause. It lacks a verb and does not express a complete thought. Often a group of words has a subject and a verb but is still a dependent clause.

incorrect

Although he enjoyed working with computers.

Here, the subordinate conjunction "Although" makes the words a dependent clause. Solve the problem with fragments by adding a subject, adding a verb, adding both a subject and a verb, adding an independent clause to the dependent clause, or removing the subordinate conjunction.

correct

Joe found that working with computers allowed him to use his educational training.
[Note: we added a subject (Joe) and a verb (found).]

Dangling and Misplaced Modifiers

A modifier is a word, phrase, or dependent clause that explains or adds detail to your independent clause. Dangling modifiers and misplaced modifiers fail to modify the words they should explain because these modifiers are placed incorrectly.

To avoid either of these modification problems, you need to place the modifying word, phrase, or dependent clause next to the word modified.

incorrect dangling modifier

While working, tiredness overcame them.
(Who was working? Who was overcome by tiredness?)

correction

While working, the staff became tired.

incorrect dangling modifier

After talking on the telephone for two hours, the equipment was prepared for shipment. (Who was talking for two hours? Not the equipment!)

correction

After talking on the telephone for two hours, the warehouse manager prepared the equipment for shipment.

incorrect misplaced modifier

He had a heart attack almost every time his supervisor reviewed him.

correction

He almost had a heart attack every time his supervisor reviewed him.

Parallelism

All items in a list should be parallel in grammatical form. Avoid mixing phrases and sentences (independent clauses).

incorrect

We will discuss the following at the department meeting:

1. Entering mileage in logs (phrase)
2. All employees have to enroll in a training seminar. (sentence)
3. Purpose of quarterly reviews (phrase)
4. Some data processors will travel to job sites. (sentence)

correct

We will discuss the following at the department meeting:

1. Entering mileage in logs
2. Enrolling in training seminars
3. Reviewing employee performance quarterly
4. Traveling to job sites

Present Participle Phrases

correct

At the department meeting, you will learn how to

1. Enter mileage in logs
2. Enroll in training seminars
3. Review employee performance quarterly
4. Travel to job sites

Verb Phrases

Punctuation Rules

Commas

1. Commas are used to separate items in a series.

   - Stewart, Trump, and Turner are three famous CEOs.
   - Turner is athletic, blue-eyed, and charming to many people.
   - Turner is famous for having colorized old movies, owning the Braves, and marrying and divorcing Jane Fonda.

2. Commas are used to separate introductory phrases or clauses.

   - Before I jog, I stretch to warm up.
   - After a good run, I feel relaxed and healthy.
   - Because of such widespread interest in jogging, many hiking and bike trails are crowded with joggers.

3. Commas are used to separate two independent clauses joined by coordinate conjunctions.

   - People who jog are healthy, and they miss fewer workdays.
   - Many people like to jog, but others find the activity boring.
   - Many people have begun jogging, so athletic shoe stores have opened everywhere.

4. Commas are used to separate *medial modifiers* (clauses or phrases that split the two halves of an independent clause).

   - My boss, who is trying to pass his CPA exams, spends long hours studying.
   - John, intent on passing this exam, even studies on weekends.
   - John, after passing the CPA tests, will receive a 10 percent salary increase.

5. Commas are used to separate cities from states.

   - Chicago, Illinois, is called the Windy City.
   - One of our most congested cities is San Francisco, California.

6. Commas are used to separate the day from the year when dates are written in a mm/dd/yy sequence. Note: if dates are written dd/mm/yy (15 April 2006), no commas are used.

   - I arrived here on August 13, 2006, from Austin.
   - I left for a vacation on June 10, 2006.

Semicolons

1. Semicolons are used to separate two independent clauses.

   - Shirley dislikes surprises; she wants all employees to be where they should be and to do what they should do.
   - One day Shirley discovered that Tom took two hours for lunch; consequently, he was called in for a discussion.
   - Tom said that he had been involved in a lunch meeting; however, Shirley reminded him that work needed to be done at the office.

2. Semicolons are used to separate items in a series from additional items in a series.

   - The committee was composed of Maureen Pierce, chairperson; Perry Michelson, secretary; and Julie Schopper, treasurer.
   - They decided to order hamburgers, soft drinks, and chips; to hold the picnic at either a park, a ranch, or in the company cafeteria; and to play softball, volleyball, or bingo.

Colons

A colon is used to separate an independent clause from items in a series.

- Three departments write proposals: marketing, human resources, and research and development.
- Five employees received the awards: Megan Clem, Galen Goben, Steve Janasz, Leah Workman, and Candi Millard.

Note: The colons in the following sentences are *incorrectly* used and should be omitted:

- The three places to eat while visiting Athens, Texas, are: Catfish Palace, Down Mexico Way, and The Spanish Trace Cafe.
- Rate hikes are a good business practice because they: stimulate customer dialogue, allow opportunities for governmental involvement, and ensure upper management realignment.

Apostrophes

1. An apostrophe is used to form a contraction.

   - It's not too late to learn how to use apostrophes.
   - You're going to learn how to use apostrophes even if it hurts.

2. An apostrophe placed *before* a final *s* is used to create singular possessive words.

   - The dog's bowl is out of water.
   - Steve's hair refuses to turn gray.

3. An apostrophe placed *after* a final *s* is used to create plural possessive words.

   - The twelve dogs' bowls are all out of water.
   - The six companies' supervisors' cars were vandalized.

Periods

1. A period ends a declarative sentence (independent clause).

   Example

   I found the business trip rewarding.

2. Periods are used with abbreviations.

e.g.	A.M. or a.m.
Mr.	P.M. or p.m.
Mrs.	Dr.
Ms.	Mgr.

3. It is incorrect to use periods with abbreviations for organizations and associations.

incorrect

I.A.B.C. (International Association of Business Communicators)

correct

IABC (International Association of Business Communicators)

4. Abbreviations for states use two capital letters and no period.

incorrect

KS. (Kansas)
MO. (Missouri)
TX. (Texas)

correct

KS
MO
TX

5. If your sentence ends with an abbreviation that includes a period, do not add a second period.

incorrect

The meeting starts at 10:00 a.m..

correct

The meeting starts at 10:00 a.m.

Numbers

Write out numbers one through nine. Use numerals for numbers 10 and above.

10	12
104	2,093
536	5,550,286

Although the preceding rules cover most situations, there are exceptions.

1. Use numerals for all percentages.
   2%   18%   25 percent   79%
2. Use numerals for addresses.
   12 Elm   935 W. Harding
3. Use numerals for miles per hour.
   5 mph   225 mph

4. Use numerals for time.
   3:15 A.M.

5. Use numerals for dates.
   May 31, 2006

6. Use numerals for monetary values.
   $45 $.95 $2 million

7. Use numerals for units of measurement.
   14′ 6 1/2″ 16 mm 10 v

8. Do not use numerals to begin sentences.

   incorrect

   568 people were fired last August.

   correct

   Five hundred sixty-eight people were fired last August.

9. Do not mix numerals and words when writing a series of numbers. The larger number determines the form.
   We attended 4 meetings over a 16-day period.

10. Use numerals and words in a compound number adjective to avoid confusion.
    The worker needed six 2-inch nails.

Works Cited or References

Parenthetical source citations are an abbreviated form of documentation. In parentheses, you tell your readers only the names of your authors and the page numbers on which the information can be found. Such documentation alone would be insufficient. Your readers would not know the names of the books, the names of the periodicals, or the dates, volumes, publishing companies, or Web site URLs. This more thorough information is found on the Works Cited page or References page, a listing of research sources alphabetized either by author's name or title (if anonymous). This is the last page of your research report or presentation.

Your entries should follow MLA or APA standards or the style sheet of your choice.

MLA Works Cited

A Book with One Author

Cottrell, Robert C. *Izzy: A Biography of I. F. Stone*. New Brunswick: Rutgers University Press, 1992.

A Book with Two or Three Authors

Tibbets, Charlene, and A. M. Tibbets. *Strategies: A Rhetoric and Reader*. Glenview: Scott, Foresman and Company, 1988.

A Book with Four or More Authors

Nadell, Judith, et al. *The Macmillan Writer*. Boston: Allyn and Bacon, 1997.

A Book with a Corporate Authorship

Corporate Credit Union Network. *A Review of the Credit Union Financial System: History, Structure, Status and Financial Trends*. Kansas City: U.S. Central, 1986.

A Translated Book

Phelps, Robert, ed. *The Collected Stories of Colette*. Trans. Matthew Ward. New York: Farrar, Straus, Giroux, 1983.

An Entry in a Collection or Anthology

Irving, Washington. "Rip Van Winkle." *Once Upon a Time: The Fairy Tale World of Arthur Rackham*. Ed. Margery Darrell. New York: Viking, 1972. 13–36.

A Signed Article in a Journal

Gerson, Steven M., and Earl Eddings. "Service Learning: Internships . . . with a Conscience." *Missouri English Bulletin* 54 (Fall 1996): 70–75.

A Signed Article in a Magazine

Kroll, Jack. "T. Rex Redux." *Newsweek* 26 May 1997: 74–75.

A Signed Article in a Newspaper

Hoffman, Donald. "Bank Consigned to Vault of Gloom." *The Kansas City Star* 24 Oct. 1988: C1.

An Unsigned Article

"Business Index." *Time* 23 Sept. 2000: 29.

Encyclopedias and Almanacs

"The Internet." *The World Book Encyclopedia*. 2001 ed. Chicago: World Book.

Computer Software

PFS: Write Sampler. Computer software. Software Pub. Corp., 2001.

Internet Source

Berst, Jesse. "Berst Alert." *ZDNet* 30 Jan. 1998. 30 Jan. 1998 http://www.zdnet.com/anchordesk/story/story_1716.html.

E-mail

Schneider, Ruth. "Teaching Workplace Communication." Personal E-mail. 2 Apr. 2004.

CD-ROM

McWard, James. "Graphics Online." TW/Inform. CD-ROM. New York: EduQuest, 2003.

APA References

A Book with One Author

Cottrell, R. C. (1992). *Izzy: A biography of I. F. Stone*. New Brunswick, NJ: Rutgers University Press.

A Book with Two Authors

Tibbets, C., & Tibbets, A. M. (1988). *Strategies: A rhetoric and reader*. Glenview, IL: Scott, Foresman and Company.

A Book with Three or More Authors

Nadell, J., McNeniman, L., & Langan, J. (1997). *The Macmillan writer*. Boston: Allyn & Bacon.

A Book with a Corporate Authorship

Corporate Credit Union Network. (1986). *A review of the credit union financial system: History, structure, and status and financial trends*. Kansas City, MO: U.S. Central.

A Translated Book

Phelps, R. (Ed.). (1983). *The collected stories of Colette* (M. Ward, Trans.). New York: Farrar, Straus & Giroux.

An Entry in a Collection or Anthology

Irving, W. (1972). Rip Van Winkle. In M. Darrell (Ed.), *Once upon a time: The fairy tale world of Arthur Rackham* (pp. 13–36). New York: Viking Press.

A Signed Article in a Journal

Gerson, S. M., & Eddings, E. (1996, Fall). Service learning: Internships . . . with a conscience. *Missouri English Bulletin*, 54, 70–75.

A Signed Article in a Magazine

Kroll, J. (1997, May 26). T. rex redux. *Newsweek*, 74–75.

A Signed Article in a Newspaper

Hoffman, D. (1988, Oct. 24). Bank consigned to vault of gloom. *The Kansas City Star*, p. C1.

An Unsigned Article

Business index. (2000, Sep. 23). *Time*, 29.

Encyclopedias and Almanacs

The internet. (2001). *The world book encyclopedia*. Chicago: World Book.

Computer Software

PFS: Write Sampler [Computer software]. (2001). Software Pub. Corp.

Internet Source

Berst, J. (1998, Jan. 30). Berst alert. *ZDNet* [on-line]. Retrieved February 1, 1998 from the World Wide Web: http://www.zdnet.com/anchordesk/story/story_1716.html.

E-mail

Schneider, R. (2004, Apr. 2). Teaching workplace communicaton. Personal e-mail.

CD-ROM

McWard, J. (2003). Graphics on-line [CD-ROM]. TW/Inform. New York: EduQuest.

ALTERNATIVE STYLE SHEETS

Although MLA and APA are popular style sheets, other style sheets are favored in certain disciplines. Refer to these if you are interested or required to do so.

- *U.S. Government Printing Office Style Manual*, 29th edition. Washington, DC: Government Printing Office, 2000.
- *The Chicago Manual of Style*, 15th edition. Chicago: University of Chicago Press, 2003.

Double space throughout the Works Cited page and list sources in alphabetical order.

Indent the second and subsequent lines 5 to 7 spaces.

If you do not have an author's name, alphabetize by the article's title (not including "the," "a," or "an").

SAMPLE MLA WORKS CITED PAGE

Works Cited

Berst, Jesse. "Berst Alert." *ZDNet* 30 Jan. 1998. 30 Jan. 1998 http://www.zdnet.com/anchordesk/story/story_1716.html.

"Business Index." *Time* 23 Sep. 2000: 29.

Corporate Credit Union Network. *A Review of the Credit Union Financial System: History, Structure, and Status and Financial Trends*. Kansas City: U.S. Central, 1986.

Cottrell, Robert C. *Izzy: A Biography of I. F. Stone*. New Brunswick: Rutgers University Press, 1992.

Gerson, Steven M., and Earl Eddings. "Service Learning: Internships . . . with a Conscience." *Missouri English Bulletin* 54 (Fall 1996): 70–75.

Hoffman, Donald. "Bank Consigned to Vault of Gloom." *The Kansas City Star* 24 Oct. 1988: C1.

Kroll, Jack. "T. Rex Redux." *Newsweek* 26 May 1997: 74–75.

Nadell, Judith, et al. *The Macmillan Writer*. Boston: Allyn and Bacon, 1997.

PFS: Write Sampler. Computer software. Software Pub. Corp., 2001.

Phelps, Robert, ed. *The Collected Stories of Colette*. Trans. Matthew Ward. New York: Farrar, Straus, Giroux, 1983.

Schneider, Ruth. "Teaching Workplace Communication." Personal E-mail. 2 Apr. 2004.

Tibbets, Charlene, and A. M. Tibbets. *Strategies: A Rhetoric and Reader*. Glenview: Scott, Foresman and Company, 1988.

Sample APA References Page

References

Berst, J. (1998, Jan. 30). Berst alert. *ZDNet* [on-line]. Retrieved February 1, 1998 from the World Wide Web: http://www.zdnet.com/anchordesk/story/story_1716.html.

Business index. (2000, Sep. 23). *Time*, 29.

Corporate Credit Union Network. (1986). *A review of the credit union financial system: History, structure, and status and financial trends*. Kansas City, MO: U.S. Central.

Cottrell, R. C. (1992). *Izzy: A biography of I. F. Stone*. New Brunswick, NJ: Rutgers University Press.

Gerson, S. M., & Eddings, E. (1996, Fall). Service learning: Internships . . . with a conscience. *Missouri English Bulletin*, 54, 70–75.

Hoffman, D. (1988, Oct. 24). Bank consigned to vault of gloom. *The Kansas City Star*, p. C1.

Kroll, J. (1997, May 26). T. rex redux. *Newsweek*, 74–75.

Nadell, J., McNeniman, L., & Langan, J. (1997). *The Macmillan writer*. Boston: Allyn & Bacon.

PFS: Write Sampler [Computer software]. (2001). Software Pub. Corp.

Phelps, R. (Ed.). (1983). *The collected stories of Colette* (M. Ward, Trans.). New York: Farrar, Straus & Giroux.

Schneider, R. (2004, Apr. 2). Teaching workplace communicaton. Personal e-mail.

Tibbets, C., & Tibbets, A. M. (1988). *Strategies: A rhetoric and reader*. Glenview, IL: Scott, Foresman and Company.

Double space throughout the References page and list sources in alphabetical order.

Indent the second and subsequent lines 5 to 7 spaces.

If you do not have an author's name, alphabetize by the article's title (not including "the," "a," or "an").

Works Cited

Chapter 1 An Introduction to Workplace Communication

Black & Veatch. "About Us." 2004. http://www2.bv.com/about/values.htm.

Campbell, Kim S., et al. "Leader-Member Relations as a Function of Rapport Management." *Journal of Business Communication* 40 (2003): 170–194.

Crane & Company. "The Cost of a Letter." 9 Nov. 2002. http://www.crane.com/business/businessidentity.html.

Miller, Carolyn, et al. "Communication in the Workplace." *Center for Communication in Science, Technology, and Management*. 2 (Oct. 1996): 1–26.

"Mission and Values." Microsoft. 17 Feb. 2005. http://www.microsoft.com/mscorp/articles/mission_values.asp.

Pope, Justin. "Poor Writing Costs Taxpayers Millions." 4 July 2005. http://www.ecy.wa.gov/quality/plaintalk/resources/Poor%20Writing%20Costs%20Taxpayers%20Millions.pdf.

Quinn, R. E., et al. "A Competing Values Framework for Analyzing Presentational Communication in Management Contexts." *Journal of Business Communication* 28 (1991): 213–232.

"Writing: A Powerful Message from State Government." *Report of The National Commission on Writing for America's Families, Schools, and Colleges*. College Entrace Examination Board, 2005.

"Writing: A Ticket to Work. Or a Ticket Out." *Report of the National Commission on Writing for America's Families, Schools, and Colleges*. College Entrance Examination Board, 2004.

"Writing Skills Necessary for Employment, Says Big Business." *The National Commission on Writing*. 4 July 2005. http://www.writingcommission.org/pr/writing_for_employ.html.

Chapter 2 Essential Goals of Successful Workplace Communication

Bowman, George, and Arthur E. Walzer. "Ethics and Technical Communication." *Proceedings: 34th Annual Technical Communication Conference*. May 10–13, 1987: [MPD—93.]

"Code for Communicators." Society for Technical Communication. 10 Mar. 2004. http://stcrmc.org/resources/resource_code.htm.

"Code of Ethics." International Association of Business Communicators. 8 Oct. 2003. http://www.iabc.com/members/joining/code.htm.

"The Fog Index." *The University of Oklahoma College of Continuing Education Public and Community Services*. 24 Feb. 2004. http://region7.ou.edu/.

GNOME Documentation Style Guide. 30 July 2003. http://developer.gnome.org/documents/style-guide/index.html.

Hilts, L., and B. J. Krilyk. *Write Readable Information to Educate*. Hamilton, Ontario: Chedoke-McMaster Hospitals and Hamilton Civic Hospitals. 1991.

Horton, William. "The Almost Universal Language: Graphics for International Documents." *Technical Communication* 40 (November 1993): 686–693.

Mollison, Andrew. "U.S. Loses Its Lead in High School and College Grads." *Cox Newspapers*. 6 November 2001: 1–3.

Nethery, Kent. "Let's Talk Business." http://www.cuspomona.edu/~cljones/powerpoints/chap02/sld001.htm.

Rains, Nancy E. "Prepare Your Documents for Better Translation." *Intercom* 41.5 (December 1994): 12.

Weiss, Edmund H. "Twenty-five Tactics to 'Internationalize' Your English." *Intercom* (May 1998): 11–15.

Chapter 3 Visual Communication: Page Layout and Graphics

Christensen, G. Jay. "Talking Captions Catch the Reader." 27 Aug. 2002. 28 Feb. 2005. http://www.csun.edu/~vcecn006/talking.html.

Everson, Larry. "Recent Trends in Technical Writing." *Technical Communication* (Fourth Quarter 1990): 396–398.

"Functional Illiteracy." *Wikipedia*. 7 Dec. 2005. http://en.wikipedia.org/wiki/Functional_illiteracy.

Chapter 4 Correspondence: Memos and Letters

"Pitney Bowes Study Finds U.S. Workers Less Overwhelmed Despite Increased E-mail Volumes." *ebizChronicle.com*. 21 Aug. 2000. 13 Nov. 2002. http://www.ebizchronicle.com/spl_reports/august/pitneybowes_email2.htm.

"Writing: A Ticket to Work . . . Or a Ticket Out—A Survey of Business Leaders." *Report of the National Commission on Writing for America's Families, Schools, and Colleges*. College Entrance Examination Board, 2004.

Chapter 5 Electronic Communication

Berst, Jesse. "Seven Deadly Web Site Sins." *ZDNet*. 30 Jan. 1998. http://www.zdnet.com/anchordesk/story/story_1716.html.

Blumberg, Max. "Blogging Statistics." 3 Jan. 2005. http://maxblumberg.typepad.com/dailymusings/2005/01/usa_blog_usage_.html.

Bruner, Rick. "Blogging Statistics." *ImaginaryPlanet.net*. 23 Apr. 2004. 6 June 2005. http://www.imaginaryplanet.net/weblogs/idiotprogrammer/?p=83397841.

Cross, Jay. "Blogging for Business." *Learning Circuits*. 23 May 2005. http://www.learningcircuits.org/2003/aug2003/cross.htm.

Dorazio, Pat. Interview. June 2000.

"Email Etiquette." Yale University. 20 Jan. 1999. 26 Nov. 2002. http://www.library.yale.edu/training/netiquette.

Foremski, Tom. "IBM Is Preparing to Launch a Massive Corporate Wide Blogging Initiative as It Seeks to Extend Its Expertise Online." *SiliconValleyWatcher*. 13 May 2005. 23 May 2005. http://www.siliconvalleywatcher.com/mt/archives/2005/05/can_blogging_bo.php.

Gard, Lauren. "The Business of Blogging." *BusinessWeek Online*. 13 Dec. 2004. http://www.businessweek.com/magazine/content/04_50/b3912115_mz016.htm.

"Gates Backs Blogs for Business." *BBCNews.com.uk*. 21 May 2004. 23 May 2005. http://news.bbc.co.uk/2/hi/technology/3734981.stm.

Goldenbaum, Don, and George Calvert. Applied Communication Group. Interview. 10 Jan. 2000.

Hoffman, Jeff. "Instant Messaging in the Workplace." *Intercom* (Feb. 2004): 16–17.

Johnson, Dana R. "Copyright Issues on the Internet." *Intercom* (June 1999): 17.

Kharif, Olga. "Blogging for Business." *BusinessWeek.com*. 9 Aug. 2004. 23 May 2004. http://www.businessweek.com/technology/content/aug2004/tc2004089_3601_tco24.htm.

Li, Charlene. "Blogging: Bubble or Big Deal?" 5 Nov. 2004. 23 May 2005. http://www.forrester.com/Research/Print/Document/0,7211,35000,00.html.

McAlpine, Rachel. "Passing the Ten-second Test." Wise-Women.com. 22 Apr. 2002. 3 Jan. 2004. http://www.wise-women.org/features/tenseconds/index.shtml.

Miller, Carolyn, et al. "Communication in the Workplace." *Center for Communication in Science, Technology, and Management*. 2 (Oct. 1996): 1–26.

Missouri Department of Transportation. http://www.modot.org.

Moore, Linda E. "Serving the Electronic Reader." *Intercom* (Apr. 2003): 16–17.

Munter, Mary, et al. "Business E-mail: Guidelines for Users." *Business Communication Quarterly* 66 (Mar. 2003): 29.

"Pitney Bowes Study Finds U.S. Workers Less Overwhelmed Despite Increased E-mail Volumes." *ebizChronicle.com*. 21 Aug. 2000. 13 Nov. 2002. http://www.ebizchronicle.com/spl_reports/august/pitneybowes_email2.htm.

Pratt, Jean A. "Where Is the Instruction in Online Help Systems?" *Technical Communication* (Feb. 1998): 33–37.

Ray, Ramon. "Blogging for Business." *Inc.com*. Sept. 2004. 23 May 2005. http://www.inc.com/partners/sbc/articles/20040929-blogging.html.

Shinder, Deb. "Instant Messaging: Does It have a Place in Business Networks?" 2 Nov. 2004. 11 Nov. 2005. http://www.windowsecurity.com/articles/Instant-Messaging-Business-Networks.html.

"The Transition to General Management." Harvard Business School. 1998. 12 November 2002. http://www.hbs.edu/gm/index.html.

"Writing: A Powerful Message from State Government." *Report of the National Commission on Writing for America's Families, Schools, and Colleges*. College Entrance Examination Board, 2005.

"Writing: A Ticket to Work . . . Or a Ticket Out—A Survey of Business Leaders." *Report of the National Commission on Writing for America's Families, Schools, and Colleges*. College Entrance Examination Board, 2004.

Wuorio, Jeff. "Blogging for Business: 7 Tips for Getting Started." *Microsoft Small Business*. 23 May 2005. http://www.microsoft.com/smallbusiness/resources/marketing/online_marketing/blogging_for_business_7_tips_for_getting_started.Mspx.

Wuorio, Jeff. "5 Ways Blogging Can Help Your Business." *Microsoft Small Business Center*. 23 May 2005. http://www.microsoft.com/smallbusiness/resources/marketing/online_marketing/5_ways_blogging_can_help_your_business.mspx.

Zubak, Cheryl L. "Choosing a Windows Help Authoring Tool." *Intercom* (Jan. 1996): 10–11, 36.

Chapter 7 Proposals and Long Reports

"Code for Communicators." Society for Technical Communications. 10 Mar. 2004. http://stcrmc.org/resources/resource_code.htm.

"Code of Ethics." International Association of Business Communicators. 8 Oct. 2003. http://www.iabc.com/members/joining/code.htm.

Chapter 8 Technical Applications

"Evacuated Blood Collection Tube." McLendon Clinical Laboratories, University of North Carolina Hospital. 2005. http://www.pathology.med.unc.edu/path/labs/textfiles/tube_guide.htm.

"HP Business Inkjet 1000 Printer." Hewlett Packard. 2005. http://h10010.wwwl.hp.com/wwpc/pscmisc/vac/us/product_pdfs/448666.pdf.

Chapter 9 The Job Search

Bloch, Janel M. "Online Job Searching: Clicking Your Way to Employment." *Intercom* (September/October 2003): 11–14.

Dikel, Margaret F. "The Online Job Application: Preparing Your Resume for the Internet." Sep. 1999. 13 Jan. 2000. http://www.dbm.com/jobguide/eresume.html.

Drakeley, Caroline A. "Viral Networking: Tactics in Today's Job Market." *Intercom* (September/October 2003): 5–7.

Kallick, Rob. "Research Pays Off During Interview." *The Kansas City Star*. March 23, 2003. D1.

"Planning Job Choices." NaceWeb. 15 Jan. 2004. http://www.naceweb.org/press/display.asp?year=2004&prid=184.

Stafford, Diane. "Show Up Armed With Answers." *The Kansas City Star*. August 3, 2003: L1.

Stern, Linda. "New Rules of the Hunt." *Newsweek* (February 17, 2003): 67.

Chapter 10 Oral Communication

Buffet, Warren. "Preface." *A Plain English Handbook: How to Create Clear SEC Disclosure Documents*. Washington, DC: U.S. Securities and Exchange Commission, August 1998: 1–2.

Coyner, Dale. "Webconferencing 101: A Dozen Ideas to Improve Your Next Online Event." 5 June 2005. http://www.isquare.com/webconf.cfm.

Keller, Julia. "Is PowerPoint the Devil?" *Chicago Tribune* 20 Sep. 2004. http://www.siliconvalley.com/mld/siliconvalley/5004120.htm.

Mahin, Linda. "PowerPoint Pedagogy." *Business Communication Quarterly* 64 (June 2004): 220.

Murray, Krysta. "Web Conferencing Tips from a Pro." *Successful Meetings*. 1 July 2005. 4 June 2005. http://www.successmtgs.com/successmtgs/magazine/search_display.jsp?vnu_content_id=1000542704.

INDEX